T0134392

Lecture Notes in Computational Science and Engineering **130**

Editors:

Timothy J. Barth
Michael Griebel
David E. Keyes
Risto M. Nieminen
Dirk Roose
Tamar Schlick

More information about this series at http://www.springer.com/series/3527

Steffen Weißer

BEM-based Finite Element Approaches on Polytopal Meshes

Steffen Weißer
FR Mathematik
Universität des Saarlandes
Saarbrücken, Germany

ISSN 1439-7358 ISSN 2197-7100 (electronic)
Lecture Notes in Computational Science and Engineering
ISBN 978-3-030-20963-6 ISBN 978-3-030-20961-2 (eBook)
https://doi.org/10.1007/978-3-030-20961-2

Mathematics Subject Classification (2010): 65N15, 65N30, 65N38, 65N50, 65D05, 41A25, 41A30

© Springer Nature Switzerland AG 2019
This work is subject to copyright. All rights are reserved by the Publisher, whether the whole or part of the material is concerned, specifically the rights of translation, reprinting, reuse of illustrations, recitation, broadcasting, reproduction on microfilms or in any other physical way, and transmission or information storage and retrieval, electronic adaptation, computer software, or by similar or dissimilar methodology now known or hereafter developed.
The use of general descriptive names, registered names, trademarks, service marks, etc. in this publication does not imply, even in the absence of a specific statement, that such names are exempt from the relevant protective laws and regulations and therefore free for general use.
The publisher, the authors, and the editors are safe to assume that the advice and information in this book are believed to be true and accurate at the date of publication. Neither the publisher nor the authors or the editors give a warranty, express or implied, with respect to the material contained herein or for any errors or omissions that may have been made. The publisher remains neutral with regard to jurisdictional claims in published maps and institutional affiliations.

This Springer imprint is published by the registered company Springer Nature Switzerland AG.
The registered company address is: Gewerbestrasse 11, 6330 Cham, Switzerland

To Anna

Preface

The BEM based finite element method has been developed within the last decade, and it is one of the first methods designed for the approximation of boundary value problems on polygonal and polyhedral meshes. This is possible due to the use of implicitly defined ansatz functions which are treated locally by means of boundary integral operators and are realized with the help of boundary element methods (BEM) in the computations.

When I started my doctoral studies in 2009, it was an appealing but also a somehow abstruse idea to generalize the well-known finite element method (FEM) to polygonal and polyhedral meshes. To that time, I was not aware of any other attempts in this direction. A few years later, the virtual element method came up, and I learned about other approaches that just started in parallel to work on general meshes. The discretization of differential equations on polygonal and polyhedral meshes became a hot research area, and the amount of publications and organized conferences is continuously increasing. The beauty of these ideas caught my attention and shaped my research interests. My resulting theoretical and computational contributions of the last 10 years have found their way into this monograph which is based on my habilitation thesis.

The monograph presents an introduction, a mathematical analysis and applications of the BEM-based finite element approach. It is intended to researchers in the field of numerics for partial differential equations in the wide community of mathematicians and mathematically aware engineers using finite element methods but also to advanced graduate students who are interested to deepen their knowledge in the field.

After the discussion of polytopal meshes, the BEM-based FEM is introduced for high-order approximation spaces over general elements, and its error analysis is given on uniform meshes for smooth solutions (Chap. 2). The method is studied for uniform, adaptive and anisotropic discretizations, where several error estimation techniques and interpolation operators are derived and the notion of anisotropic polytopal elements is introduced (Chap. 3). Beside these, the numerical treatment of boundary integral equations is discussed (Chap. 4). Local error estimation techniques give rise to adaptive mesh refinement as in classical finite element strategies.

Here, the flexibility of polytopal meshes yields benefits in the refinement process and in the application of post-processing operators as in goal-oriented techniques, for instance (Chap. 5). Finally, some recent developments on mixed finite element formulations and problem-adapted approximation spaces are highlighted (Chap. 6). Throughout the monograph, all theoretical results are motivated and validated by numerous numerical experiments and tests, which demonstrate the applicability and the flexibility of the BEM-based FEM.

This monograph and the involved research could not have been realized without the encouragement and support of various persons. Among all these friends in the scientific community, I thank, in particular, Sergej Rjasanow and his research group for their open-mindedness and helpfulness which created a great and stimulating working atmosphere. My deepest gratitude, however, is reserved for my wife, Anna Benz-Weißer, whose logical mind, sound and thoughtful advice, sympathetic ear and unselfish support are of indispensable value for me.

Saarbrücken, Germany Steffen Weißer
April 2019

Contents

Acronyms

List of abbreviations which are sorted according to their context

Boundary Value Problem

Γ	Boundary of the domain Ω
Γ_D	Dirichlet boundary with $\Gamma_D \subset \Gamma$, $\lvert\Gamma_D\rvert > 0$
Γ_N	Neumann boundary with $\Gamma_N \subset \Gamma$
Ω	Domain of the boundary value problem with $\Omega \subset \mathbb{R}^d$, which is open and bounded
a	Scalar valued diffusion coefficient with $0 < a_{\min} \leq a \leq a_{\max}$
d	Space dimension with $d = 2, 3$
f	Source function
g_D	Dirichlet data
g_N	Neumann data
$\mathbf{n}$	Outward unit normal vector on Γ
$\mathbf{x}, \mathbf{y}$	Points in $\mathbb{R}^d$ with $\mathbf{x} = (x_1, \ldots, x_d)^\top$

Function Spaces

Let $\omega \subset \overline{\Omega}$ be either a domain or a lower-dimensional manifold.

$\mathscr{P}^k(\omega)$	Space of polynomials with degree smaller or equal $k \in \mathbb{N}_0$ with the convention $\mathscr{P}^{-1}(\omega) = \{0\}$
$\mathscr{P}^k_{\text{pw}}(\mathscr{T}_h(K))$	Space of continuous functions on K which are polynomials of degree smaller or equal k on each triangle ($d = 2$) or tetrahedron ($d = 3$) of the auxiliary discretization $\mathscr{T}_h(K)$ of the polytopal element K

$\mathscr{P}^k_{\text{pw}}(\partial K)$	Space of continuous functions on ∂K which are polynomials of degree smaller or equal k on each edge ($d = 2$) or triangular face ($d = 3$) of the polytopal element K; $\mathscr{P}^k_{\text{pw}}(\mathscr{B}_h)$ for a general boundary mesh $\mathscr{B}_h$
$\mathscr{P}^k_{\text{pw,d}}(\partial K)$	Space of $L_2(\partial K)$ functions which are polynomials of degree smaller or equal k on each edge ($d = 2$) or triangular face ($d = 3$) of the polytopal element K; $\mathscr{P}^k_{\text{pw,d}}(\mathscr{B}_h)$ for a general boundary mesh $\mathscr{B}_h$
$C^0(\omega)$	Space of continuous functions on ω
$C^k(\omega)$	Space of $k \in \mathbb{N}$ times continuously differentiable functions on ω
$C^{k,\kappa}(\omega)$	Space of Hölder continuous functions with $k \in \mathbb{N}_0$ and $\kappa \in (0, 1]$ on ω and in particular the space of Lipschitz continuous functions $C^{0,1}(\omega)$
$H^s(\omega)$	Sobolev space of order $s \in \mathbb{R}$ over ω (see Sect. 1.3)
$H^1_0(\omega)$	Space of $H^1(\omega)$ functions with vanishing trace on $\partial\omega$
$H^1_*(K)$	Space of $H^1(K)$ functions with vanishing mean values on ∂K
$H^{1/2}_*(\partial K)$	Space of $H^{1/2}(\partial K)$ functions with vanishing mean values
$H^1_\Delta(\mathscr{K}_h)$	Space of $H^1(\Omega)$ functions which are weakly harmonic on each $K \in \mathscr{K}_h$
$H^1_D(\Omega)$	Space of $H^1(\Omega)$ functions with vanishing trace on the Dirichlet boundary Γ_D
$\mathbf{H}(\text{div}, \Omega)$	Space of vector valued L_2-functions with divergence in $L_2(\Omega)$
$\mathbf{H}_N(\text{div}, \Omega)$	Space of $\mathbf{H}(\text{div}, \Omega)$ functions with vanishing normal trace on Γ_N
$L_2(\omega)$	Lebesgue space of square integrable functions on ω
$L_\infty(\omega)$	Lebesgue space of measurable functions which are bounded almost everywhere on ω
$W^2_\infty(\omega)$	Sobolev space of functions which are, together with their first and second derivatives, measurable and bounded almost everywhere on ω

Polytopal Mesh

$c_{\mathscr{K}}, c_{\mathscr{F}}$	Regularity parameter for stable polytopal meshes (see Definitions 2.2, 2.12 and 2.17)
$\mathscr{E}_h$	Set of all edges
$\mathscr{E}_{h,\Omega}$	Set of all edges in the interior of Ω
$\mathscr{E}_{h,D}$	Set of all edges on the Dirichlet boundary Γ_D
$\mathscr{E}_{h,N}$	Set of all edges on the interior of the Neumann boundary Γ_N
$\mathscr{F}_h$	Set of all faces for 3D, might also refer to the set of edges
$\mathscr{F}_{h,\Omega}$	Set of all faces in the interior of Ω
$\mathscr{F}_{h,D}$	Set of all faces on the Dirichlet boundary Γ_D
$\mathscr{F}_{h,N}$	Set of all faces on the interior of the Neumann boundary Γ_N
$\mathscr{E}(K)$	Set of edges which belong to the element K

$\mathscr{E}(F)$	Set of edges which belong to the face F
$\mathscr{F}(K)$	Set of faces which belong to the element K
Γ_S	Skeleton of the discretization
$\mathscr{K}_h$	Set of all polygonal or polyhedral elements in the mesh
$\mathscr{N}_h$	Set of all nodes
$\mathscr{N}_{h,\Omega}$	Set of all nodes in the interior of Ω
$\mathscr{N}_{h,D}$	Set of all nodes on the Dirichlet boundary Γ_D
$\mathscr{N}_{h,N}$	Set of all nodes on the interior of the Neumann boundary Γ_N
$\mathscr{N}(K)$	Set of nodes which belong to the element K
$\mathscr{N}(E)$	Set of nodes which belong to the edge E
$\mathscr{N}(F)$	Set of nodes which belong to the face F
ρ_K, ρ_F	Radius of inscribed ball/circle of an element or face see (Definitions 2.1, 2.10 and 2.11)
$\sigma_{\mathscr{K}}, \sigma_{\mathscr{F}}$	Uniform bound of aspect ratio in regular meshes (see Definitions 2.1, 2.10 and 2.11)
E	Edge in the mesh with $E = \overline{\mathbf{z}_b\mathbf{z}_e}$
F	Polygonal face of a polyhedral element, might also be an edge depending on the context
h_E	Length of an edge
h_F	Diameter of a face
h_K	Diameter of an element
K	Polygonal or polyhedral element
$\mathbf{z}$	Node in the mesh
$\mathbf{z}_K, \mathbf{z}_F$	Centre of inscribed ball/circle of an element or face (see Definitions 2.1, 2.10 and 2.11)

Approximation Spaces, Interpolation and Projection

$\psi_{\mathbf{z}}$	Nodal basis function
ψ_E	Edge basis function
ψ_F	Face basis function
ψ_K	Element basis function
Ψ_h^k	Basis of V_h^k with $\Psi_h^k = \Psi_{h,H}^K \cup \Psi_{h,B}^k$
V_h^k	Approximation space of order k over polytopal mesh with the decomposition $V_h^k = V_{h,H}^k \oplus V_{h,B}^k$
$V_{h,D}^k$	Subspace of V_h^k with functions vanishing on the Dirichlet boundary Γ_D
$V_{h,H,D}^k$	Subspace of $V_{h,H}^k$ with functions vanishing on the Dirichlet boundary Γ_D
$\mathfrak{I}_h^k$	Interpolation operator from $H^2(\Omega)$ into V_h^k with $\mathfrak{I}_h^k = \mathfrak{I}_{h,H}^k + \mathfrak{I}_{h,B}^k$
$\mathfrak{I}_C$	Clément interpolation operator
$\mathfrak{I}_{SZ}$	Scott–Zhang interpolation operator
Π_ω	L_2-projection over ω into the space of constants
$\omega_{\mathbf{z}}$	Neighbourhood of all elements adjacent to the node $\mathbf{z}$

ω_E	Neighbourhood of all elements adjacent to the edge E
ω_F	Neighbourhood of all elements adjacent to the face F
ω_K	Neighbourhood of all elements adjacent to the element K
ω_E^*	Neighbourhood of all elements whose closure contain the edge E
ω_F^*	Neighbourhood of all elements whose closure contain the face F
ω_K^*	Neighbourhood of all elements that share an edge ($d = 2$) or face ($d = 3$) with K
$C_P(\omega)$	Poincaré constant for the domain ω

Boundary Integral Equations

Φ_D	Basis of $\mathscr{P}^k_{\mathrm{pw}}(\mathscr{B}_h)$
$\Phi_{D,F}$	Set of basis functions from Φ_D with support in $F \in \mathscr{F}_h$
Φ_N	Basis of $\mathscr{P}^{k-1}_{\mathrm{pw,d}}(\mathscr{B}_h)$
$\mathscr{P}^k_{\mathrm{pw}}(\mathscr{B}_h)$	Space of continuous functions which are piecewise polynomials of degree smaller or equal k over the boundary mesh $\mathscr{B}_h$ with basis Φ_D
$\mathscr{P}^{k-1}_{\mathrm{pw,d}}(\mathscr{B}_h)$	Space of $L_2(\partial K)$ functions which are piecewise polynomials of degree smaller or equal $k-1$ over the boundary mesh $\mathscr{B}_h$ with basis Φ_N
$\varsigma(\mathbf{x})$	Scalar term on ∂K (see (4.5) and Remark 4.1)
γ_0^K	Trace operator, which gives the trace of a function defined on K on the boundary ∂K
γ_1^K	Conormal derivative operator, which gives the outer normal derivative on ∂K for sufficiently smooth functions defined on K (might be scaled by diffusion matrix A)
$\widetilde{\gamma_1^K}$	Modified conormal derivative operator on ∂K for convection-diffusion-reaction equation (see (6.34))
$\mathbf{D}_K$	Hypersingular integral operator
$\mathbf{D}_{K,h}$	Discretization of the hypersingular integral operator
$\widetilde{\mathbf{D}}_{K,h}$	Regularized discretization of the hypersingular integral operator
$\mathbf{K}_K$	Double-layer potential operator
$\mathbf{K}'_K$	Adjoint double-layer potential operator
$\mathbf{K}_{K,h}$	Discretization of the double-layer potential operator
$\mathbf{M}_{K,h}$	Massmatrix for Φ_N and Φ_D
$\mathbf{M}^{DD}_{K,h}$	Massmatrix for Φ_D and Φ_D
$\mathbf{M}^{NN}_{K,h}$	Massmatrix for Φ_N and Φ_N
$\mathbf{n}_K(\mathbf{x})$	Outward unit normal vector on ∂K at the point $\mathbf{x} \in \partial K$
$\mathbf{P}_K$	Poincaré–Steklov operator (Neumann-to-Dirichlet map)
$\mathbf{S}_K$	Steklov–Poincaré operator (Dirichlet-to-Neumann map)
$\mathbf{S}_{K,h}$	Symmetric discretization of the Steklov–Poincaré operator
$\mathbf{S}^{\mathrm{unsym}}_{K,h}$	Unsymmetric discretization of the Steklov–Poincaré operator
$\widetilde{\mathbf{S}}_K$	Approximation of the Steklov–Poincaré operator

$\mathbf{V}_{K,h}$ Discretization of the single-layer potential operator
$\mathbf{V}_K$ Single-layer potential operator

Adaptivity and Error Estimates

$\|\cdot\|_b$ Energy norm induced by the bilinear form of the variational formulation
$\|\cdot\|_{b,\omega}$ Energy norm with restricted integration domain ω
$[\![u_h]\!]_F$ Jump of the conormal derivative of u_h over F, where F is either an edge ($d = 2$) or a face ($d = 3$)
R_F Edge/face residual
R_K Element residual

Miscellaneous

δ_{ij} Kronecker symbol
$\|\cdot\|_F$ Frobenius norm of a matrix
$\sigma_{\mathcal{T}}$ Uniform bound of aspect ratio in auxiliary discretization $\mathcal{T}_h(\mathcal{K}_h)$
$\mathcal{T}_h(K)$ Auxiliary discretization of K with triangles or tetrahedrons (see Sects. 2.2.1 and 2.2.2); we also use $\mathcal{T}_h(\cdot) = \mathcal{T}_0(\cdot)$
$\mathcal{T}_h(\mathcal{K}_h)$ Admissible auxiliary discretization of Ω with triangles or tetrahedrons constructed by $\mathcal{T}_h(K)$, $K \in \mathcal{K}_h$
$\mathcal{T}_l(F)$ Auxiliary triangulation of a face with mesh level l
$\mathcal{T}_l(\partial K)$ Admissible auxiliary triangulation of the surface of an element constructed by $\mathcal{T}_l(F)$, $F \in \mathcal{F}(K)$
$\mathcal{T}_l(K)$ Auxiliary discretization into tetrahedra of element constructed by $\mathcal{T}_l(\partial K)$
$\mathcal{T}_l(\mathcal{K}_h)$ Admissible auxiliary discretization of Ω into tetrahedra constructed by $\mathcal{T}_l(K)$, $K \in \mathcal{K}_h$
$\mathfrak{F}_K$ Affine mapping which serves for the scaling and depends on the context
$M_{\mathrm{Cov}}(K)$ Covariance matrix of element K
$\overline{\mathbf{x}}_K$ Barycentre of element K

Chapter 1
Introduction

The numerical solution of boundary value problems on polygonal and polyhedral discretizations is an emerging topic, and the interest on it increased continuously during the last few years. This work presents a self-contained and systematic introduction, study and application of the BEM-based FEM with high-order approximation spaces on general polytopal meshes in two and three space dimensions. This approach makes use of local boundary integral formulations that will also be discussed in more detail. The study of interpolation and approximation properties over isotropic and anisoptropic polytopal discretizations build fundamental concepts for the analysis and the development. The BEM-based FEM is applied to adaptive FEM strategies, yielding locally refined meshes with naturally included hanging nodes, and to convection-diffusion-reaction problems, showing stabilizing effects while incorporating the differential equation into the approximation space. All theoretical considerations are substantiated with several numerical tests and examples.

1.1 Overview

Computer simulations play a crucial role in modern research institutes and development departments of all areas. The simulations rely on physical models which describe the underlying principles and the interplay of all components. Such properties are often modelled by differential equations and in particular by boundary value problems in the mathematical framework. In order to realise these formulations on a computer, discretizations have to be introduced such that the unknown quantities of interest can be approximated accurately and efficiently. The probably most accepted and most successful discretization strategy for boundary value problems in science and engineering is the finite element method (FEM). Furthermore, there are the discontinuous Galerkin (DG), the finite volume (FV) and the boundary element methods (BEM), which all have their advantages in certain application areas.

© Springer Nature Switzerland AG 2019

S. Weißer, *BEM-based Finite Element Approaches on Polytopal Meshes*,
Lecture Notes in Computational Science and Engineering 130,
https://doi.org/10.1007/978-3-030-20961-2_1

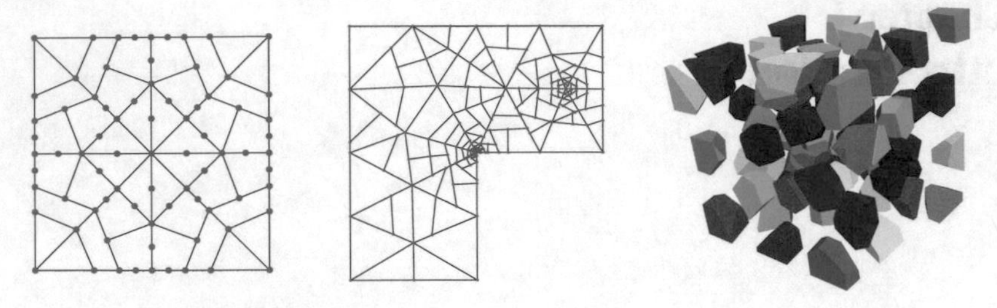

Fig. 1.1 Polygonal and polyhedral meshes

Most of these approaches, and in particular the finite element method, rely on the decomposition of the computational domain into simplicial elements, which include triangles as well as quadrangles in two-dimensions and tetrahedra as well as hexahedra in three-dimensions. These elements form the mesh of the domain that is utilized in the discretization of the corresponding function spaces containing the sought quantities. In high-performance computing and in applications needing accurate and efficient results, the use of problem adapted meshes is essential. Such meshes might be constructed using a priori knowledge or adaptive finite element strategies, which successively and automatically refine the meshes in appropriate regions. But in all these cases the meshes have to be admissible and satisfy some regularity properties. Nowadays, there are efficient approaches, which deal with these mesh requirements. Nevertheless, they still have to be considered in the implementations. Such issues appear in contact problems and when some parts of the computational domain are meshed independently. Obviously, the restrictive nature is caused by the small variety of supported element shapes. This also results in difficulties, when meshing complex geometries.

An enormous gain on flexibility is achieved by the use of polygonal (2D) and polyhedral (3D) meshes in the discretizations, cf. Fig. 1.1. General polygonal and polyhedral element shapes adapt more easily to complex situations. They naturally include so called hanging nodes, for example, which cause additional effort in classical meshes when dealing with adaptivity or non-matching meshes. Already in 1975, Wachspress [172] proposed the construction of conforming rational basis functions on convex polygons with any number of sides for the finite element method. In that time, however, the construction was not attractive for the realization in efficient computer codes, since the processing power was too low. The advent of mean value coordinates [75] in 2003 was a turning point in the sustained interest and further development of finite element methods on polygonal meshes. Only recently, these meshes received a lot of attentions. Several improved basis functions on polygonal elements have been introduced and applied in linear elasticity for example, see [164, 165]. They are often referred to generalized barycentric coordinates and polygonal finite elements. Beside of mean value coordinates [75, 101], there are maximum entropy coordinates [12, 102, 163] and several others as described in the book *Generalized Barycentric Coordinates in Computer Graphics and Computational Mechanics*, see [103]. These coordinates are applied in computer graphics for character articulation [112, 126], for instance. A mathematical discussion of properties and applications as well as an error analysis can be found in [77, 84]. An up-to-date survey of barycentric coordinates is given in [76, 103].

Beside of the polygonal finite element methods, the finite volume methods are successfully applied on polygonal and polyhedral meshes [57]. These methods produce non-conforming approximations and they are popular in computational fluid dynamics (CFD), where polyhedral meshes often yield more accurate results as structured grids. Due to this advantage, the polyhedral meshes for CFD simulations with finite volume approximations were integrated in software packages like OpenFOAM, ANSYS Fluent and STAR-CD from CD-adapco. The mimetic finite difference methods [124] are a related methodology, which have been initially stated on orthogonal meshes and have then be transferred to general polyhedral discretizations. A mathematical analysis on general meshes has been performed in [46] and only a few years ago new insights enabled conforming and arbitrary order approximations within mimetic discretizations [24]. A detailed discussion and introduction can be found in the monograph [27]. The newly derived concept gave rise to the development of the virtual element method (VEM), see [25]. The analysis of VEM is performed in the finite element framework, which offers a rich set of tools. Since 2013, the development of VEM spread fast into several areas including linear elasticity [26, 81], the Helmholtz [138] and the Navier–Stokes problem [32], mixed formulations [42], stabilizations for convection problems [33], adaptivity [23, 31, 35, 50] and many more. Further non-conforming discretization techniques, that are applied and analysed on polygonal and polyhedral meshes, are the discontinuous Galerkin [66] and the recently introduced hybrid higher-order [65] and weak Galerkin [173] methods.

Another conforming approximation scheme came up in parallel to VEM when D. Copeland, U. Langer and D. Pusch proposed to study the boundary element domain decomposition methods [106] in the framework of finite element methods in [60]. This class of discretization methods uses PDE-harmonic shape functions in every element of a polytopal mesh. Therefore, these methods can be considered as local Trefftz FEM following the early work [168]. In order to generate the local stiffness matrices efficiently, boundary element techniques are employed locally. This is the reason why these non-standard finite element methods are called BEM-based FEM. The papers [95] and [93] provide the a priori discretization error analysis with respect to the energy and L_2 norms, respectively, where homogeneous diffusion problems serve as model problems. Fast FETI-type solvers for solving the large linear systems arising from the BEM-based FEM discretization of diffusion problems are studied in [97, 98]. Residual-type a posteriori discretization error estimates are derived in a sequence of papers for adaptive versions of the BEM-based FEM [174, 178–180, 182] and anisotropic polytopal elements are studied in [181]. Additionally, high-order trial functions are introduced in [145, 146, 175, 177], which open the development towards fully *hp*-adaptive strategies. Furthermore, the ideas of BEM-based FEM are transferred into several other application areas. There are, for instance, first results on vector valued, $H(\mathrm{div})$-conforming approximations [73] and on time dependent problems [176]. Additionally, the notion of anisotropic polytopal elements has been applied to VEM [9]. The construction of PDE-harmonic trial functions seems to be especially appropriate for convection-diffusion-reaction problems. First results are presented in [96] and extended in [99] utilizing the hierarchical construction discussed in [147].

One of the probably most attractive features of polygonal and polyhedral meshes is their enormous flexibility, which has not been fully exploited in the literature so far. In modern high-performance computations the use of problem adapted meshes is one of the key ingredients for their success. The finite element method is often combined with an adaptive strategy, where the mesh is successively adapted to the problem. Error indicators gauge the approximation quality and steer a local mesh refinement procedure. Refining classical elements like triangles and tetrahedra affect their neighbouring elements in the mesh. After subdividing several elements, the neighbours are not correctly aligned any more. Therefore, some kind of post-processing is mandatory in order to maintain the mesh admissibility. When polygonal and polyhedral elements are used, local refinements might affect the neighbouring elements, but these elements are still polygonal and polyhedral and are thus naturally supported. This effect solves the handling of so called hanging nodes and it has been demonstrated in [174]. Although this is a fruitful topic, there are only few results available on a posteriori error analysis and adaptivity for conforming approximation methods on polygonal and polyhedral meshes. For the virtual element method see [23, 31, 35, 50] and for the mimetic discretization technique there are also only several references which are limited to low order methods, compare the recent work [7]. Further analysis on quasi-interpolation operators, residual-based error estimates and local mesh refinement for polygonal elements has been performed in [174, 178] with applications to the BEM-based FEM. Additionally, an extension to non-convex elements and high-order approximations with upper and lower bounds for the residual-based error estimator is derived in [180]. Beside of the classical residual-based error estimation techniques, there exist goal-oriented error estimation [22]. Instead of considering the energy error, goal-oriented strategies allow for adaptive refinement steered by some quantity of interest. Thus, these methods are practical in engineering simulations. First results on polygonal meshes have been obtained in [182].

Problem adapted meshes are also utilized in computations, where sharp layers in the solution are expected. This appears in convection-dominated problems, for instance. If the unknown solution changes rapidly in one direction but rather slowly orthogonal to it, anisotropic meshes are beneficial in finite element computations. These meshes contain anisotropic tetrahedral and hexahedral elements which are stretched in one or two directions but thin in the others. In contrast to the usual, isotropic meshes these stretched elements need special care in their analysis. The anisotropy of the mesh has to be aligned with the anisotropy of the approximated function in order to obtain satisfactory results [10]. The quality of the alignment is reflected in a posteriori error estimates [119]. Such estimates can be used to steer an adaptive refinement strategy with anisotropic elements [11, 79]. In comparison to adaptive, isotropic mesh refinement, less elements are required and the efficiency is increased. Simplicial meshes with triangles and tetrahedra or rectangles and hexahedra have often restrictions on their possible anisotropic refinements. Therefore, the initial mesh should be aligned already with the sought function. The use of polygonal and polyhedral elements simplifies the anisotropic refinement, since they do not rely on any restricted direction for subdividing the elements. These new opportunities have been exploited in [9, 181].

The BEM-based FEM has its advantageous not only in the treatment of general meshes. It can be considered a local Trefftz method. This means that the shape functions in the approximation space are build with accordance to the differential equation in the underlying problem. These shape functions satisfy the homogeneous differential equation locally and thus build in some features of the problem into the approximation space. This behaviour has been studied numerically for convection-dominated diffusion problems in [96]. Where conventional approaches without any stabilization lead to oscillations in the solution, the BEM-based FEM remains stable for increased Péclet numbers, i.e., in the convection-dominated regime. The results have been improved in [99], where the idea of Trefftz approximations has been build in on the level of polyhedral elements, their polygonal faces and on the edges of the discretization. Comparisons with a stabilized FEM, the Streamline Upwind/Petrov-Galerkin (SUPG) method [48], have shown an improved resolution of exponential layers at outflow boundaries.

The use of local Trefftz-type approximation spaces is also studied in other areas. One example is the plane wave approximation for the Helmholtz equation [91] or the Trefftz-DG method for time-harmonic Maxwell equations [92]. The combination and coupling of such innovative approaches is quite appealing in order to combine the flexibility of polygonal and polyhedral meshes with problem adapted approximation spaces. Just recently, plane waves have been combined with the virtual element method [138]. All these quite new developments have a great potential and might benefit from each other. Their interplay has been studied rarely and opens opportunities for future developments.

1.2 Outline

The aim of this book is to give a systematic introduction, study and application of the BEM-based FEM. The topics range from high-order approximation spaces on isotropic as well as anisotropic polytopal meshes over a posteriori error estimation and adaptive mesh refinement to specialized adaptations of approximation spaces and interpolation operators. The chapters are organized as follows.

Chapter 2 contains a discussion of polygonal as well as polyhedral meshes including regularity properties and their treatment in mesh refinement. Furthermore, the construction of basis functions is carried out for an approximation space over these general meshes. They are applied in the formulation of the BEM-based FEM, which is obtained by means of a Galerkin formulation. Its convergence and approximation properties are analysed with the help of introduced interpolation operators.

In Chap. 3, best approximation results and trace inequalities are given for polytopal elements. By their application, quasi-interpolation operators for non-smooth functions over polytopal meshes are introduced and analysed. In particular, operators of Clément- and Scott–Zhang-type are studied. Furthermore, the notion of anisotropic meshes is established for polytopal discretizations. These meshes do not satisfy the classical regularity properties introduced in Chap. 2. Consequently, they have to be treated in a special way.

The local problems in the definition of basis functions for the BEM-based FEM are handled by means of boundary integral equations. Chapter 4 gives a short introduction into this topic with a special emphasis on its application in the BEM-based FEM. Therefore, the boundary integral operators for the Laplace problem are reviewed in two- and three-dimensions and corresponding boundary integral equations are derived. Their discretization is realized by a Galerkin boundary element method and by an alternative approach that relies on the Nyström method.

In Chap. 5, adaptive mesh refinement strategies are applied to polytopal meshes in the presence of singularities. In particular, a posteriori error estimates are derived which are used to drive the adaptive procedure. For the error estimation, the classical residual based error estimator as well as goal-oriented techniques are covered on general polytopal meshes. Whereas, the reliability and efficiency estimates for the first mentioned estimator are proved, the benefits and potentials of the second one are discussed for general meshes.

In Chap. 6, some further developments and extensions are taken up. The introduction of $H(\mathrm{div})$-conforming approximation spaces in the sense of the BEM-based FEM is highlighted. Additionally, a hierarchical construction of basis functions in three-dimensions is discussed and applied to convection-diffusion-reaction problems. The presented strategy integrates the underlying differential equation into the approximation space and yields therefore stabilizing properties.

Throughout the book, there are numerical examples, tests and experiments that illustrate and substantiate the theoretical findings.

1.3 Mathematical Preliminaries

In the following, we summarize some mathematical preliminaries on variational formulations and we give the definition of certain function spaces. For the precise definitions, however, we refer to the specialized literature. The classical results on Sobolev spaces can be found in Adams [1], and for Sobolev spaces of rational exponent we refer to Grisvard [87]. A detailed discussion on Sobolev spaces on manifolds and their application to boundary integral equations is given in

McLean [128]. The experienced reader might skip the following sections and come back to them if needed. This section serves as reference only and does not contain all mathematical details.

1.3.1 Function Spaces and Trace Operators

In the study of boundary value problems, the solutions have to be specified in proper function spaces. In the following, we give definitions of several spaces. For this reason, let Ω be any measurable subset of $\mathbb{R}^d$, $d \in \mathbb{N}$ with strictly positive Lebesgue measure. The Banach spaces $L_1(\Omega)$ and $L_2(\Omega)$ are defined in the usual way with the corresponding norms

$$\|u\|_{L_1(\Omega)} = \int_\Omega |u(\mathbf{x})| \,\mathrm{d}\mathbf{x} \quad \text{and} \quad \|u\|_{L_2(\Omega)} = \left(\int_\Omega |u(\mathbf{x})|^2 \,\mathrm{d}\mathbf{x}\right)^{1/2},$$

respectively. Here, the symbol $|\cdot|$ denotes the absolute value. But in other contexts, it might denote the Euclidean norm, the d or $d-1$ dimensional measure or even the cardinality of a discrete set. Furthermore, let the space of locally integrable functions be labeled by

$$L_1^{loc}(\Omega) = \{u : u \in L_1(K) \text{ for any compact } K \subset \Omega\}.$$

The space $L_2(\Omega)$ together with the inner product

$$(u, v)_{L_2(\Omega)} = \int_\Omega u(\mathbf{x})v(\mathbf{x}) \,\mathrm{d}\mathbf{x}$$

becomes a Hilbert space. Additionally, we denote by $L_\infty(\Omega)$ the space of measurable and almost everywhere bounded functions. It is equipped with the norm

$$\|u\|_{L_\infty(\Omega)} = \operatorname*{ess\,sup}_{\mathbf{x}\in\Omega} |u(\mathbf{x})| = \inf_{K\subset\Omega, |K|=0} \sup_{\mathbf{x}\in\Omega\setminus K} |u(\mathbf{x})|,$$

where $|K|$ is the d dimensional Lebesgue measure of K. For a $d-1$ dimensional manifold Γ, the space $L_2(\Gamma)$ is defined in an analog way. Here, the surface measure is used instead of the volume measure.

The space of continuous functions over Ω is denoted by $C^0(\Omega)$ and equipped with the supremum norm

$$\|u\|_{C^0(\Omega)} = \sup_{\mathbf{x}\in\Omega} |u(\mathbf{x})|.$$

Let $\alpha = (\alpha_1, \ldots, \alpha_d) \in \mathbb{N}_0^d$ be a multi-index, i.e., a d-tuple with non-negative entries, and set

$$|\alpha| = \alpha_1 + \cdots + \alpha_d \quad \text{as well as} \quad \partial^\alpha = \left(\frac{\partial}{\partial x_1}\right)^{\alpha_1} \cdots \left(\frac{\partial}{\partial x_d}\right)^{\alpha_d} .$$

The order of the partial derivative ∂^α is the number $|\alpha|$. For any integer $k \geq 0$ and Ω open, we define

$$C^k(\Omega) = \{u : \partial^\alpha u \text{ exists and is continuous on } \Omega \text{ for } |\alpha| \leq k\} .$$

In the special case that $k = 0$, the space of continuous functions over Ω is recovered. Furthermore, we define

$$C_0^k(\Omega) = \{u \in C^k(\Omega) : \operatorname{supp} u \subset \Omega\} ,$$

where

$$\operatorname{supp} u = \overline{\{\mathbf{x} \in \Omega : u(\mathbf{x}) \neq 0\}} ,$$

and set

$$C^\infty(\Omega) = \bigcap_{k \geq 0} C^k(\Omega) \quad \text{as well as} \quad C_0^\infty(\Omega) = \bigcap_{k \geq 0} C_0^k(\Omega) .$$

Finally, we review the space of Lipschitz functions

$$C^{0,1}(\Omega) = \{u \in C^0(\Omega) : \exists L > 0 : |u(\mathbf{x}) - u(\mathbf{y})| \leq L|\mathbf{x} - \mathbf{y}| \text{ for } \mathbf{x}, \mathbf{y} \in \Omega\}$$

and

$$C^{k,1}(\Omega) = \{u \in C^k(\Omega) : \partial^\alpha u \in C^{0,1}(\Omega) \text{ for } |\alpha| = k\}$$

for $k \in \mathbb{N}$. The space of Hölder continuous functions is a straightforward generalization. For $\kappa \in (0, 1]$, it is

$$C^{0,\kappa}(\Omega) = \{u \in C^0(\Omega) : \exists C > 0 : |u(\mathbf{x}) - u(\mathbf{y})| \leq C|\mathbf{x} - \mathbf{y}|^\kappa \text{ for } \mathbf{x}, \mathbf{y} \in \Omega\}$$

and

$$C^{k,\kappa}(\Omega) = \{u \in C^k(\Omega) : \partial^\alpha u \in C^{0,\kappa}(\Omega) \text{ for } |\alpha| = k\}$$

for $k \in \mathbb{N}$.

1.3.1.1 Sobolev Spaces

Let Ω be a non-empty open subset of $\mathbb{R}^d$, $d \in \mathbb{N}$. The Sobolev space $H^k(\Omega)$ of order $k \in \mathbb{N}_0$ is defined by

$$H^k(\Omega) = \{u \in L_2(\Omega) : \partial^\alpha u \in L_2(\Omega) \text{ for } |\alpha| \le k\} \tag{1.1}$$

with the norm $\|\cdot\|_{H^k(\Omega)}$ and the semi-norm $|\cdot|_{H^k(\Omega)}$, where

$$\|u\|_{H^k(\Omega)} = \left(\sum_{|\alpha| \le k} \|\partial^\alpha u\|^2_{L_2(\Omega)} \right)^{1/2} \quad \text{and} \quad |u|_{H^k(\Omega)} = \left(\sum_{|\alpha| = k} \|\partial^\alpha u\|^2_{L_2(\Omega)} \right)^{1/2}.$$

Here, the partial derivative $\partial^\alpha u$ has to be understood in the weak sense. More precisely, let the functional $g_\alpha : C_0^\infty(\Omega) \to \mathbb{R}$ be the distributional derivative of u with index α, i.e., g_α satisfies

$$(u, \partial^\alpha \varphi)_{L_2(\Omega)} = (-1)^{|\alpha|} g_\alpha(\varphi)$$

for all $\varphi \in C_0^\infty(\Omega)$. Furthermore, let g_α have the representation

$$g_\alpha(\varphi) = \int_\Omega \varphi(\mathbf{x})\, \partial^\alpha u(\mathbf{x})\, \mathrm{d}\mathbf{x}$$

for all $\varphi \in C_0^\infty(\Omega)$ with some function $\partial^\alpha u \in L_1^{loc}(\Omega)$ which is defined uniquely up to an equivalence class. In this case, $\partial^\alpha u$ is called the weak derivative of u with index α. The additional condition $\partial^\alpha u \in L_2(\Omega)$ in (1.1) ensures that the weak derivative can be chosen such that it is square integrable.

For the definition of Sobolev spaces with fractional order $s \ge 0$, let $s = k + \mu$ with $k \in \mathbb{N}_0$ and $\mu \in [0, 1)$. The Sobolev–Slobodekij norm is given by

$$\|u\|_{H^s(\Omega)} = \left(\|u\|^2_{H^k(\Omega)} + \sum_{|\alpha| = k} |\partial^\alpha u|^2_{H^\mu(\Omega)} \right)^{1/2},$$

where

$$|u|_{H^\mu(\Omega)} = \left(\int_\Omega \int_\Omega \frac{|u(\mathbf{x}) - u(\mathbf{y})|^2}{|\mathbf{x} - \mathbf{y}|^{d+2\mu}}\, \mathrm{d}\mathbf{x}\, \mathrm{d}\mathbf{y} \right)^{1/2}.$$

Therefore, we define

$$H^s(\Omega) = \{u \in H^k(\Omega) : |\partial^\alpha u|_{H^\mu(\Omega)} < \infty \text{ for } |\alpha| = k\}.$$

The Sobolev norm $\|\cdot\|_{H^s(\Omega)}$ for arbitrary real $s \geq 0$ is induced by the inner product

$$(u, v)_{H^s(\Omega)} = (u, v)_{H^k(\Omega)} + \sum_{|\alpha|=k} (\partial^\alpha u, \partial^\alpha v)_{H^\mu(\Omega)}$$

with

$$(u, v)_{H^k(\Omega)} = \sum_{|\alpha|\leq k} (\partial^\alpha u, \partial^\alpha v)_{L_2(\Omega)}$$

and

$$(u, v)_{H^\mu(\Omega)} = \int_\Omega \int_\Omega \frac{\big(u(\mathbf{x}) - u(\mathbf{y})\big)\big(v(\mathbf{x}) - v(\mathbf{y})\big)}{|\mathbf{x} - \mathbf{y}|^{d+2\mu}} \, \mathrm{d}\mathbf{x} \, \mathrm{d}\mathbf{y} \, .$$

Hence, $H^s(\Omega)$ is a Hilbert space for all $s \geq 0$.

1.3.1.2 Sobolev Spaces on the Boundary

For the definition of Sobolev spaces on the boundary of a domain, we have to restrict the class of admitted domains. Therefore, let $\Omega \subset \mathbb{R}^d$, $d \in \mathbb{N}$ be a bounded open set with boundary Γ. Additionally, we assume that Γ is non-empty and has an overlapping cover that can be parametrized in the way

$$\Gamma = \bigcup_{i=1}^{p} \Gamma_i \, , \qquad \Gamma_i = \left\{ \mathbf{x} \in \mathbb{R}^d : \mathbf{x} = \chi_i(\xi) \text{ for } \xi \in K_i \subset \mathbb{R}^{d-1} \right\} \, . \tag{1.2}$$

With regard to the decomposition of Γ, let $\{\varphi_i\}_{i=1}^p$ be a partition of unity with non-negative cut off functions $\varphi_i \in C_0^\infty(\mathbb{R}^d)$ such that

$$\sum_{i=1}^{p} \varphi_i(\mathbf{x}) = 1 \quad \text{for } \mathbf{x} \in \Gamma \, , \qquad \varphi_i(\mathbf{x}) = 0 \quad \text{for } \mathbf{x} \in \Gamma \setminus \Gamma_i \, .$$

For a function u defined on Γ, we write

$$u(\mathbf{x}) = \sum_{i=1}^{p} u(\mathbf{x})\varphi_i(\mathbf{x}) = \sum_{i=1}^{p} u_i(\mathbf{x}) \quad \text{for } \mathbf{x} \in \Gamma \, ,$$

where $u_i(\mathbf{x}) = u(\mathbf{x})\varphi_i(\mathbf{x})$. In the next step, $\mathbf{x}$ is replaced by the parametrisation from (1.2) and we obtain

$$u_i(\mathbf{x}) = u(\mathbf{x})\varphi_i(\mathbf{x}) = u(\chi_i(\xi))\varphi_i(\chi_i(\xi)) \quad \text{for } \xi \in K_i \subset \mathbb{R}^{d-1}, \ i = 1, \ldots, p \, .$$

The last expression is abbreviated to $\widetilde{u}_i(\xi)$. These functions are defined on bounded subsets of $\mathbb{R}^{d-1}$, and thus the Sobolev spaces from Sect. 1.3.1.1 can be used. To satisfy $u_i \in H^s(K_i)$ for $s > 0$, the corresponding derivatives of the parametrisation χ_i have to exist. For the definition of these derivatives of order up to $s \leq k$, we have to assume $\chi_i \in C^{k-1,1}(K_i)$.

For $0 \leq s \leq k$, the Sobolev norm

$$\|u\|_{H^s(\Gamma),\chi} = \left(\sum_{i=1}^{p} \|u_i\|^2_{H^s(K_i)}\right)^{1/2},$$

which depends on the parametrisation of Γ, is defined. By the use of this norm the Sobolev spaces $H^s(\Gamma)$ can be introduced. For a Lipschitz domain Ω and $s \in (0, 1)$, the Sobolev–Slobodekij norm

$$\|u\|_{H^s(\Gamma)} = \left(\|u\|^2_{L_2(\Gamma)} + \int_\Gamma \int_\Gamma \frac{|u(\mathbf{x}) - u(\mathbf{y})|^2}{|\mathbf{x} - \mathbf{y}|^{d-1+2s}} \,\mathrm{d}s_{\mathbf{x}} \,\mathrm{d}s_{\mathbf{y}}\right)^{1/2}$$

is equivalent to $\|\cdot\|_{H^s(\Gamma),\chi}$, and thus, the space $H^s(\Gamma)$ is independent of the parametrisation chosen in (1.2). For $s < 0$, we define $H^s(\Gamma)$ as the dual space of $H^{-s}(\Gamma)$ and equip it with the norm

$$\|u\|_{H^s(\Gamma)} = \sup_{0 \neq v \in H^{-s}(\Gamma)} \frac{|u(v)|}{\|v\|_{H^{-s}(\Gamma)}}.$$

Additionally, we need some spaces which are only defined on a part of the boundary. Let Γ_0 be an open part of the sufficiently smooth boundary Γ. For $s \geq 0$, we set the Sobolev space

$$H^s(\Gamma_0) = \left\{u = \widetilde{u}|_{\Gamma_0} : \widetilde{u} \in H^s(\Gamma)\right\}$$

with the norm

$$\|u\|_{H^s(\Gamma_0)} = \inf_{\widetilde{u} \in H^s(\Gamma) : \widetilde{u}|_{\Gamma_0} = u} \|\widetilde{u}\|_{H^s(\Gamma)}.$$

Furthermore, let

$$\widetilde{H}^s(\Gamma_0) = \left\{u = \widetilde{u}|_{\Gamma_0} : \widetilde{u} \in H^s(\Gamma),\ \operatorname{supp} \widetilde{u} \subset \Gamma_0\right\},$$

and for $s < 0$, we set $H^s(\Gamma_0)$ as the dual space of $\widetilde{H}^s(\Gamma_0)$. Finally, we define a Sobolev space over the boundary with piecewise regularity. Therefore, let

$$\Gamma = \bigcup_{i=1}^{p} \overline{\Gamma}_i, \qquad \Gamma_i \cap \Gamma_j = \varnothing \quad \text{for } i \neq j,$$

and define

$$H^s_{\mathrm{pw}}(\Gamma) = \left\{ u \in H^{\min\{\frac{d-1}{2}, s\}}(\Gamma) : u|_{\Gamma_i} \in H^s(\Gamma_i),\ i = 1, \ldots, p \right\} .$$

This space is equipped with the norm

$$\|u\|_{H^s_{\mathrm{pw}}(\Gamma)} = \left(\sum_{i=1}^{p} \|u|_{\Gamma_i}\|^2_{H^s(\Gamma_i)} \right)^{1/2} .$$

1.3.1.3 Properties of Sobolev Spaces

To state some properties of Sobolev spaces, we have to guarantee certain regularities of the domain Ω and its boundary Γ. Therefore, we take from [87] the following definition.

Definition 1.1 Let Ω be an open subset of $\mathbb{R}^d$. We say that its boundary Γ is continuous (respectively Lipschitz, continuously differentiable, of class $C^{k,1}$, k times differentiable) if for every $\mathbf{x} \in \Gamma$ there exists a neighbourhood U of $\mathbf{x}$ in $\mathbb{R}^d$ and new orthogonal coordinates $\{\xi_1, \ldots, \xi_d\}$ such that

1. U is an hypercube in the new coordinates:

$$U = \{(\xi_1, \ldots, \xi_d) : -c_i < \xi_i < c_i,\ i = 1, \ldots, d\} ,$$

2. there exists a continuous (respectively Lipschitz, continuous differentiable, of class $C^{k,1}$, k times continuously differentiable) function f, defined in

$$U' = \{(\xi_1, \ldots, \xi_{d-1}) : -c_i < \xi_i < c_i,\ i = 1, \ldots, d-1\} ,$$

and such that

$$|f(\xi')| \le c_d/2 \quad \text{for every } \xi' = (\xi_1, \ldots, \xi_{d-1}) \in U' ,$$
$$\Omega \cap U = \{\xi = (\xi', \xi_d) \in U : \xi_d < f(\xi')\} ,$$
$$\Gamma \cap U = \{\xi = (\xi', \xi_d) \in U : \xi_d = f(\xi')\} .$$

If Ω has a Lipschitz boundary, we call Ω a Lipschitz domain. From now on, we restrict ourselves to bounded domains Ω. So, the boundary Γ is compact, and thus there is a finite cover of Γ, which can be used to construct a parametrisation as given in (1.2). We state the famous Sobolev embedding theorem, see, e.g., [1, 49].

Theorem 1.2 (Sobolev Embedding) *Let $\Omega \subset \mathbb{R}^d$, $d \in \mathbb{N}$ be a bounded domain with Lipschitz boundary and let $2k > d$ with $k \in \mathbb{N}$. For $u \in H^k(\Omega)$, it is*

$u \in C^0(\overline{\Omega})$ and there exists a constant $C_S > 0$ such that

$$\|u\|_{C^0(\overline{\Omega})} \le C_S \|u\|_{H^k(\Omega)} \quad \text{for } u \in H^k(\Omega)\,.$$

Remark 1.3 In [49], it is shown that for convex domains Ω with diameter smaller or equal to one, the constant in Theorem 1.2 has the form

$$C_S = c\,|\Omega|^{-1/2}$$

with a constant $c > 0$ which only depends on d and k.

Next, we give some results for traces of functions in Sobolev spaces. For sufficiently smooth functions u over $\overline{\Omega}$, we set the trace operator γ_0 as restriction of u to the boundary Γ, i.e.

$$\gamma_0 u = u\big|_{\Gamma}\,.$$

This operator has continuous extensions such that the following theorems taken from [61] and [128] are valid.

Theorem 1.4 *If the bounded subset Ω of $\mathbb{R}^d$ has a boundary Γ of class $C^{k-1,1}$ and if $1/2 < s \le k$, then*

$$\gamma_0 : H^s(\Omega) \to H^{s-1/2}(\Gamma)$$

is a bounded linear operator, i.e.

$$\|\gamma_0 u\|_{H^{s-1/2}(\Gamma)} \le c_T\, \|u\|_{H^s(\Omega)} \quad \text{for } u \in H^s(\Omega)\,.$$

This operator has a continuous right inverse

$$\mathfrak{E} : H^{s-1/2}(\Gamma) \to H^s(\Omega)$$

with $\gamma_0 \mathfrak{E} v = v$ for all $v \in H^{s-1/2}(\Gamma)$ and

$$\|\mathfrak{E} v\|_{H^s(\Omega)} \le c_{IT}\, \|v\|_{H^{s-1/2}(\Gamma)} \quad \text{for } v \in H^{s-1}(\Gamma)\,.$$

Theorem 1.5 *If $\Omega \subset \mathbb{R}^d$ is a bounded domain with Lipschitz boundary Γ, then the trace operator γ_0 is bounded for $\frac{1}{2} < s < \frac{3}{2}$.*

1.3.2 Galerkin Formulations

At several places in this book, we are concerned with operator equations and in particularly with weak formulations of differential equations. These are treated by means of Galerkin formulations in the continuous as well as in the discretized versions. We also call these formulations variational problems. In the following, we give a summary on this topic.

Let V be a Hilbert space with inner product $(\cdot,\cdot)_V$ and corresponding induced norm $\|\cdot\|_V = \sqrt{(\cdot,\cdot)_V}$. The abstract setting of a Galerkin formulation is

$$\text{Find } u \in V: \quad a(u,v) = \ell(v) \quad \forall v \in V\ , \tag{1.3}$$

where $a(\cdot,\cdot) : V \times V \to \mathbb{R}$ denotes a bilinear and $\ell(\cdot) : V \to \mathbb{R}$ a linear form. The bilinear form is said to be continuous or bounded on V if there exists a constant $c_1 > 0$ such that

$$|a(u,v)| \le c_1 \|u\|_V \|v\|_V \quad \text{for } u, v \in V\ .$$

Furthermore, $a(\cdot,\cdot)$ is called V-elliptic if there is another constant $c_2 > 0$ such that

$$a(v,v) \ge c_2 \|v\|_V^2 \quad \text{for } v \in V\ .$$

Analogously, $\ell(\cdot)$ is said to be continuous if

$$|\ell(v)| \le c_\ell \|v\|_V \quad \text{for } v \in V\ ,$$

for a constant $c_\ell > 0$. Hence, a continuous linear form is a bounded functional on V and therefore, it belongs to the dual space of V.

Theorem 1.6 (Lax–Milgram Lemma) *Let V be a Hilbert space, $a(\cdot,\cdot) : V \times V \to \mathbb{R}$ be a continuous V-elliptic bilinear form, and let $\ell : V \to \mathbb{R}$ be a continuous linear form. The abstract variational problem* (1.3) *has one and only one solution.*

In the proof of the Lax–Milgram Lemma, the Riesz representation theorem is utilized, see, e.g., [58] or the original work [121].

Theorem 1.7 (Riesz Representation Theorem) *Let V' be the dual space of V equipped with the norm*

$$\|\ell\|_{V'} = \sup_{0 \ne v \in V} \frac{|\ell(v)|}{\|v\|_V}\ .$$

For each $\ell \in V'$, there exists a unique $u \in V$ such that

$$(u,v)_V = \ell(v) \quad \textit{for } v \in V$$

and

$$\|u\|_V = \|\ell\|_{V'}\ .$$

In the numerics, it is not possible to work with the space V directly. Therefore, a finite dimensional subspace V_h of V is introduced and the discrete Galerkin formulation

$$\text{Find } u_h \in V_h : \quad a(u_h, v_h) = \ell(v_h) \quad \forall v_h \in V_h \tag{1.4}$$

is considered. Since $V_h \subset V$, the method is said to be conforming. Due to the finite dimension of V_h, we can introduce a basis Ψ with $V_h = \operatorname{span} \Psi$ and $\dim V_h = n$ for some $n \in \mathbb{N}$. Next, we express u_h as linear combination of basis functions

$$u_h = \sum_{\psi \in \Psi} u_\psi \psi ,$$

and we have to test (1.4) only with $v_h = \varphi$ for all $\varphi \in \Psi$. Consequently, we end up with a system of linear equations to compute the unknown coefficients u_ψ of u_h. More precisely, let $\underline{u}_h$ be the vector with components u_ψ, i.e. $\underline{u}_h = \left(u_\psi\right)_{\psi \in \Psi}$. We obtain

$$A\underline{u}_h = b \tag{1.5}$$

with

$$A = \big(a(\psi, \varphi)\big)_{\varphi, \psi \in \Psi} \in \mathbb{R}^{n \times n} \quad \text{and} \quad b = \big(\ell(\varphi)\big)_{\varphi \in \Psi} \in \mathbb{R}^n .$$

The system matrix A is positive definite because of the V-ellipticity of the bilinear form $a(\cdot, \cdot)$. Therefore, the $n \times n$ system of linear equations admits a unique solution. If the system (1.5) of linear equations is small, we use an efficient direct solver of LAPACK [6]. In case of large systems, however, iterative solvers are preferable. For symmetric matrices we apply the conjugate gradient method (CG) [90] and for non-symmetric matrices we utilize GMRES [150].

Nevertheless, the question remains how the Galerkin formulations (1.3) and (1.4) are related to each other. Céa's Lemma gives the answer. The discrete Galerkin formulation (1.4) yields the quasi-best approximation of the solution of (1.3).

Lemma 1.8 (Céa's Lemma) *Let V be a Hilbert space and $V_h \subset V$ a finite dimensional subspace of V, let $a(\cdot, \cdot) : V \times V \to \mathbb{R}$ be a continuous V-elliptic bilinear form, and let $\ell : V \to \mathbb{R}$ be a continuous linear form. Furthermore, let $u \in V$ be the solution of* (1.3) *and $u_h \in V_h$ be the solution of* (1.4)*. The abstract error estimate*

$$\|u - u_h\|_V \leq \frac{c_1}{c_2} \min_{v_h \in V_h} \|u - v_h\|_V \tag{1.6}$$

holds.

Consequently, we can estimate the error of the Galerkin approximation by studying interpolation properties of the finite dimensional subspace. More precisely, we can estimate the minimum on the right hand side of (1.6) by inserting an interpolation of u in the space V_h. Thus, we have to introduce interpolation operators and to prove interpolation error estimates. This yields error estimates for the Galerkin approximation of the form

$$\|u - u_h\| \le c\,\mathfrak{h}^s\,|||u|||$$

with certain norms $\|\cdot\|$ and $|||\cdot|||$, where $s \in \mathbb{R}$ and $\mathfrak{h}$ corresponds either to the characteristic mesh size h, defined later, or to the number of degrees of freedom (DoF) in the system of linear Eq. (1.5). We say in this case that the error in the norm $\|\cdot\|$ converges with order s with respect to $\mathfrak{h}$. In our computational tests, we verify the theoretical orders of convergence. Therefore, let V_h and V_{h_*} be two approximation spaces with corresponding $\mathfrak{h}$ and $\mathfrak{h}_*$. We compute the numerical order of convergence (noc) as

$$\frac{\log(\|u - u_h\|) - \log(\|u - u_{h_*}\|)}{\log(\mathfrak{h}) - \log(\mathfrak{h}_*)}, \tag{1.7}$$

which is an approximation on s in the error model

$$\|u - u_h\| \approx c\,\mathfrak{h}^s\,|||u|||\,.$$

An important analytical tool in order to prove interpolation error estimates is the Bramble–Hilbert Lemma, see [58] and below. Beside of this, we extensively apply the triangle and reverse triangle inequality,

$$\|x + y\| \le \|x\| + \|y\| \qquad \text{and} \qquad \big|\,\|x\| - \|y\|\,\big| \le \|x + y\|\,,$$

for all kinds of norms, as well as the Cauchy–Schwarz inequality

$$|(x, y)| \le \|x\|\|y\|\,,$$

where the norm $\|\cdot\|$ is induced by the inner product $(\cdot,\cdot)$, i.e. $\|\cdot\| = \sqrt{(\cdot,\cdot)}$. Here, x and y might refer to vectors, functions or vector valued functions depending on the context of the inequality.

Theorem 1.9 (Bramble–Hilbert Lemma) *Let $\Omega \subset \mathbb{R}^d$ be a Lipschitz domain. For some integer $k \ge 0$, let $\mathbf{f}$ be a continuous linear form on the space $H^{k+1}(\Omega)$ with the property that*

$$\mathbf{f}(p) = 0 \quad \forall p \in \mathscr{P}^k(\Omega)\,.$$

There exists a constant $C(\Omega)$ such that

$$|\mathbf{f}(v)| \le C(\Omega)\,\|\mathbf{f}\|_*\,|v|_{H^{k+1}(\Omega)}\,,$$

where $\|\cdot\|_*$ is the norm in the dual space of $H^{k+1}(\Omega)$.

Chapter 2
Finite Element Method on Polytopal Meshes

The finite element method (FEM) is a powerful tool for the approximation of boundary value problems, which is widely applied and accepted in science and engineering. The approach relies on the decomposition of the underlying domain into elements and the construction of a discrete approximation space over the given discretization. The BEM-based finite element method can be seen as a generalization in order to handle more general elements in the mesh. This chapter contains a discussion of polygonal as well as polyhedral meshes and the construction of basis functions for the approximation space over these general meshes. The formulation of the BEM-based FEM is obtained by means of a Galerkin formulation and its convergence and approximation properties are analysed with the help of introduced interpolation operators. Numerical experiments confirm the theoretical findings.

2.1 Preliminaries

The approximation space in the BEM-based finite element method is defined in accordance with the underlying differential equation of the considered boundary value problem. For this presentation, we choose the diffusion problem with mixed boundary conditions on a bounded polygonal/polyhedral domain $\Omega \subset \mathbb{R}^d, d = 2, 3$. Its boundary $\Gamma = \Gamma_D \cup \Gamma_N$ is split into a Dirichlet and a Neumann part, where we assume $|\Gamma_D| > 0$. Given a source term $f \in L_2(\Omega)$, a Dirichlet datum $g_D \in H^{1/2}(\Gamma_D)$ as well as a Neumann datum $g_N \in L_2(\Gamma_N)$, the problem reads

$$\begin{aligned} -\operatorname{div}(a\nabla u) &= f \quad \text{in } \Omega\,, \\ u &= g_D \quad \text{on } \Gamma_D\,, \\ a\nabla u \cdot \mathbf{n} &= g_N \quad \text{on } \Gamma_N\,, \end{aligned} \tag{2.1}$$

© Springer Nature Switzerland AG 2019

S. Weißer, *BEM-based Finite Element Approaches on Polytopal Meshes*,
Lecture Notes in Computational Science and Engineering 130,
https://doi.org/10.1007/978-3-030-20961-2_2

where $a \in L_\infty(\Omega)$ with $0 < a_{\min} \leq a \leq a_{\max}$ almost everywhere in Ω and $\mathbf{n}$ is the outward unit normal vector on Γ. This boundary value problem is considered in the week sense with the help of a Galerkin formulation. Thus, we seek a solution $u \in H^1(\Omega)$, where we denote, as usual, the Sobolev spaces of order $s \in \mathbb{R}$ with $H^s(D)$ for some domain $D \subset \Omega$, cf. Sect. 1.3. Furthermore, we utilize the space of polynomials $\mathscr{P}^k(D)$ with degree smaller or equal $k \in \mathbb{N}_0$ with the convention that $\mathscr{P}^{-1}(D) = \{0\}$. Here, D might also be a one- or two-dimensional submanifold of Ω. For simplicity, we assume in the first part that the diffusion coefficient a is piecewise constant and its jumps are resolved by the meshes later on. Nevertheless, we will also give a strategy for the more general situation of continuously varying diffusion coefficients. Our goal is to introduce a H^1-conforming approximation space of arbitrary order $k \in \mathbb{N}$ which yields optimal rates of convergence in the finite element framework. In all estimates, c denotes a generic constant that depends on the mesh regularity and stability, the space dimension d and the approximation order k. The following discrete approximation of $H^1(\Omega)$ is constructed but not limited to the diffusion equation. It can also be applied to other boundary value problems where H^1-conforming approximations are desirable.

2.2 Polygonal and Polyhedral Meshes

For the finite element method, we have to introduce a discretization $\mathscr{K}_h$ of Ω. In this section, we distinguish the two- and three-dimensional case $\Omega \subset \mathbb{R}^d$, $d = 2, 3$. In contrast to classical conforming finite element methods, we allow meshes with general polygonal and polyhedral elements which are bounded. Examples of such meshes are given in Fig. 2.1. If we do not distinguish between the space dimension d, we call the meshes and the elements polytopal. The elements $K \in \mathscr{K}_h$

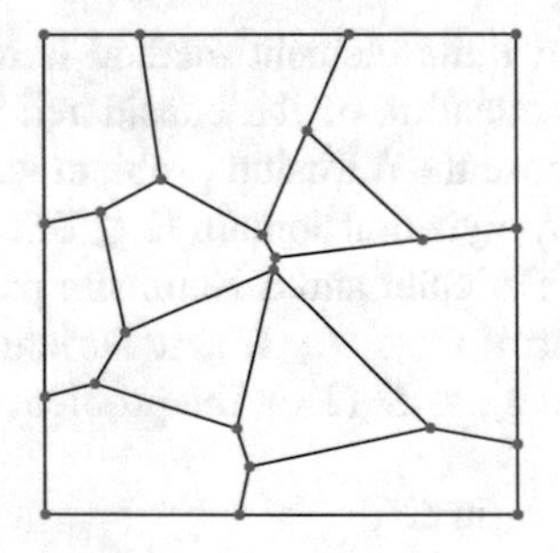
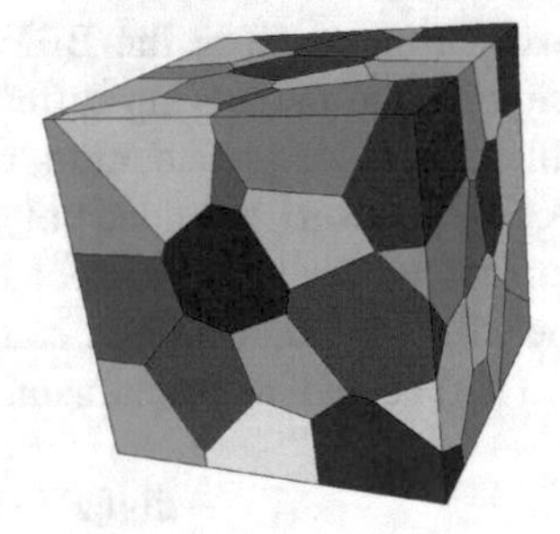

Fig. 2.1 Two examples for meshes with polygonal and polyhedral elements

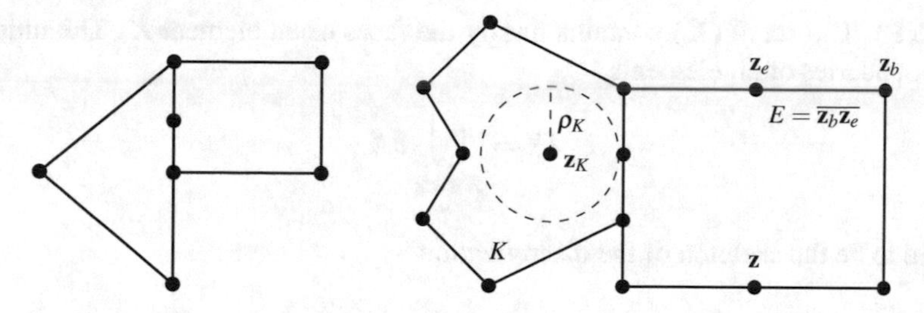

Fig. 2.2 Two examples of neighbouring elements with additional nodes on the straight boundary

are non-overlapping open sets such that

$$\overline{\Omega} = \bigcup_{K \in \mathscr{K}_h} \overline{K}\,.$$

The boundaries of the elements consist of nodes and edges in 2D as well as of faces in 3D. An edge $E = \overline{\mathbf{z}_b\mathbf{z}_e}$ is always located between two nodes, the one at the beginning $\mathbf{z}_b$ and the one at the end $\mathbf{z}_e$. These points are fixed once per edge and they are the only nodes on E. In each corner of an element K, a node is located, but in 2D there could also be some nodes on straight lines of the polygonal boundary ∂K, cf. Fig. 2.2. We stress this fact more carefully. If we have a triangle with three nodes and we add some nodes on the boundary, this triangle turns formally into a polygon. These additional nodes enrich the approximation space in the finite element method. In this context, nodes on straight lines are natural since they are just ordinary nodes for polygons. In triangular or quadrilateral meshes these nodes appear as hanging nodes which are undesirable and do not influence the accuracy of the approximation. In classical finite element implementations, such hanging nodes have to be treated in a special way as conditional nodes or by removing them. Methods working on polygonal meshes include such nodes naturally. In 3D, hanging nodes appear naturally on edges of the polyhedral elements and one may have hanging edges on the polygonal faces. The polygonal faces are assumed to be flat and they are surrounded by edges which are coplanar.

For the later analysis, we need some notation. $\mathscr{N}_h$ denotes the set of all nodes in the mesh $\mathscr{K}_h$. It is $\mathscr{N}_h = \mathscr{N}_{h,\Omega} \cup \mathscr{N}_{h,D} \cup \mathscr{N}_{h,N}$, where $\mathscr{N}_{h,\Omega}$, $\mathscr{N}_{h,D}$, $\mathscr{N}_{h,N}$ contain the nodes in the interior of Ω, on the Dirichlet boundary Γ_D and on the interior of the Neumann boundary Γ_N, respectively. The transition points between Γ_D and Γ_N belong to $\mathscr{N}_{h,D}$. We denote the set of all edges of the mesh with $\mathscr{E}_h$. In analogy to the set of nodes, we decompose $\mathscr{E}_h = \mathscr{E}_{h,\Omega} \cup \mathscr{E}_{h,D} \cup \mathscr{E}_{h,N}$, where $\mathscr{E}_{h,\Omega}$, $\mathscr{E}_{h,D}$ and $\mathscr{E}_{h,N}$ contain all edges in the interior of Ω, on the Dirichlet boundary Γ_D and on the Neumann boundary Γ_N, respectively. In 3D, we additionally have the set of all faces $\mathscr{F}_h = \mathscr{F}_{h,\Omega} \cup \mathscr{F}_{h,D} \cup \mathscr{F}_{h,N}$ with subsets analogous as before. Moreover, the sets $\mathscr{N}(K)$, $\mathscr{N}(E)$ and $\mathscr{N}(F)$ contain all nodes which belong to the element $K \in \mathscr{K}_h$, the edge $E \in \mathscr{E}_h$ and the face $F \in \mathscr{F}_h$, respectively. We denote the set of edges which belong to the element K by $\mathscr{E}(K)$ and those which belong to a face F

by $\mathscr{E}(F)$. The set $\mathscr{F}(K)$ contains finally the faces of an element K. The union of the boundaries of all elements

$$\Gamma_S = \bigcup_{K \in \mathscr{K}_h} \partial K$$

is said to be the skeleton of the discretization.

2.2.1 Mesh Regularity and Properties in 2D

The length of an edge E and the diameter of an element K are denoted by h_E and $h_K = \sup\{|x - y| : x, y \in \partial K\}$, respectively.

Definition 2.1 (Regular Mesh in 2D) The family of meshes $\mathscr{K}_h$ is called regular if it satisfies:

1. Each element $K \in \mathscr{K}_h$ is a star-shaped polygon with respect to a circle of radius ρ_K and midpoint $\mathbf{z}_K$.
2. The aspect ratio is uniformly bounded from above by $\sigma_{\mathscr{K}}$, i.e. $h_K / \rho_K < \sigma_{\mathscr{K}}$ for all $K \in \mathscr{K}_h$.

The circle in the definition is chosen in such a way that its radius is maximal, cf. Fig. 2.2. If the position of the circle is not unique, its midpoint $\mathbf{z}_K$ is fixed once per element. Additionally, we assume that $h_K < 1$ for all elements $K \in \mathscr{K}_h$. This condition is no grievous restriction on the mesh since $h_K < 1$ can always be satisfied by scaling Ω. Nevertheless, it is necessary in the forthcoming local boundary integral formulations in 2D.

For the analysis of local boundary element methods used in the BEM-based FEM and some proofs in Chap. 5, the regularity of a mesh is not enough. Another important property is that the diameter of an element is comparable to the length of its shortest edge. This is ensured by the following definition.

Definition 2.2 (Stable Mesh in 2D) The family of meshes $\mathscr{K}_h$ is called stable if there is a constant $c_{\mathscr{K}} > 0$ such that for all elements $K \in \mathscr{K}_h$ and all its edges $E \in \mathscr{E}(K)$ it holds

$$h_K \le c_{\mathscr{K}} h_E.$$

When we consider convergence or error estimates with respect to the mesh size $h = \max\{h_K : K \in \mathscr{K}_h\}$, it is important that the constants in the definitions above hold uniformly for the whole family of meshes. For convenience we only write mesh and mean a whole family for $h \to 0$.

In the following, we give some useful properties of regular meshes. An important analytical tool is an auxiliary triangulation $\mathscr{T}_h(K)$ of the elements $K \in \mathscr{K}_h$. This triangulation is constructed by connecting the nodes on the boundary of K with

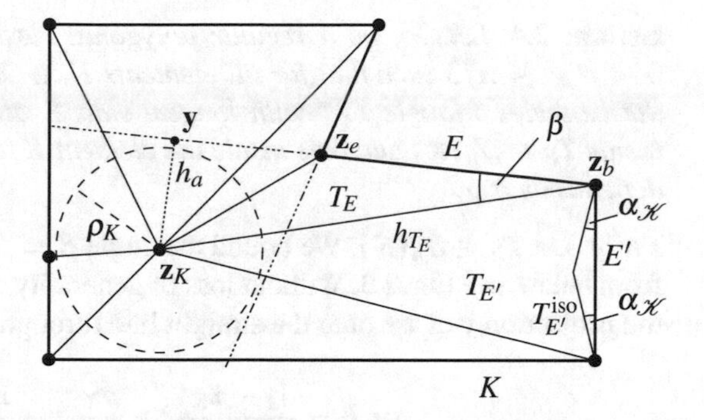

Fig. 2.3 Auxiliary triangulation $\mathcal{T}_h(K)$ of star-shaped element K, altitude $h_a = \text{alt}(T_E, E)$ of triangle $T_E \in \mathcal{T}_h(K)$ perpendicular to E and angle $\beta = \angle \mathbf{z}_K \mathbf{z}_b \mathbf{z}_e$ as well as triangle $T_{E'} \in \mathcal{T}_h(K)$ with isosceles triangle $T_{E'}^{\text{iso}}$

the point $\mathbf{z}_K$ of Definition 2.1, see Fig. 2.3. Consequently, $\mathcal{T}_h(K)$ consists of the triangles T_E for $E = \overline{\mathbf{z}_b\mathbf{z}_e} \in \mathcal{E}(K)$, which are defined by the points $\mathbf{z}_b$, $\mathbf{z}_e$ and $\mathbf{z}_K$.

Lemma 2.3 *Let K be a polygonal element of a regular and stable mesh $\mathcal{K}_h$. The auxiliary triangulation $\mathcal{T}_h(K)$ is shape-regular in the sense of Ciarlet [58], i.e., neighbouring triangles share either a common node or edge and the aspect ratio of each triangle is uniformly bounded by some constant $\sigma_{\mathcal{T}}$, which only depends on $\sigma_{\mathcal{K}}$ and $c_{\mathcal{K}}$.*

Proof Let $T_E \in \mathcal{T}_h(K)$ be a triangle with diameter h_{T_E} and let ρ_{T_E} be the radius of the incircle. It is known that the area of T_E is given by $|T_E| = \frac{1}{2}|\partial T_E|\rho_{T_E}$, where $|\partial T_E|$ is the perimeter of T_E. Obviously, it is $|\partial T_E| \le 3h_{T_E}$. On the other hand, we have the formula $|T_E| = \frac{1}{2}h_E h_a$, where $h_a = \text{alt}(T_E, E)$ is the altitude of the triangle perpendicular to E, see Fig. 2.3. Since the element K is star-shaped with respect to a circle of radius ρ_K, the line through the side $E \in \mathcal{E}_h$ of the triangle does not intersect this circle. Thus, $h_a \ge \rho_K$ and we have the estimate $|T_E| \ge \frac{1}{2}h_E\rho_K$. Together with Definition 2.1, we obtain

$$\frac{h_{T_E}}{\rho_{T_E}} = \frac{|\partial T_E|h_{T_E}}{2|T_E|} \le \frac{3h_{T_E}^2}{h_E\rho_K} \le 3c_{\mathcal{K}}\sigma_{\mathcal{K}}\frac{h_{T_E}^2}{h_K^2} \le 3c_{\mathcal{K}}\sigma_{\mathcal{K}} = \sigma_{\mathcal{T}} .$$

□

In the previous proof, we discovered and applied the estimate

$$|T_E| \ge \tfrac{1}{2}h_E\rho_K \tag{2.2}$$

for the area of the auxiliary triangle. This inequality will be of importance once more. We may also consider the auxiliary triangulation $\mathcal{T}_h(\mathcal{K}_h)$ of the whole domain Ω which is constructed by gluing the local triangulations $\mathcal{T}_h(K)$. Obviously, $\mathcal{T}_h(\mathcal{K}_h)$ is also shape-regular in the sense of Ciarlet. Furthermore, the angles in the auxiliary triangulation $\mathcal{T}_h(K)$ next to ∂K can be bounded from below. This gives rise to the following result.

Lemma 2.4 *Let $\mathcal{K}_h$ be a regular polygonal mesh. There is an angle $\alpha_{\mathcal{K}}$ with $0 < \alpha_{\mathcal{K}} \leq \pi/3$ such that for all elements $K \in \mathcal{K}_h$ and all its edges $E \in \mathcal{E}(K)$ the isosceles triangle T_E^{iso} with longest side E and two interior angles $\alpha_{\mathcal{K}}$ lies inside $T_E \in \mathcal{T}_h(K)$ and thus inside the element K, see Fig. 2.3. The angle $\alpha_{\mathcal{K}}$ only depends on $\sigma_{\mathcal{K}}$.*

Proof Let $T_E \in \mathcal{T}_h(K)$. We bound the angle $\beta = \angle \mathbf{z}_K \mathbf{z}_b \mathbf{z}_e$ in T_E next to $E = \overline{\mathbf{z}_b \mathbf{z}_e}$ from below, see Fig. 2.3. Without loss of generality, we assume that $\beta < \pi/2$. Using the projection $\mathbf{y}$ of $\mathbf{z}_K$ onto the straight line through the edge E, we recognize

$$\sin\beta = \frac{|\mathbf{y} - \mathbf{z}_K|}{|\mathbf{z}_b - \mathbf{z}_K|} \geq \frac{\rho_K}{h_K} \geq \frac{1}{\sigma_{\mathcal{K}}} \in (0, 1)\,. \tag{2.3}$$

Consequently, it is $\beta \geq \arcsin \sigma_{\mathcal{K}}^{-1}$. Since this estimate is valid for all angles next to ∂K of the auxiliary triangulation, the isosceles triangles T_E^{iso}, $E \in \mathcal{E}(K)$ with common angle $\alpha_{\mathcal{K}} = \min\{\pi/3, \arcsin \sigma_{\mathcal{K}}^{-1}\}$ lie inside the auxiliary triangles T_E and therefore inside K. □

Remark 2.5 The upper bound of $\alpha_{\mathcal{K}}$ is chosen in such a way that the longest side of the isosceles triangle T_E^{iso} is always the edge E. This fact is not important in the previous lemma, but it simplifies forthcoming proofs.

Corollary 2.6 *Let $\mathcal{K}_h$ be a regular mesh. Every node belongs to a uniformly bounded number of elements, i.e. $|\{K \in \mathcal{K}_h : \mathbf{z} \in \mathcal{N}(K)\}| \leq c$, $\forall \mathbf{z} \in \mathcal{N}_h$. The constant $c > 0$ only depends on $\sigma_{\mathcal{K}}$.*

Proof Due to the regularity of $\mathcal{K}_h$, every interior angle of an element is bounded from below by $\alpha_{\mathcal{K}}$ as we have seen in Lemma 2.4. This angle only depends on $\sigma_{\mathcal{K}}$. Therefore, we have

$$|\{K \in \mathcal{K}_h : \mathbf{z} \in \mathcal{N}(K)\}| \leq \left\lfloor \frac{2\pi}{\alpha_{\mathcal{K}}} \right\rfloor,$$

where the term on the right hand side denotes the biggest integer smaller than or equal to $2\pi/\alpha_{\mathcal{K}}$. □

Conversely, we have a more restrictive result, which additionally assumes the stability of the mesh. Without the stability, the lengths of the edges might degenerate and thus a regular polygonal element can have arbitrary many nodes on its boundary.

Lemma 2.7 *Let $\mathcal{K}_h$ be regular and stable. Every element contains a uniformly bounded number of nodes and edges, i.e. $|\mathcal{N}(K)| = |\mathcal{E}(K)| \leq c$, $\forall K \in \mathcal{K}_h$. The constant $c > 0$ only depends on $\sigma_{\mathcal{K}}$ and $c_{\mathcal{K}}$.*

Proof We exploit the regularity of the mesh. Let $K \in \mathcal{K}_h$. In 2D it is obviously $|\mathcal{N}(K)| = |\mathcal{E}(K)|$. With the help of (2.2), we obtain

$$\begin{aligned} h_K^2 |\mathcal{N}(K)| &\le \sigma_{\mathcal{K}} \rho_{\mathcal{K}}\, h_K |\mathcal{E}(K)| \\ &\le \sigma_{\mathcal{K}} \rho_{\mathcal{K}} \sum_{E \in \mathcal{E}(K)} c_{\mathcal{K}} h_E \\ &\le \sigma_{\mathcal{K}} c_{\mathcal{K}} \sum_{E \in \mathcal{E}(K)} 2|T_E| \\ &= 2\sigma_{\mathcal{K}} c_{\mathcal{K}}\, |K| \\ &\le 2\sigma_{\mathcal{K}} c_{\mathcal{K}}\, h_K^2 \,. \end{aligned}$$

Consequently, we have $|\mathcal{N}(K)| \le 2\sigma_{\mathcal{K}} c_{\mathcal{K}}$. □

The isosceles triangles and the auxiliary triangulation play an important role in the analysis of error estimates later on. They are used in order to handle polygonal elements and, in particular, to apply some results on interpolation of functions over triangulations. Such results are applicable, if the polygonal mesh is regular and stable, since then, the auxiliary triangulation is regular in the sense of Ciarlet according to Lemma 2.3. However, in certain situations, we can weaken the assumptions on the polygonal mesh and remove the stability. In [84], the following result is proven with similar considerations as in the proof of Lemma 2.4 for convex elements. However, the result is also valid in our more general case.

Lemma 2.8 *For a regular mesh* $\mathcal{K}_h$*, all angles of all triangles in the auxiliary triangulation* $\mathcal{T}_h(\mathcal{K}_h)$ *are less than* $\pi - \arcsin \sigma_{\mathcal{K}}^{-1}$ *and, in particular, they are strictly less than* π.

Proof We proceed similar as in the proof of Lemma 2.4. Therefore, let $K \in \mathcal{K}_h$ be an element with edge $E = \overline{\mathbf{z}_b\mathbf{z}_e}$ and we consider the triangle $T_E \in \mathcal{T}_h(K)$. It is sufficient to bound the angle $\angle \mathbf{z}_b\mathbf{z}_K\mathbf{z}_e$ and the larger angle of the others adjacent to E, lets say $\angle \mathbf{z}_b\mathbf{z}_e\mathbf{z}_K$. It is easily seen form (2.3) that

$$\angle \mathbf{z}_b\mathbf{z}_K\mathbf{z}_e \le \pi - 2 \arcsin \sigma_{\mathcal{K}}^{-1} \,.$$

In order to bound $\angle \mathbf{z}_b\mathbf{z}_e\mathbf{z}_K$ we employ the point $\mathbf{y}$ once more which is the projection of $\mathbf{z}_K$ onto the line through E, see Fig. 2.3. Without loss of generality we assume $\angle \mathbf{z}_b\mathbf{z}_e\mathbf{z}_K > \pi/2$ and thus $\mathbf{y} \notin \overline{\mathbf{z}_b\mathbf{z}_e}$. It is

$$\sin(\pi - \angle \mathbf{z}_b\mathbf{z}_e\mathbf{z}_K) = \sin(\angle \mathbf{z}_K\mathbf{z}_e\mathbf{y}) = \frac{|\mathbf{y} - \mathbf{z}_K|}{|\mathbf{z}_e - \mathbf{z}_K|} \ge \frac{\rho_K}{h_K} \ge \frac{1}{\sigma_{\mathcal{K}}} \,.$$

Applying arcsin yields $\angle \mathbf{z}_b\mathbf{z}_e\mathbf{z}_K \le \pi - \arcsin \sigma_{\mathcal{K}}^{-1}$ and the result follows because of $\arcsin \sigma_{\mathcal{K}}^{-1} > 0$ due to $\sigma_{\mathcal{K}} > 0$. □

An important consequence of this proposition is the following corollary.

Corollary 2.9 *Let $K \in \mathcal{K}_h$ be an element of a regular polygonal mesh $\mathcal{K}_h$. The auxiliary triangulations $\mathcal{T}_h(K)$ and $\mathcal{T}_h(\mathcal{K}_h)$ satisfy a maximum angle condition, i.e., every angle in the triangles of the mesh is uniformly bounded from above by a constant which is strictly less than π. The maximum angle only depends on $\sigma_{\mathcal{K}}$.*

Therefore, several approximation properties of finite element interpolation for linear as well as for higher order basis functions are valid on this auxiliary discretization, see [14, 114]. The constants appearing in those estimates depend on the maximum angle and thus, on the aspect ratio $\sigma_{\mathcal{K}}$ of the mesh $\mathcal{K}_h$, but not on the stability parameter $c_{\mathcal{K}}$.

2.2.2 Mesh Regularity and Properties in 3D

In addition to the diameter h_K of an element $K \in \mathcal{K}_h$ and the edge length h_E of $E \in \mathcal{E}_h$, we use the diameter h_F of polygonal faces $F \in \mathcal{F}_h$ in the following.

Definition 2.10 (Regular Faces) A set of faces $\mathcal{F}_h$ is called regular if all faces are flat polygons which are regular in the sense of Definition 2.1 with regularity parameter $\sigma_{\mathcal{F}}$. The radius of the inscribed circle of $F \in \mathcal{F}_h$ is denoted by ρ_F and its center by $\mathbf{z}_F$.

Definition 2.11 (Regular Mesh in 3D) The family of meshes $\mathcal{K}_h$ is called regular if it satisfies:

1. The associated set of faces $\mathcal{F}_h$ is regular.
2. Each element $K \in \mathcal{K}_h$ is a star-shaped polyhedron with respect to a ball of radius ρ_K and midpoint $\mathbf{z}_K$.
3. The aspect ratio is uniformly bounded from above by $\sigma_{\mathcal{K}}$, i.e. $h_K/\rho_K < \sigma_{\mathcal{K}}$ for all $K \in \mathcal{K}_h$.

The ball in the definition is chosen in such a way that its radius is maximal and, if its position is not unique, the midpoint $\mathbf{z}_K$ is fixed once per element. In contrast to the two-dimensional case, we do not impose the restriction on the diameter of the elements.

Definition 2.12 (Stable Mesh in 3D) The family of meshes $\mathcal{K}_h$ is called stable if there is a constant $c_{\mathcal{K}} > 0$ such that for all elements $K \in \mathcal{K}_h$ and all its edges $E \in \mathcal{E}(K)$ it holds

$$h_K \le c_{\mathcal{K}} h_E.$$

When we consider convergence or error estimates with respect to the mesh size $h = \max\{h_K : K \in \mathcal{K}_h\}$, it is important that the constants in the definitions above hold uniformly for the whole family of meshes. As in the two-dimensional case, we only write mesh in the following and mean a whole family for $h \to 0$. The stability ensures that for an element the lengths of its edges, the diameters of its faces and the diameter of itself are comparable. It yields

$$h_E \leq h_F \leq h_K \leq c_{\mathscr{K}} h_E \leq c_{\mathscr{K}} h_F$$

for $K \in \mathscr{K}_h$ and all $F \in \mathscr{F}(K)$ and $E \in \mathscr{E}(F)$.

Remark 2.13 For a regular and stable mesh $\mathscr{K}_h$, it holds

$$h_K^{d-1} \leq c\,|F|\,, \tag{2.4}$$

for $K \in \mathscr{K}_h$ with $F \in \mathscr{F}(K)$. This is a direct generalization of the stability condition in two-dimensions, cf. Definition 2.2. Thus, (2.4) is valid for $d = 2, 3$ if F is interpreted as edge and face, respectively. This inequality follows by

$$|F| \geq \pi\rho_F^2 \geq \pi\frac{h_F^2}{\sigma_{\mathscr{F}}^2} \geq \pi\frac{h_K^2}{c_{\mathscr{K}}^2\sigma_{\mathscr{F}}^2}\,.$$

In the derivation of interpolation and error estimates, an auxiliary discretization into tetrahedra will be the counterpart to the constructed triangulation in 2D. We employ the introduced auxiliary triangulation from Sect. 2.2.1 in order to discretize the polygonal faces and denote it by $\mathscr{T}_0(F)$ for $F \in \mathscr{F}_h$. Note, that we have chosen an index 0 instead of h. We introduce a family $\mathscr{T}_l(F)$ of triangulations, where the meshes of level $l \geq 1$ are defined recursively by splitting each triangle of the previous level into four similar triangles, see Fig. 2.4. The set of nodes in the triangular mesh is denoted by $\mathscr{M}_l(F)$. Obviously, the discretizations of the faces can be combined to a whole conforming surface mesh of an element $K \in \mathscr{K}_h$ by setting

$$\mathscr{T}_l(\partial K) = \bigcup_{F\in\mathscr{F}(K)} \mathscr{T}_l(F) \quad \text{and} \quad \mathscr{M}_l(\partial K) = \bigcup_{F\in\mathscr{F}(K)} \mathscr{M}_h(F)\,.$$

Finally, the auxiliary tetrahedral mesh $\mathscr{T}_l(K)$ of the polyhedral element $K \in \mathscr{K}_h$ is constructed by connecting the nodes of the triangular surface mesh $\mathscr{T}_l(\partial K)$ with

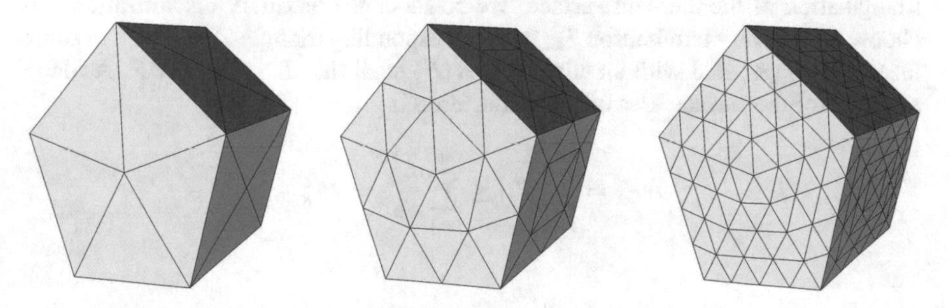

Fig. 2.4 Polyhedral element with surface triangulations of level $l = 0, 1, 2$

the point $\mathbf{z}_K$. The tetrahedra constructed this way are denoted by T_{tet}. Although, this auxiliary discretization may contain needle-like tetrahedra, their regularity can be controlled by the mesh regularity and stability of the polyhedral mesh. Combining the tetrahedral discretizations of the polyhedral elements forms an auxiliary discretization $\mathscr{T}_l(\mathscr{K}_h)$ for the whole domain Ω. If the mesh level l is not important in later proofs and $l = 0$ is sufficient, we also write $\mathscr{T}_h(\cdot)$ for $\mathscr{T}_0(\cdot)$ in order to treat the two- and three-dimensional cases simultaneously.

Lemma 2.14 *Let K be a polyhedral element of a regular and stable mesh $\mathscr{K}_h$. The auxiliary discretizations $\mathscr{T}_l(K)$ and $\mathscr{T}_l(\mathscr{K}_h)$, $l \geq 0$ are shape-regular in the sense of Ciarlet [58], i.e., neighbouring tetrahedra share either a common node, edge or triangular face and the aspect ratio of each tetrahedra is uniformly bounded by some constant $\sigma_{\mathscr{T}}$, which only depends on $\sigma_{\mathscr{K}}$, $\sigma_{\mathscr{F}}$, $c_{\mathscr{K}}$ and the mesh level l in the face discretization.*

Proof The conformity of the auxiliary mesh is rather obvious. Thus, we only have to bound the aspect ratio of the tetrahedra $T_{\text{tet}} \in \mathscr{T}_l(K)$, i.e., the ratio of the diameter $h_{T_{\text{tet}}}$ and the radius $\rho_{T_{\text{tet}}}$ of their insphere. For an arbitrary tetrahedron, we have the relation

$$\rho_{T_{\text{tet}}} = \frac{3V_{T_{\text{tet}}}}{A_{T_{\text{tet}}}} ,$$

where $V_{T_{\text{tet}}}$ is the volume and $A_{T_{\text{tet}}}$ is the surface area of the tetrahedron. This relation is seen as follows. $V_{T_{\text{tet}}}$ is equal to the sum of the volumes $V_{T_{\text{tet},i}}$, $i = 1, \ldots, 4$, of the four tetrahedra $T_{\text{tet},i}$ obtained by connecting the vertexes with the center of the insphere. Each volume is computed as $V_{T_{\text{tet},i}} = \frac{1}{3}\rho_{T_{\text{tet}}}|T_i|$, where T_i is the triangle on the surface of the initial tetrahedron T_{tet} and $\rho_{T_{\text{tet}}}$ corresponds to the hight of $T_{\text{tet},i}$ over T_i. Consequently, it holds

$$V_{T_{\text{tet}}} = \sum_{i=1}^{4} V_{T_{\text{tet},i}} = \sum_{i=1}^{4} \frac{1}{3}\rho_{T_{\text{tet}}}|T_i| = \frac{1}{3}\rho_{T_{\text{tet}}} A_{T_{\text{tet}}} .$$

First, we study the case $l = 0$, where only one node per face is added for the triangulation of the element surface. We consider the auxiliary discretization and choose an arbitrary tetrahedron T_{tet} with corresponding triangle $T \in \mathscr{T}_l(F)$ in some face $F \in \mathscr{F}(K)$ and with an edge $E \in \mathscr{E}(F)$ such that $E \subset \partial T \cap \partial F$. A rough estimate for the surface area of this tetrahedron is

$$A_{T_{\text{tet}}} = \sum_{i=1}^{4} |T_i| \leq \sum_{i=1}^{4} \frac{h_K^2}{2} = 2h_K^2 .$$

Let $\operatorname{alt}(T_{\text{tet}}, T)$ be the altitude of the tetrahedron T_{tet} over the side T and let $\operatorname{alt}(T, E)$ be the altitude of the triangle T over the edge E. For the volume of T_{tet}, we have

$$V_{T_{\text{tet}}} = \frac{1}{3} \operatorname{alt}(T_{\text{tet}}, T)|T| = \frac{1}{6} \operatorname{alt}(T_{\text{tet}}, T) \operatorname{alt}(T, E)\, h_E \,.$$

Since the faces of the element K and the element itself are star-shaped with respect to circles and a ball according to Definitions 2.10 and 2.11, it holds $\rho_F \leq \operatorname{alt}(T, E)$ as well as $\rho_K \leq \operatorname{alt}(T_{\text{tet}}, T)$ due to the construction of T_{tet} and T. Consequently, we obtain

$$V_{T_{\text{tet}}} \geq \frac{1}{6} \rho_K \rho_F h_E \geq \frac{1}{6\sigma_{\mathcal{K}}\sigma_{\mathcal{F}}} h_K h_F h_E \geq \frac{1}{6\sigma_{\mathcal{K}}\sigma_{\mathcal{F}}} h_E^3 \,.$$

This yields together with the stability, see Definition 2.12,

$$\frac{h_{T_{\text{tet}}}}{\rho_{T_{\text{tet}}}} = \frac{h_{T_{\text{tet}}} A_{T_{\text{tet}}}}{3 V_{T_{\text{tet}}}} \leq \frac{4\sigma_{\mathcal{K}}\sigma_{\mathcal{F}} h_K^3}{h_E^3} \leq 4\sigma_{\mathcal{K}}\sigma_{\mathcal{F}} c_{\mathcal{K}}^3 \,.$$

In the case $l \geq 1$, the volume $V_{T_{\text{tet}}}$ gets smaller. The triangle $T \subset F \in \mathcal{F}(K)$ is obtained by successive splitting of an initial triangle T_0 of the mesh with level zero. Due to the construction, these triangles are similar and the relation $|T| = |T_0|/4^l$ holds. Taking into account this relation in the considerations above gives the general estimate

$$\frac{h_{T_{\text{tet}}}}{\rho_{T_{\text{tet}}}} \leq \sigma_{\text{tet}} \quad \text{with} \quad \sigma_{\text{tet}} = 4^{l+1} \sigma_{\mathcal{K}}\sigma_{\mathcal{F}} c_{\mathcal{K}}^3 \,.$$

□

Similar to the two-dimensional case we obtain the following two results on the object counts. In the corollary for three space dimensions, however, we additionally assume the stability of the mesh in contrast to the lower dimensional setting.

Corollary 2.15 *Let $\mathcal{K}_h$ be a regular and stable mesh. Every node belongs to a uniformly bounded number of elements, i.e. $|\{K \in \mathcal{K}_h : \mathbf{z} \in \mathcal{N}(K)\}| \leq c$, $\forall \mathbf{z} \in \mathcal{N}_h$. The constant $c > 0$ only depends on $\sigma_{\mathcal{K}}$, $\sigma_{\mathcal{F}}$ and $c_{\mathcal{K}}$.*

Proof According to the previous Lemma 2.14, the auxiliary discretization $\mathcal{T}_0(\mathcal{K}_h)$ is shape-regular in the sense of Ciarlet. Therefore, each node in the mesh $\mathcal{K}_h$ belongs to a uniformly bounded number of auxiliary tetrahedra, and consequently to a probably smaller uniformly bounded number of polyhedral elements. □

Lemma 2.16 *Let $\mathcal{K}_h$ be a regular and stable mesh. Every element contains a uniformly bounded number of nodes, edges and faces, i.e.*

$$|\mathcal{N}(K)| \leq c \,, \quad |\mathcal{E}(K)| \leq c \,, \quad |\mathcal{F}(K)| \leq c \,, \quad \forall K \in \mathcal{K}_h \,.$$

The constants $c > 0$ only depend on $\sigma_{\mathcal{K}}$, $\sigma_{\mathcal{F}}$ and $c_{\mathcal{K}}$.

Proof For the surface area of a polytopal element $K \in \mathcal{K}_h$, we have due to the regularity and stability of the mesh

$$|\partial K| = \sum_{F\in\mathcal{F}(K)} |F| \geq |\mathcal{F}(K)|\,\pi\,\frac{h_K^2}{c_{\mathcal{K}}^2\sigma_{\mathcal{F}}^2}\,,$$

see Remark 2.13. Using the auxiliary discretization into tetrahedra T_{tet} with corresponding triangles $T \in \mathcal{T}_0(\partial K)$, for which $|T| \leq 3V_{\text{tet}}/\rho_K$ since $V_{\text{tet}} = \frac{1}{3}\,\text{alt}(T_{\text{tet}}, T)|T|$, we obtain on the other hand

$$|\partial K| = \sum_{T\in\mathcal{T}_0(\partial K)} |T| \leq \sum_{T\in\mathcal{T}_0(\partial K)} \frac{3V_{\text{tet}}}{\rho_K} = \frac{3|K|}{\rho_K} \leq 3\sigma_{\mathcal{K}}h_K^2\,,$$

due to $|K| \leq h_K^3$ and the regularity. Thus, the number of faces is uniformly bounded, namely $|\mathcal{F}(K)| \leq 3\sigma_{\mathcal{K}}c_{\mathcal{K}}^2\sigma_{\mathcal{F}}^2/\pi$. According to Lemma 2.7, each of these faces has a uniformly bounded number of nodes and edges. Consequently, the number of nodes and edges of the element K is also uniformly bounded. □

If only the regularity of a polyhedral mesh is assumed, the auxiliary discretization of tetrahedra is not necessarily regular. The edges might degenerate without the stability and thus, the condition on the aspect ratio for the tetrahedra does not hold anymore. But, the stability can be weakend such that the tetrahedral mesh still satisfies a maximum angle condition.

Definition 2.17 (Weakly Stable Mesh in 3D) The family of meshes $\mathcal{K}_h$ is called weakly stable if there is a constant $c_{\mathcal{F}} > 0$ such that for all polygonal faces $F \in \mathcal{F}_h$ in the mesh and all its edges $E \in \mathcal{E}(F)$ it holds

$$h_F \leq c_{\mathcal{F}}h_E\,.$$

In contrast to stable meshes, the edges of elements in weakly stable meshes might degenerate with respect to the element diameter. But, due to the weak stability, small edges involve that adjacent faces are also small in their size. Thus, if an edge degenerates to a point, the adjacent faces will degenerate to this point, too. A consequence of this definition is, that the polygonal faces in a regular and weakly stable mesh are regular and stable in the two-dimensional sense.

Lemma 2.18 *Let $K \in \mathcal{K}_h$ be an element of a regular and weakly stable polyhedral mesh $\mathcal{K}_h$. The auxiliary discretization of tetrahedra $\mathcal{T}_l(K)$ and $\mathcal{T}_l(\mathcal{K}_h)$ satisfy a maximum angle condition, i.e., all dihedral angles between faces and all angles within a triangular face are uniformly bounded from above by a constant which is strictly less than π. The maximum angle only depends on $\sigma_{\mathcal{K}}$, $\sigma_{\mathcal{F}}$, $c_{\mathcal{F}}$ and the mesh level l in the face discretization.*

Proof Similar as in the proof of Lemma 2.14, we only consider $l = 0$. The general case $l > 0$ follows due to the fact that the triangles in the face triangulation $\mathcal{T}_l(F)$

are similar to those in $\mathcal{T}_0(F)$. Thus, the arguments turn over and the dependence on l enters the constants. In order to prove the maximum angle condition for the tetrahedral mesh, we distinguish several cases. First we show that the angles in the surface triangles of the tetrahedra are bounded uniformly by a constant strictly less than π. Afterwards, we bound the dihedral angles.

Let $T_{\text{tet}} \in \mathcal{T}_0(K)$ be a tetrahedra and T one of its triangular faces. If $T \in \mathcal{T}_0(F)$ for a face $F \in \mathcal{F}(K)$, then all angles of T are bounded uniformly from above by a constant strictly less than π depending only on $\sigma_{\mathcal{F}}$ according to Lemma 2.8, since F is a regular polygon. On the other hand, if $T \subset K$, we consider the intersection of the polyhedral element K with the plane in which T lies. The intersection is obviously a polygon and we denote it by P. Since K is star-shaped with respect to a ball of radius ρ_K and center $\mathbf{z}_K$, we easily see that P is star-shaped with respect to the enclosed circle of radius ρ_K and center $\mathbf{z}_K$. Thus, P is a regular polygon because of

$$\frac{h_P}{\rho_P} \leq \frac{h_K}{\rho_K} \leq \sigma_{\mathcal{K}} .$$

Consequently, T is part of an auxiliary triangulation of a regular polygonal element and thus its angles are bounded from above according to Lemma 2.8 by a constant depending only on $\sigma_{\mathcal{K}}$.

Next, we consider the dihedral angles of T_{tet}. Let T_i, $i = 1, 2, 3, 4$ be the triangular faces of T_{tet} and $E_{ij} = \overline{T_i} \cap \overline{T_j}$ be the edge shared by the triangles T_i and T_j for $i \neq j$. Furthermore, let the triangles be numbered such that $T_1 \in \mathcal{T}_0(F)$ for some face $F \in \mathcal{F}(K)$ and $T_i \subset K$ for $i = 2, 3, 4$. We distinguish again two cases. First, consider the dihedral angle between $T_1 \subset \partial K$ and $T_i \subset K, i = 2, 3, 4$. We denote the dihedral angle by δ. It is given in the plane orthogonal to E_{1i} as the angle between the two planes spanned by T_1 and T_i. Without loss of generality let $\mathbf{z}_K$ lie on the plane orthogonal to E_{1i} and denote by $\mathbf{y}$ the intersection of all three planes, see Fig. 2.5. For $\delta \geq \pi/2$, it is

$$\delta = \frac{\pi}{2} + \arccos\left(\frac{h_a}{|\mathbf{y} - \mathbf{z}_K|}\right),$$

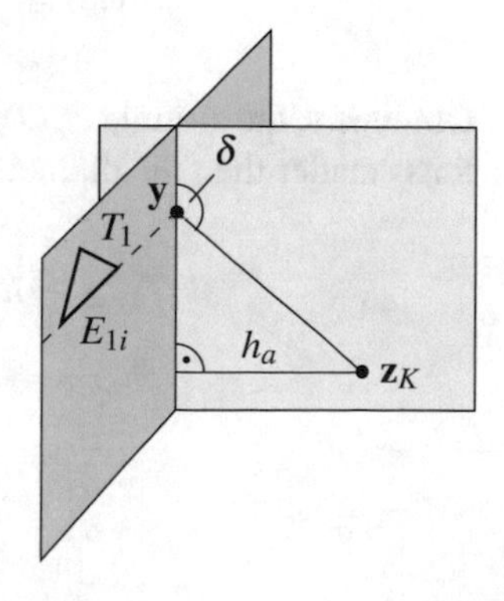

Fig. 2.5 The altitude $h_a = \text{alt}(T_{\text{tet}}, T_1)$, intersection point $\mathbf{y}$ and demonstration of dihedral angle δ in plane orthogonal to the edge E_{ij} through $\mathbf{z}_K$ between the triangles T_1 and T_i as described in the proof of Lemma 2.18

where $h_a = \operatorname{alt}(T_{\text{tet}}, T_1)$ is the altitude of the tetrahedron T_{tet} with respect to the side T_1, which corresponds to the distance of $\mathbf{z}_K$ to the plane through T_1. Consequently, it is $h_a \geq \rho_K$ due to the regularity of the polyhedral element K. Furthermore, K is enclosed by a sphere of radius h_K and center $\mathbf{z}_K$. Since $\mathbf{y}$ is the orthogonal projection of $\mathbf{z}_K$ onto the line through E_{1i}, its distance to $\mathbf{z}_K$ is smaller than the distance of z_K to the edge, thus $|\mathbf{y} - \mathbf{z}_K| \leq h_K$. This yields

$$\frac{h_a}{|\mathbf{y} - \mathbf{z}_K|} \geq \frac{\rho_K}{h_K} \geq \frac{1}{\sigma_{\mathcal{K}}} ,$$

because of the regularity. Since arc cosine is monotonically decreasing, we obtain

$$\delta \leq \frac{\pi}{2} + \arccos\left(\frac{1}{\sigma_{\mathcal{K}}}\right) < \pi .$$

It remains to bound the dihedral angle between triangular faces of the tetrahedra with $T_i, T_j \subset K$. We denote the angle again by δ. According to Proposition 3.1 in [122], the volume $V_{T_{\text{tet}}}$ of T_{tet} satisfy the relation

$$V_{T_{\text{tet}}} = \frac{2}{3h_{E_{ij}}} |T_i|\,|T_j| \sin\delta .$$

On the other hand it is

$$V_{T_{\text{tet}}} = \frac{1}{3} \operatorname{alt}(T_{\text{tet}}, T_1)\,|T_1| .$$

If we assume $\pi/2 \leq \delta \leq \pi$, this yields

$$\delta = \frac{\pi}{2} + \arccos\left(\frac{h_{E_{ij}} \operatorname{alt}(T_{\text{tet}}, T_1)\,|T_1|}{2|T_i|\,|T_j|}\right) . \tag{2.5}$$

The areas of the triangles are given by

$$|T_\ell| = \frac{1}{2} h_{E_{ij}} \operatorname{alt}(T_\ell, E_{ij}) \quad \text{for } \ell = i, j .$$

Obviously, the altitude $\operatorname{alt}(T_\ell, E_{ij})$ is smaller than the edge shared by T_ℓ and T_1 and thus smaller than the diameter of T_1. This yields

$$|T_\ell| \leq \frac{1}{2} h_{E_{ij}} h_{E_{1\ell}} \leq \frac{1}{2} h_{E_{ij}} h_{T_1} \quad \text{for } \ell = i, j .$$

Furthermore, it is $\mathrm{alt}(T_{\mathrm{tet}}, T_1) \geq \rho_K$ and $h_{E_{ij}} \leq h_K$. Consequently, we obtain for the argument in the arc cosine in (2.5)

$$\frac{h_{E_{ij}}\ \mathrm{alt}(T_{\mathrm{tet}}, T_1)\,|T_1|}{2|T_i|\,|T_j|} \geq \frac{2\ \mathrm{alt}(T_{\mathrm{tet}}, T_1)\,|T_1|}{h_{E_{ij}}\, h_{T_1}^2} \geq \frac{2\,\rho_K\,|T_1|}{h_K\, h_{T_1}^2} \geq \frac{2\pi\rho_{T_1}^2}{\sigma_{\mathcal{K}}\, h_{T_1}^2}\,,$$

where we used the regularity of the mesh and the incircle of T_1 with radius ρ_{T_1}, which gives $|T_1| \geq \pi\rho_{T_1}^2$, in the last step. Finally, we employ the weak stability of the mesh, which ensures that the polygonal faces are regular and stable in the two-dimensional sense. Therefore, the auxiliary triangulation of the polygonal faces is regular in the sense of Ciarlet and it is $h_{T_1}/\rho_{T_1} \leq \sigma_{\mathcal{T}}$, where $\sigma_{\mathcal{T}}$ only depends on $\sigma_{\mathcal{F}}$ and $c_{\mathcal{F}}$, see Lemma 2.3. Since the arc cosine is monotonically decreasing, Eq. (2.5) yields with the previous considerations

$$\delta \leq \frac{\pi}{2} + \arccos\left(\frac{2\pi\rho_{T_1}^2}{\sigma_{\mathcal{K}}\, h_{T_1}^2}\right) \leq \frac{\pi}{2} + \arccos\left(\frac{2\pi}{\sigma_{\mathcal{K}}\,\sigma_{\mathcal{T}}^2}\right) < \pi\,.$$

In summary, all angles in the surface triangles of the tetrahedra and all dihedral angles between faces are bounded by constants that are strictly less than π. Taking the maximum of them proves the maximal angle condition for the auxiliary discretization of tetrahedra. □

2.2.3 Mesh Refinement

Although the use of polygonal and polyhedral meshes is quite interesting for practical applications, only a few commercial mesh generators are able to create and refine such general meshes. For the two-dimensional case there is the free MATLAB tool PolyMesher available, see [167], and in three-dimensions one often exploits either Voronoi meshes, see [70], or dual meshes to given tetrahedral discretizations. In the following, we assume that a polygonal or polyhedral mesh is given and we address the refinement of such meshes. We may perform uniform refinement, where all elements of a mesh are refined, or adaptive refinement, where only a few elements are refined according to some criterion. For polygonal and polyhedral meshes, there is a great flexibility for the refinement process. We do not have to take care on hanging nodes and edges, since they are naturally included in such meshes.

For the refinement process, we choose the bisection of elements. For the description of the procedure, we focus on a single polygonal or polyhedral element $K \subset \mathbb{R}^d$, $d = 2, 3$. Furthermore, we assume that K is convex. The method might be adapted to non-convex, star-shaped elements, but this would yield several special cases which shall be omitted here. In order to obtain some geometrical information

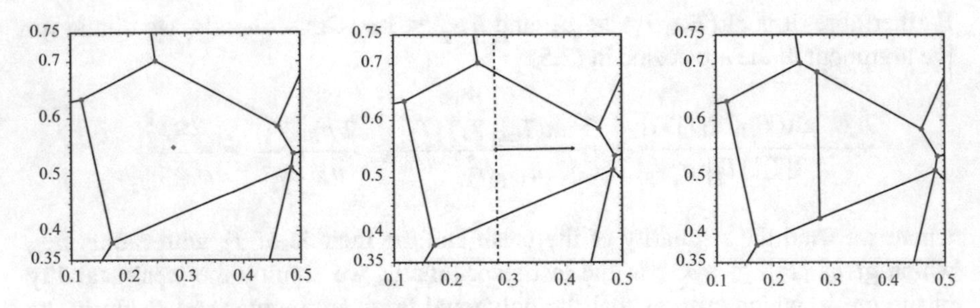

Fig. 2.6 Refinement of an element: element with center $\bar{\mathbf{x}}_K$ (left), element with eigenvector (middle), two new elements (right)

of the element shape, we first compute the covariance matrix

$$M_{\mathrm{Cov}}(K) = \frac{1}{|K|}\int_K (\mathbf{x}-\bar{\mathbf{x}}_K)(\mathbf{x}-\bar{\mathbf{x}}_K)^\top \,\mathrm{d}\mathbf{x}\,,$$

where

$$\bar{\mathbf{x}}_K = \frac{1}{|K|}\int_K \mathbf{x}\,\mathrm{d}\mathbf{x}$$

is the barycenter of the element. The matrix $M_{\mathrm{Cov}}(K) \in \mathbb{R}^{d\times d}$ is symmetric and positive definite due to construction. We compute its eigenvalues and the corresponding eigenvectors. This principle component analysis provides some information on the dimensions of the element. The square root of the eigenvalues give the standard deviation in the direction of the corresponding eigenvector. Thus, the eigenvector which belongs to the biggest eigenvalue points into the direction of the longest extend of the element K. Consequently, we split the element orthogonal to this eigenvector through the barycenter $\bar{\mathbf{x}}_K$ of K, see Fig. 2.6. Afterwards, two new elements are obtained. This strategy actually works in any dimension $d \in \mathbb{N}$. Similar ideas are used in [144] to cluster point clouds which are used for matrix approximation in fast boundary element methods.

Figure 2.7 shows the uniform refinement starting from a triangle. The meshes are obtained after one, three, five and seven refinement steps. We recognize that even a refinement of a triangle results in an unstructured polygonal mesh. Nevertheless, the resulting sequence of meshes has a uniform character. A big advantage of the introduced strategy can be seen in an adaptive context. It is possible to perform local refinements within a few elements. Classical mesh refinement techniques for triangular meshes, for example, suffer from the fact that local refinement propagates into neighbouring regions. This behaviour is necessary since the resulting meshes have to be admissible and thus the use of hanging nodes is very restricted or even avoided.

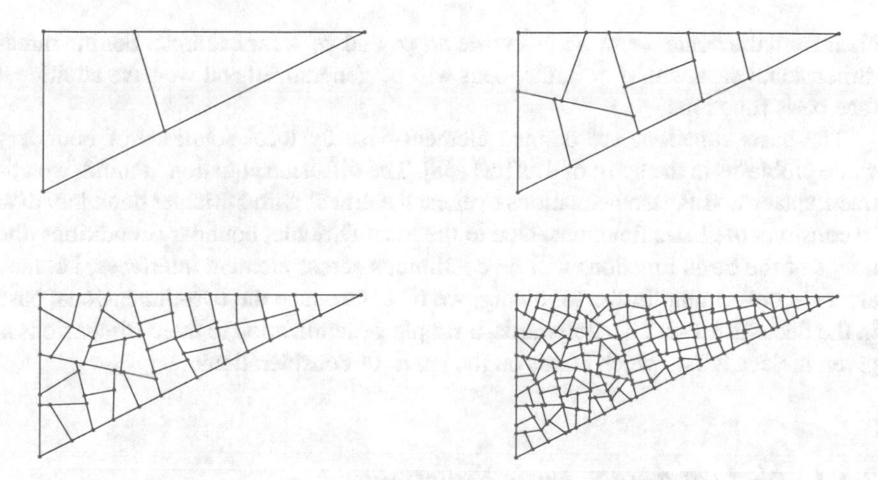

Fig. 2.7 Uniform refinement of a triangle after one, three, five and seven refinement steps

Finally, the question arises whether the regularity and the stability of a mesh is preserved during the refinement. In general this is not possible with the prescribed procedure. During the bisection of elements, small edges and thin faces might occur. However, the aspect ratios for convex elements stay bounded, since the algorithm tries to equilibrate the extend of the element in its characteristic directions. Thus, the regularity is preserved for convex elements in two-dimensions. The stability, however, has to be enforced during the refinement process if it is needed. The introduced bisection strategy for mesh refinement is applied in most of the numerical experiments presented in this book.

2.3 Trefftz-Like Basis Functions

Our goal is to introduce finite dimensional spaces V_h^k over polygonal and polyhedral discretizations of the domain $\Omega \subset \mathbb{R}^d$, $d = 2, 3$, which approximate the Sobolev space $H^1(\Omega)$. The index $k \in \mathbb{N}$ denotes the order of the approximation space. In this section, a more general strategy is presented which extends the original idea in [146] to arbitrary order. The approximation space $V_h^k = \operatorname{span} \Psi_h^k$ is constructed as span of some basis Ψ_h^k. For $d = 2$, this basis is specified in the following and consists of nodal, edge and element basis functions. These functions are indicated by $\psi_{\mathbf{z}}$, ψ_E and ψ_K, respectively. All of them have certain degrees and thus they are marked and numbered by indices like $\psi_{E,i}$ and $\psi_{K,i,j}$ for some i, j. However, for shorter notation, we will skip sometimes parts of the indices if the meaning is clear from the context and we just write ψ, ψ_i and $\psi_{i,j}$, for example. For the three-dimensional case with $d = 3$, the ideas will be generalized and we have additional face basis functions.

The basis functions are defined element-wise by local solutions of boundary value problems in the spirit of Trefftz [168]. The diffusion equation in mind, we utilize Laplace and Poisson equations over each element with Dirichlet boundary data to construct the basis functions. Due to the local Dirichlet boundary conditions, the traces of the basis functions will be continuous across element interfaces, i.e. they are H^1-conforming. In the following, we first introduce the two-dimensional case in the Sects. 2.3.1–2.3.3. Afterwards, a simple generalization to three-dimensions is given in Sect. 2.3.4, which builds on the previous considerations.

2.3.1 Node and Edge Basis Functions

Let $\mathcal{K}_h$ be a polygonal mesh of a bounded domain $\Omega \subset \mathbb{R}^2$. The functions $\psi_{\mathbf{z}}$ and ψ_E, which are assigned to nodes and edges, are defined to satisfy the Laplace equation on each element. Their Dirichlet trace on the element boundaries is chosen to be continuous and piecewise polynomial. Thus, we define for each node $\mathbf{z} \in \mathcal{N}_h$ the basis function $\psi_{\mathbf{z}}$ as unique solution of

$$
\begin{aligned}
-\Delta\psi_{\mathbf{z}} = 0 \quad &\text{in } K \quad \text{for all } K \in \mathcal{K}_h\,, \\
\psi_{\mathbf{z}}(\mathbf{x}) &= \begin{cases} 1 & \text{for } \mathbf{x} = \mathbf{z}\,, \\ 0 & \text{for } \mathbf{x} \in \mathcal{N}_h \setminus \{\mathbf{z}\}\,, \end{cases} \\
\psi_{\mathbf{z}} \text{ is linear on each edge of the mesh}\,.
\end{aligned}
\tag{2.6}
$$

So, the function $\psi_{\mathbf{z}}$ is locally defined as solution of a boundary value problem over each element. If the element $K \in \mathcal{K}_h$ is convex, the boundary value problem can be understood in the classical sense and it is $\psi_{\mathbf{z}} \in C^2(K) \cap C^0(\overline{K})$, see [82, 87]. However, we explicitly allow star-shaped elements within the discretization $\mathcal{K}_h$ of the domain Ω. In this case, the boundary value problem is understood in the weak sense and we obtain $\psi_{\mathbf{z}} \in H^1(K)$. Since the Dirichlet trace is continuous across element interfaces, the local regularity of $\psi_{\mathbf{z}}$ yields $\psi_{\mathbf{z}} \in H^1(\Omega)$. This will also be true for the edge and element basis functions. In the following, the local problems for the definition of basis functions are always understood in the classical or weak sense depending on the shape of the elements. In contrast to [146], we only make use of the fact that the nodal, edge and element basis functions satisfy $\psi \in H^1(K)$ for $K \in \mathcal{K}_h$ and we do not use a maximum principle [82, 140] for harmonic functions which would require convex elements.

In the case that $\mathcal{K}_h$ is an admissible triangulation without hanging nodes, the basis functions turn out to be the standard hat functions of classical finite element methods. This relation is quite obvious since the lowest order linear basis functions satisfy the data on the boundary of each element and they are harmonic because of their linearity. According to the unique solvability of the Dirichlet problem for the Laplace equation the hat functions coincide with the basis functions defined here. In this sense, the BEM-based FEM can be seen as a generalization of standard finite element methods.

If $\mathcal{K}_h$ is a polygonal mesh containing only convex elements, another connection can be recognized. For the model problem, we rediscover the so called harmonic coordinates mentioned in several articles like [77, 84, 112, 126]. These harmonic coordinates restricted to one element $K \in \mathcal{K}_h$ are a special type of barycentric coordinates, i.e., they satisfy

$$\psi_{\mathbf{z}}(\mathbf{x}) \geq 0 \quad \text{on } \overline{K} \tag{2.7}$$

for $\mathbf{z} \in \mathcal{N}(K)$ and it is

$$v = \sum_{\mathbf{z} \in \mathcal{N}(K)} v(\mathbf{z}) \psi_{\mathbf{z}} \tag{2.8}$$

for any linear function v on $\overline{K}$ according to [84]. Condition (2.7) follows directly from the minimum-maximum principle [82, 140]. To verify (2.8), we observe that both sides of the equation are harmonic and coincide on the boundary of K. Therefore, the difference of both sides is harmonic and identical to zero on the boundary. Using the minimum-maximal principle again shows that Eq. (2.8) is valid in the whole element. In [76, 77], the authors have proven for any set of barycentric coordinates and especially for the harmonic coordinates, which are considered in this section, that they satisfy the estimate

$$0 \leq L_{\mathbf{z}}^{\text{low}} \leq \psi_{\mathbf{z}} \leq L_{\mathbf{z}}^{\text{up}} \leq 1 \quad \text{on } \overline{K}$$

for $\mathbf{z} \in \mathcal{N}(K)$. Here, $L_{\mathbf{z}}^{\text{low}}$ and $L_{\mathbf{z}}^{\text{up}}$ are piecewise linear functions defined as follows. Both functions are equal to one at the node $\mathbf{z}$ and they are equal to zero at every other node on the boundary of K. Additionally, $L_{\mathbf{z}}^{\text{low}}$ is linear on the triangle constructed by connecting the node before and after $\mathbf{z}$ on the boundary, and zero else, see Fig. 2.8. The function $L_{\mathbf{z}}^{\text{up}}$ is linear on each triangle that is obtained by connecting $\mathbf{z}$ with all other nodes on the boundary of K.

To introduce the edge basis functions ψ_E, polynomial data is prescribed on the element boundaries. Therefore, we first review a hierarchical polynomial basis over the interval $[0, 1]$. We set

$$p_0(t) = t \qquad \text{and} \qquad p_1(t) = 1 - t \qquad \text{for } t \in [0, 1]\,,$$

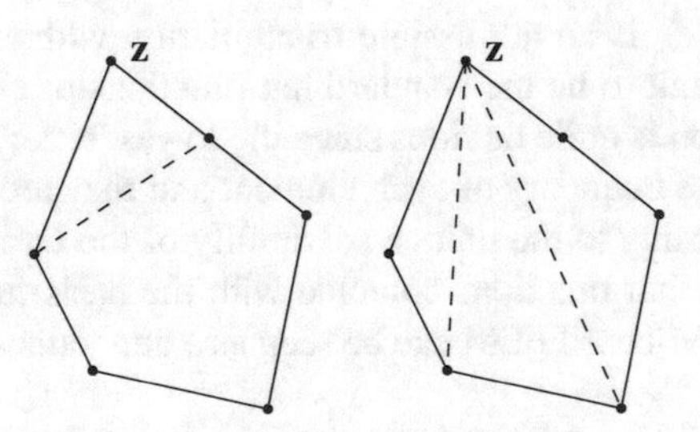

Fig. 2.8 Triangles for construction of $L_{\mathbf{z}}^{\text{low}}$ (left) and $L_{\mathbf{z}}^{\text{up}}$ (right)

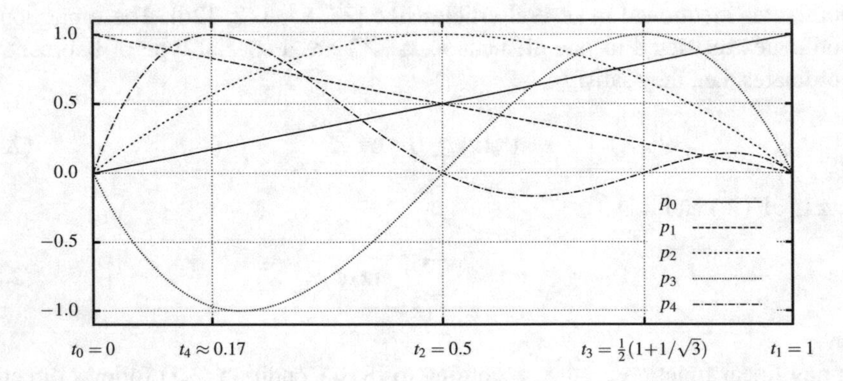

Fig. 2.9 Visualization of p_i for $i = 0, \ldots, 4$

and assign these functions to the points $t_0 = 0$ and $t_1 = 1$, respectively. Afterwards, we define $p_i \in \mathscr{P}^i([0, 1])$, $i \geq 2$ with exact degree i recursively as

$$p_i = \frac{\widetilde{p}_i}{\widetilde{p}_i(t_i)} ,$$

where $\widetilde{p}_i \in \mathscr{P}^i([0, 1]) \setminus \{0\}$ is a polynomial with $\widetilde{p}_i(t_j) = 0$ for $j = 0, \ldots, i - 1$ and

$$t_i = \max\{\arg \max_{t\in[0,1]} |\widetilde{p}_i(t)|\} .$$

The polynomial p_i is well defined since $\widetilde{p}_i$ is unique up to a multiplicative constant and we obviously have $t_i \neq t_j$ for $j < i$. In Fig. 2.9, the first polynomials are visualized. One easily sees that these polynomials are linearly independent and that for $k \geq 1$

$$\mathscr{P}^k([0, 1]) = \operatorname{span}\{p_i : i = 0, \ldots, k\} .$$

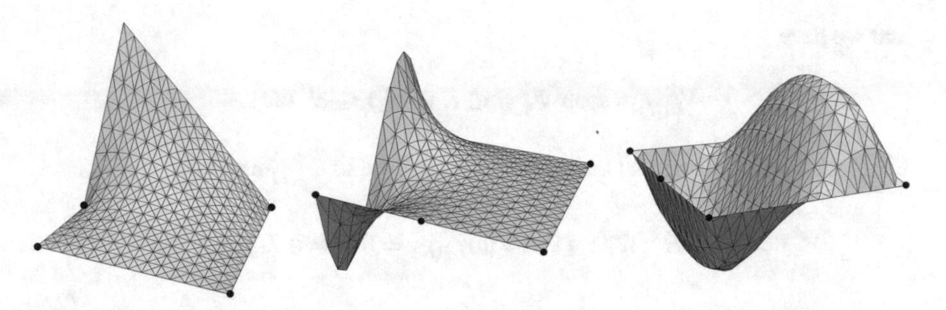

Fig. 2.10 Visualization of $\psi_{\mathbf{z}}$, $\psi_{E,3}$ and $\psi_{K,1,0}$ over rectangular element with additional node on straight line, nodes are marked with black dots

For the definition of edge basis functions ψ_E, we make use of a linear parametrization of the corresponding edge. Let $E \in \mathcal{E}_h$ with $E = \overline{\mathbf{z}_b\mathbf{z}_e}$ and

$$E : [0,1] \ni t \mapsto \mathfrak{F}_E(t) = \mathbf{z}_b + t(\mathbf{z}_e - \mathbf{z}_b) .$$

In contrast to nodal basis functions, we have more than one basis function per edge. We define $\psi_{E,i}$ for $i = 2, \ldots, k$ as unique solution of

$$-\Delta\psi_{E,i} = 0 \quad \text{in } K \quad \text{for all } K \in \mathcal{K}_h ,$$

$$\psi_{E,i} = \begin{cases} p_i \circ \mathfrak{F}_E^{-1} & \text{on } E , \\ 0 & \text{on } \mathcal{E}_h \setminus \{E\} , \end{cases}$$

and we assign these functions to the points $\mathbf{z}_{E,i} = \mathfrak{F}_E(t_i)$. In Fig. 2.10, an approximation of such a function is visualized over one rectangular element. As in the case of nodal basis functions, we observe that the Dirichlet trace is continuous along element boundaries. Thus, we have $\psi_{E,i} \in H^1(K)$ for $K \in \mathcal{K}_h$ which yields $\psi_{E,i} \in H^1(\Omega)$. With the conventions

$$\psi_{E,0} = \psi_{\mathbf{z}_b} \qquad \text{and} \qquad \psi_{E,1} = \psi_{\mathbf{z}_e} ,$$

we find that

$$\mathcal{P}^k(E) = \operatorname{span}\{\psi_{E,i}\big|_E : i = 0, \ldots, k\}$$

and

$$\psi_{E,i}(\mathbf{z}_{E,j}) = \delta_{ij} \quad \text{for } j = 0, \ldots, i ,$$

where δ_{ij} is the Kronecker symbol. According to the last property, the functions $\psi_{\mathbf{z}}$ and $\psi_{E,i}$ are linearly independent. So, we collect them in the basis

$$\Psi_{h,H}^k = \{\psi_{\mathbf{z}}, \psi_{E,i} : \mathbf{z} \in \mathcal{N}_h, E \in \mathcal{E}_h, i = 2, \ldots, k\} ,$$

and we have

$$V_{h,H}^k = \text{span}\, \Psi_{h,H}^k \subset H_\Delta^1(\mathcal{K}_h) \subset H^1(\Omega)\,.$$

Here, for $k = 1$, only nodal basis functions are used in $\Psi_{h,H}^k$ and for $k \in \mathbb{N}$,

$$H_\Delta^k(\mathcal{K}_h) = \left\{ v \in H^k(\Omega) : \; (\nabla v, \nabla w)_{L_2(K)} = 0 \;\; \forall w \in H_0^1(K), \; \forall K \in \mathcal{K}_h \right\} \tag{2.9}$$

is the space of piecewise weakly harmonic functions.

2.3.2 Element Basis Functions

Next, we address the definition of element basis functions over the polygonal mesh $\mathcal{K}_h$ of a domain $\Omega \subset \mathbb{R}^2$. To motivate the procedure, we remember that the nodal and edge basis functions satisfy the Laplace equation inside the elements and are polynomial on the edges. The nodal functions $\psi_{\mathbf{z}}$ are linear on edges, and thus they satisfy the one dimensional Laplace equation along edges: $\Delta_1 \psi_{\mathbf{z}} = 0$ on $E \in \mathcal{E}_h$. If we compute the 1D-Laplacian of the edge functions ψ_E along the edge E, we observe that $\Delta_1 \psi_{E,i} \in \mathcal{P}^{i-2}(E)$, $i \geq 2$, and thus the edge basis functions satisfy the Poisson equation with polynomial right hand side on each edge. Additionally, it is easy to check that

$$\mathcal{P}^{k-2}(E) = \text{span}\,\{\Delta_1 \psi_{E,i} : i = 2, \ldots, k\}$$

for $k \geq 2$. From this point of view, we exchanged the Laplace equation for the Poisson equation on the edges as we have made the step from nodal to edge basis functions. The same is done for the element basis functions. Here, we exchange the Laplace for the Poisson equation in the elements and we prescribe right hand sides such that they form a basis of $\mathcal{P}^{k-2}(K)$. Thus, we define $\psi_{K,i,j}$ for $K \in \mathcal{K}_h$, $i = 0, \ldots, k-2$ and $j = 0, \ldots, i$ as unique solution of

$$\begin{aligned} -\Delta \psi_{K,i,j} &= p_{K,i,j} && \text{in } K\,, \\ \psi_{K,i,j} &= 0 && \text{else}\,, \end{aligned} \tag{2.10}$$

where

$$\mathcal{P}^{k-2}(K) = \text{span}\,\{p_{K,i,j} : i = 0, \ldots, k-2 \text{ and } j = 0, \ldots, i\}\,. \tag{2.11}$$

Consequently, we have $\frac{1}{2}k(k-1)$ element basis functions per element. The support of such a function is limited to one element, i.e. supp $\psi_{K,i,j} = \overline{K}$, and the function itself belongs to $H_0^1(K)$. Due to the local regularity, we obtain $\psi_{K,i,j} \in H^1(\Omega)$. See Fig. 2.10 for a visualization of such an element basis function.

Remark 2.19 In the numerical experiments we will choose the polynomial basis as shifted monomials, namely as

$$p_{K,i,j}(\mathbf{x}) = \left(x_1 - z_{K,1}\right)^{i-j} \left(x_2 - z_{K,2}\right)^j , \quad \mathbf{x} = \left(x_1, x_2\right)^\top \in K ,$$

where $\mathbf{z}_K = \left(z_{K,1}, z_{K,2}\right)^\top$ is given in Definition 2.1. For $i, j = 0$, the element bubble function from [146] is recovered, since $p_{K,0,0} = 1$.

We define the set of functions

$$\Psi_{h,B}^k = \{\psi_{K,i,j} : K \in \mathcal{K}_h, i = 0, \ldots, k-2 \text{ and } j = 0, \ldots, i\}$$

and the space

$$V_{h,B}^k = \operatorname{span} \Psi_{h,B}^k \subset H^1(\Omega) ,$$

which consists of element bubble functions that vanish on the skeleton of the mesh. For $k = 1$, this means $\Psi_{h,B}^k = \varnothing$. Furthermore, we point out that the definition of element basis functions $\psi_{K,i,j} \in \Psi_{h,B}^k$ is equivalent to the variational formulation

$$\begin{aligned} &\text{Find } \psi_{K,i,j} \in H_0^1(K) : \\ &\left(\nabla\psi_{K,i,j}, \nabla w\right)_{L_2(K)} = \left(p_{K,i,j}, w\right)_{L_2(K)} \quad \forall w \subset H_0^1(K). \end{aligned} \tag{2.12}$$

Lemma 2.20 *The functions in* $\Psi_{h,B}^k$ *are linearly independent.*

Proof Since the support of an element basis function is restricted to one element, the functions belonging to different elements are independent. Therefore, it is sufficient to consider just functions over one element in this proof. Let $\alpha_{i,j} \in \mathbb{R}$ for $i = 0, \ldots, k-2$ and $j = 0, \ldots, i$ and let $\sum_{i,j} \alpha_{i,j}\psi_{i,j} = 0$. Consequently, we have $\sum_{i,j} \alpha_{i,j}\nabla\psi_{i,j} = 0$. Due to this and since the element basis functions $\psi_{i,j} = \psi_{K,i,j}$ satisfy (2.12), we obtain

$$\Big(\sum_{i,j} \alpha_{i,j} p_{i,j}, w\Big)_{L_2(K)} = \Big(\sum_{i,j} \alpha_{i,j}\nabla\psi_{i,j}, \nabla w\Big)_{L_2(K)} = 0 \quad \text{for } w \in H_0^1(K) .$$

The function space $C_0^\infty(K)$ is dense in $H_0^1(K)$ and thus the fundamental lemma of the calculus of variations yields $\sum_{i,j} \alpha_{i,j} p_{i,j} = 0$. Because of the choice of $p_{i,j}$ as basis of $\mathcal{P}^{k-2}(K)$, it follows that $\alpha_{i,j} = 0$ for $i = 0, \ldots, k-2$ and $j = 0, \ldots, i$. □

2.3.3 Final Approximation Space

The final basis for the approximation space of $H^1(\Omega)$ is now defined as

$$\Psi_h^k = \Psi_{h,H}^k \cup \Psi_{h,B}^k ,$$

and combines the nodal, edge and element basis functions. All functions in $\Psi_{h,H}^k$ locally satisfy the Laplace equation on each element and so, they are piecewise harmonic in a weak sense. Different from the functions in $\Psi_{h,H}^k$, the functions in $\Psi_{h,B}^k$ are exactly those which are not locally harmonic. They obviously serve the approximation of non-harmonic functions. Furthermore, we observe that

$$(\nabla\psi, \nabla\varphi)_{L_2(K)} = 0 \quad \text{for } \psi \in \Psi_{h,H}^k,\ \varphi \in \Psi_{h,B}^k , \tag{2.13}$$

since $\psi \in H^1_\Delta(\mathscr{K}_h)$ and $\varphi \in H^1_0(K)$, cf. (2.9). Sometimes, we will consider the basis functions restricted to a single element. For this reason, we define for $K \in \mathscr{K}_h$

$$\Psi_h^k\big|_K = \left\{\psi\big|_K : \psi \in \Psi_h^k\right\}$$

and $\Psi_{h,H}^k\big|_K$ as well as $\Psi_{h,B}^k\big|_K$ accordingly. The final approximation space is conforming, i.e.

$$V_h^k = \operatorname{span} \Psi_h^k \subset H^1(\Omega) ,$$

and can be written as a direct sum of piecewise weakly harmonic functions and element bubble functions. The space of element bubble functions can be further decomposed into its contributions from the single elements, because of the zero traces on the element boundaries. Thus, it is

$$V_h^k = V_{h,H}^k \oplus V_{h,B}^k \quad \text{with} \quad V_{h,B}^k = \bigoplus_{K\in\mathscr{K}_h} V_{h,B}^k\big|_K ,$$

where the same notation holds for the restriction to a single element as above. A simple counting argument shows that

$$\dim V_h^k\big|_K = k\,|\mathscr{N}(K)| + \tfrac{1}{2}k(k-1) ,$$

since

$$\dim V_{h,H}^k\big|_K = k\,|\mathscr{N}(K)| \quad \text{and} \quad \dim V_{h,B}^k\big|_K = \binom{d+k-2}{d} = \tfrac{1}{2}k(k-1) .$$

Due to the construction of the basis, it is easily seen that the approximation space can be written in the following form

$$V_h^k = \left\{ v \in H^1(\Omega) : \Delta v \in \mathscr{P}^{k-2}(K) \text{ and } v|_{\partial K} \in \mathscr{P}_{\mathrm{pw}}^k(\partial K)\ \forall K \in \mathscr{K}_h \right\}$$

with the convention $\mathscr{P}^{-1}(K) = \{0\}$. Thus, the functions in V_h^k are polynomials of degree k over each edge and their Laplacian over each element is a polynomial of degree $k-2$.

The virtual element method (VEM) in [25] also uses this approximation space. Therefore, the BEM-based FEM and the VEM seek the approximation of the solution of the boundary-value problem for the diffusion Eq. (2.1) in the same discrete space. The VEM reduces all computations to carefully chosen degrees of freedom and to local projections into polynomial spaces. The BEM-based FEM in contrary makes use of the explicit knowledge of the basis functions and thus enables the evaluation of the approximation inside the elements. Both methods rely on clever reformulations to avoid volume integration. Since the BEM-based FEM applies Trefftz-like basis functions, which are related to the differential equation of the global problem, the discrete space for the BEM-based FEM and the VEM differ as soon as more general boundary-value problems are considered.

2.3.4 *Simple Generalization to 3D*

This section gives a straight forward generalization to the three-dimensional case. A more involved one is postponed to a later chapter. Let $\mathscr{K}_h$ be a polyhedral mesh of a bounded domain $\Omega \subset \mathbb{R}^3$. We restrict ourselves here to polyhedral elements that have triangular faces. This can be always achieved by triangulating the polygonal faces of general polyhedra. For this purpose we may use the auxiliary triangulation $\mathscr{T}_0(\partial K)$ introduced in Sect. 2.2.2 and reinterpret K as element with triangular faces. Consequently, one additional node per face is introduced on the surface of the polyhedral element K. Several constructed triangular faces meet in this node and lie on a flat part of ∂K. However, the notion of polyhedral elements allows for such degenerations. A more direct approach for the treatment of polygonal faces will be discussed in Sect. 6.2.

Turning to the construction of the approximation space V_h^k in three-dimensions, we may recognize that it can be written down immediately as

$$V_h^k = \left\{ v \in H^1(\Omega) : \Delta v \in \mathscr{P}^{k-2}(K) \text{ and } v|_{\partial K} \in \mathscr{P}_{\mathrm{pw}}^k(\partial K)\ \forall K \in \mathscr{K}_h \right\} .$$

Thus, the only difference to the two-dimensional case is that the functions in V_h^k are now piecewise polynomial of degree k over the triangular faces of polyhedra instead of piecewise polynomials over the edges of polygonal elements. Consequently, the considerations from the previous sections can be directly generalized to polyhedral meshes $\mathscr{K}_h$, for which the set of faces $\mathscr{F}_h$ consists of triangles only.

The space V_h^k is again constructed as a direct sum of piecewise weakly harmonic functions with polynomial traces on the faces of the mesh and element bubble functions that vanish on the skeleton but have a polynomial Laplacian inside the elements, i.e.

$$V_h^k = V_{h,H}^k \oplus V_{h,B}^k \quad \text{and} \quad V_{h,B}^k = \bigoplus_{K\in\mathscr{K}_h} V_{h,B}^k\big|_K .$$

Let $v_h = v_{h,H} + v_{h,B} \in V_h^k$ with $v_{h,H} \in V_{h,H}^k$ and $v_{h,B} \in V_{h,B}^k$. For each element $K \in \mathscr{K}_h$, it holds

$$-\Delta v_{h,H} = 0 \quad \text{in } K \quad \text{and} \quad v_{h,H} = p_{\partial K} \quad \text{on } \partial K , \tag{2.14}$$

as well as

$$-\Delta v_{h,B} = p_K \quad \text{in } K \quad \text{and} \quad v_{h,B} = 0 \quad \text{on } \partial K , \tag{2.15}$$

for some $p_{\partial K} \in \mathscr{P}_{\text{pw}}^k(\partial K)$ and $p_K \in \mathscr{P}^{k-2}(K)$. Thus, $v_{h,H}$ and $v_{h,B}$ are uniquely defined by the polynomial data $p_{\partial K}$ and p_K, respectively. Consequently, the basis Ψ_h^k of V_h^k is constructed in an element-wise fashion respecting the direct sum, such that

$$\Psi_h^k\big|_K = \Psi_{h,H}^k\big|_K \cup \Psi_{h,B}^k\big|_K \quad \text{for} \quad V_h^k\big|_K = V_{h,H}^k\big|_K \oplus V_{h,B}^k\big|_K .$$

We choose a basis for $\mathscr{P}_{\text{pw}}^k(\partial K)$ and $\mathscr{P}^{k-2}(K)$. For each function in these sets a harmonic basis function and an element basis function are obtained by (2.14) and (2.15), respectively. Due to this construction, a simple counting argument shows that

$$\dim V_h^k\big|_K = |\mathscr{N}(K)| + (k-1)|\mathscr{E}(K)| + \tfrac{1}{2}(k-1)(k-2)|\mathscr{F}(K)| + \tfrac{1}{6}k(k-1)(k+1) ,$$

since

$$\dim V_{h,H}^k\big|_K = |\mathscr{N}(K)| + (k-1)|\mathscr{E}(K)| + \tfrac{1}{2}(k-1)(k-2)|\mathscr{F}(K)|$$

and

$$\dim V_{h,B}^k\big|_K = \binom{d+k-2}{d} = \tfrac{1}{6}k(k-1)(k+1) .$$

In the previous sections on the two-dimensional case, this construction has been carried out in more detail and we have given a precise choice of basis functions. For the three-dimensional case we are content with the abstract setting and pass a detailed presentation. We point out, however, the important orthogonality property given in (2.13), which still holds on $K \in \mathscr{K}_h$, namely

$$(\nabla v_{h,H}, \nabla v_{h,B})_{L_2(K)} = 0 \quad \text{for } v_{h,H} \in V^k_{h,H}, v_{h,B} \in V^k_{h,B} .$$

This is a consequence of the weakly harmonic functions, cf. (2.9), and the element bubble functions that satisfy

$$V^k_{h,H} \subset H^1_\Delta(\mathscr{K}_h) \quad \text{and} \quad V^k_{h,B} \subset \bigoplus_{K \in \mathscr{K}_h} H^1_0(K) .$$

Remark 2.21 For the implementation and the numerical experiments it is important to specify the choice of basis functions. As discussed above, the sets $\Psi^k_{h,H}$ and $\Psi^k_{h,B}$ are constructed by choosing a basis for $\mathscr{P}^k_{\text{pw}}(\partial K)$ and $\mathscr{P}^{k-2}(K)$, respectively. It is convenient to choose the classic Lagrange elements over triangles, cf. [40, 151], for the basis of $\mathscr{P}^k_{\text{pw}}(\partial K)$, whereas the basis of $\mathscr{P}^{k-2}(K)$ might be chosen according to the two-dimensional case as shifted monomials, for instance.

2.4 Interpolation Operators

In this section, we are concerned with the interpolation of function in $H^2(\Omega)$ by functions in $V^k_h = V^k_{h,H} \oplus V^k_{h,B}$. Due to the Sobolev embedding theorem it holds that $H^2(\Omega) \subset C^0(\overline{\Omega})$, see [1], and the pointwise evaluation of such functions is well defined. Thus, we may exploit nodal interpolation to some extend. The interpolation of non-smooth functions in $H^1(\Omega)$ is postponed to later considerations, see Chap. 3.

Since V^k_h is given as a direct sum of weakly harmonic and element bubble functions, it is natural to decompose the interpolation into two corresponding operators. Therefore, we study

$$\mathfrak{I}^k_h = \mathfrak{I}^k_{h,H} + \mathfrak{I}^k_{h,B} : H^2(\Omega) \to V^k_h \subset H^1(\Omega)$$

with

$$\mathfrak{I}^k_{h,H} : H^2(\Omega) \to V^k_{h,H} \subset H^1_\Delta(\mathscr{K}_h) \quad \text{and} \quad \mathfrak{I}^k_{h,B} : H^2(\Omega) \to V^k_{h,B} \subset \bigoplus_{K \in \mathscr{K}_h} H^1_0(K) .$$

The interpolation operators $\mathfrak{I}^k_{h,H}$ and $\mathfrak{I}^k_{h,B}$ are discussed in the following. Furthermore, it is sufficient to introduce them over a single element, since the local nature of the operators directly extend to their global definition. Thus, we restrict ourselves to a single element of a regular polytopal mesh and denote the restrictions of the operators with the same symbols.

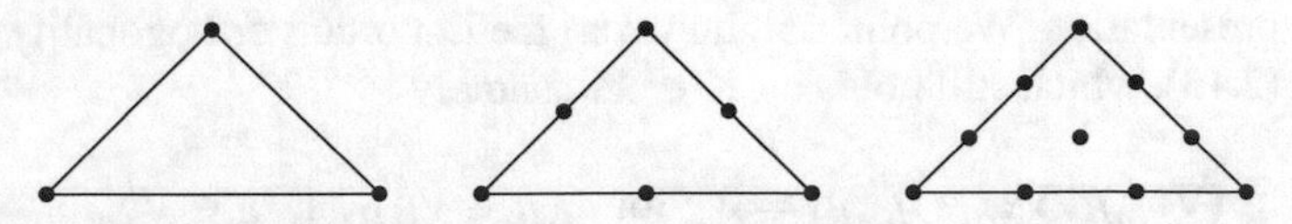

Fig. 2.11 Points for Lagrange interpolation into $\mathscr{P}^k(\partial K)$ on triangles for $k = 1, 2, 3$

We start by the interpolation into the space of weakly harmonic functions. For $v_h \in V^k_{h,H}$, it is $\Delta v_h = 0$ in K and $v|_{\partial K} \in \mathscr{P}^k_{\mathrm{pw}}(\partial K)$. Thus, for the definition of

$$\mathfrak{I}^k_{h,H} : H^2(K) \to V^k_{h,H}\big|_K \subset H^1_\Delta(K)\,,$$

we prescribe $\mathfrak{I}^k_{h,H} v|_{\partial K}$ to be equal to a standard nodal interpolation operator into the space $\mathscr{P}^k_{\mathrm{pw}}(\partial K)$ on the boundary of the polytopal element K. Afterwards, we extend this boundary data harmonically into K. By fixing a standard interpolation operator on ∂K, the operator $\mathfrak{I}^k_{h,H}$ is uniquely defined. For this purpose, we exploit the classical Lagrange interpolation on equidistant points along the edges in the boundary of polygonal elements ($d = 2$) and on equidistributed points, see Fig. 2.11, in the triangular faces in the boundary of polyhedral elements ($d = 3$). Consequently, $\mathfrak{I}^k_{h,H} v$ is constructed in such a way that it coincides in $k + 1$ points on each edge and in $\frac{1}{2}(k + 1)(k + 2)$ points on each triangular face of the elements.

Remark 2.22 In 2D, we can alternatively follow the idea from [175] and choose a different interpolation operator for $k \geq 2$ on the boundary of the polygonal elements. The introduced points $\mathbf{z}_{E,i}$ from Sect. 2.3.1 can be used for the pointwise interpolation. For $v \in H^2(\Omega)$, this yields

$$\mathfrak{I}^k_{h,H} v = \sum_{\mathbf{z}\in\mathscr{N}_h} v_{\mathbf{z}}\psi_{\mathbf{z}} + \sum_{E\in\mathscr{E}_h}\sum_{i=2}^{k} v_{E,i}\psi_{E,i}\,,$$

where the coefficients are given as

$$v_{\mathbf{z}} = v(\mathbf{z}) \quad \text{for } \mathbf{z} \in \mathscr{N}_h$$

and

$$v_{E,i} = v(\mathbf{z}_{E,i}) - \sum_{j=0}^{i-1} v_{E,j}\psi_{E,j}(\mathbf{z}_{E,j}) \quad \text{for } E \in \mathscr{E}_h,\ i = 2, \ldots, k\,.$$

Next, we consider the definition of the interpolation operator into the space of element bubble functions, namely

$$\mathfrak{I}^k_{h,B} : H^2(K) \to V^k_{h,B}\big|_K \subset H^1_0(K)\,.$$

Therefore, let

$$\Psi^k_{h,B}\big|_K = \{\psi_{K,i} : i = 1, \ldots, n(k)\} \quad \text{and} \quad \{p_{K,i} = -\Delta\psi_{K,i} : i = 1, \ldots, n(k)\} \tag{2.16}$$

be the basis of $V^k_{h,B}\big|_K$ and the corresponding basis of $\mathscr{P}^{k-2}(K)$, respectively, where $n(k) = \dim V^k_{h,B}\big|_K$ is the number of basis functions. Compare the former definition (2.10)–(2.11) in 2D and the construction (2.15) in 3D. For $v \in H^2(K)$, we have

$$\mathfrak{I}^k_{h,B} v = \sum_{i=1}^{n(k)} v_{K,i}\psi_{K,i} \in V^k_{h,B}\big|_K ,$$

where the coefficients $v_{K,i}$ are defined such that $\mathfrak{I}^k_{h,B} v$ is the orthogonal projection of $v - \mathfrak{I}^k_{h,H} v$ into $V^k_{h,B}\big|_K$ with respect to the weighted inner product

$$(u, v)_{hH^1(K)} = (u, v)_{L_2(K)} + h_K^2 (\nabla u, \nabla v)_{L_2(K)} . \tag{2.17}$$

Thus, $\mathfrak{I}^k_{h,B} v$ is uniquely defined by

$$\left(\mathfrak{I}^k_{h,B} v, w\right)_{hH^1(K)} = \left(v - \mathfrak{I}^k_{h,H} v, w\right)_{hH^1(K)} \qquad \forall w \in V^k_{h,B}\big|_K . \tag{2.18}$$

The properties of the orthogonal projection yield

$$\|\mathfrak{I}^k_{h,B} v\|_{hH^1(K)} \le \|v - \mathfrak{I}^k_{h,H} v\|_{hH^1(K)} , \tag{2.19}$$

where the weighted norm is given as $\|\cdot\|^2_{hH^1(K)} = (\cdot,\cdot)_{hH^1(K)}$. If $h_K = 1$ the weighted inner product and the weighted norm coincide with the usual ones in $H^1(K)$, which are denoted by $(\cdot,\cdot)_{H^1(K)}$ and $\|\cdot\|_{H^1(K)}$, respectively.

In the following, we investigate the properties of the interpolation operators in more details. For this purpose, let $\mathscr{K}_h$ be a regular polytopal mesh.

Lemma 2.23 *The restrictions of the interpolation operators $\mathfrak{I}^k_{h,H}$ and $\mathfrak{I}^k_h$ onto an element $K \in \mathscr{K}_h$ satisfy*

$$\mathfrak{I}^k_{h,H}\, p = p \quad \text{for } p \in \mathscr{P}^k(K) \text{ with } \Delta p = 0 \text{ in } K ,$$

and

$$\mathfrak{I}^k_h\, p = p \quad \text{for } p \in \mathscr{P}^k(K) .$$

Proof Let $p \in \mathscr{P}^k(K)$ with $\Delta p = 0$. According to the definition, $\mathfrak{I}_{h,H}^k\, p$ is given as a classical, nodal interpolation into the space $\mathscr{P}_{\mathrm{pw}}^k(\partial K)$ on the boundary of the element K. Since $p \in \mathscr{P}_{\mathrm{pw}}^k(\partial K)$ and polynomials are preserved by the classical interpolation operators, p and $\mathfrak{I}_{h,H}^k p$ are identical on the boundary of the element K. Furthermore, both functions satisfy the Laplace equation inside K. Thus, the unique solvability of the Dirichlet problem for the Laplace equation yields $\mathfrak{I}_{h,H}^k\, p = p$, the first statement of the lemma.

Next, let $p \in \mathscr{P}^k(K)$ and therefore $-\Delta p \in \mathscr{P}^{k-2}(K)$. Since the polynomials $p_{K,i}$ form a basis of $\mathscr{P}^{k-2}(K)$, see (2.16), there are unique coefficients $\beta_{K,i} \in \mathbb{R}$ such that

$$-\Delta p = \sum_{i=1}^{n(k)} \beta_{K,i}\, p_{K,i} \,.$$

Furthermore, we define

$$\widetilde{p} = \mathfrak{I}_{h,H}^k\, p + \sum_{i=1}^{n(k)} \beta_{K,i}\, \psi_{K,i} \,. \tag{2.20}$$

We observe that p as well as $\widetilde{p}$ satisfy the boundary value problem

$$-\Delta u = \sum_{i=1}^{n(k)} \beta_{K,i}\, p_{K,i} \quad \text{in } K \,,$$

$$u = p \quad \text{on } \partial K \,,$$

at least in the weak sense, due to construction. Because of the unique solvability of this problem, we conclude that $p = \widetilde{p}$. By (2.20), we obtain

$$p - \mathfrak{I}_{h,H}^k\, p = \sum_{i=1}^{n(k)} \beta_{K,i}\, \psi_{K,i} \in V_{h,B}^k\big|_K \,.$$

Since $\mathfrak{I}_{h,B}^k\, p$ is defined as orthogonal projection of $p - \mathfrak{I}_{h,H}^k\, p$ into $V_{h,B}^k\big|_K$, it is $\mathfrak{I}_{h,B}^k\, p = p - \mathfrak{I}_{h,H}^k\, p$ and the second statement of the lemma follows. □

A consequence of this lemma is that

$$\mathscr{P}^k(K) \subset V_h^k\big|_K \,,$$

i.e., the space of polynomials of degree k is locally embedded in the approximation space over each element. Obviously, the element basis functions are essential to capture the non-harmonic polynomials.

Lemma 2.24 *The restrictions of the interpolation operators* $\mathfrak{I}_{h,H}^k$ *and* $\mathfrak{I}_h^k$ *onto an element* $K \in \mathcal{K}_h$ *of a regular and stable polytopal mesh* $\mathcal{K}_h$ *with* $h_K = 1$ *are linear and continuous. Furthermore, there are constants* c_1 *and* c_2*, which only depend on the regularity and stability parameters of the mesh, on* k *and on the dimension* d*, such that*

$$\|\mathfrak{I}_{h,H}^k v\|_{H^1(K)} \leq c_1 \|v\|_{H^2(K)} \quad \text{and} \quad \|\mathfrak{I}_h^k v\|_{H^1(K)} \leq c_2 \|v\|_{H^2(K)}$$

for all $v \in H^2(K)$.

Proof The linearity of the operators is obvious, so we only have to prove the given estimates which also ensure the continuity. Therefore, we make use of an auxiliary discretization $\mathcal{T}_h(K)$ of K into simplicial elements, i.e., into triangles ($d = 2$) and tetrahedra ($d = 3$). In two-dimensions, we connect the nodes on the boundary of K with the point $\mathbf{z}_K$ and in three-dimensions we exploit $\mathcal{T}_0(K)$ from Sect. 2.2.2. Since $\mathcal{K}_h$ is regular and stable, these auxiliary meshes are shape-regular in the sense of Ciarlet [58] according to Lemmata 2.3 and 2.14, respectively. Thus, neighbouring simplices share either a common node, edge or face and the aspect ratio of each simplex is uniformly bounded by some constant $\sigma_{\mathcal{T}}$. Because the auxiliary mesh is regular, we can use classical interpolation operators, see e.g. [58]. Let

$$\mathfrak{I}_{\mathcal{T}} : H^2(K) \to \mathcal{P}_{\mathrm{pw}}^k(\mathcal{T}_h(K))$$

be such a classical operator with

$$\|v - \mathfrak{I}_{\mathcal{T}} v\|_{H^1(K)} \leq C_{\mathcal{T},1} h_{\mathcal{T}} |v|_{H^2(K)} \quad \text{and} \quad \|\mathfrak{I}_{\mathcal{T}} v\|_{H^1(K)} \leq C_{\mathcal{T},2} \|v\|_{H^2(K)} \tag{2.21}$$

for $v \in H^2(K)$, where $h_{\mathcal{T}} = \max\{h_T : T \in \mathcal{T}_h(K)\}$ and

$$\mathcal{P}_{\mathrm{pw}}^k(\mathcal{T}_h(K)) = \left\{ p \in C^0(K) : \, p\big|_T \in \mathcal{P}^k(T) \;\; \forall T \in \mathcal{T}_h(K) \right\} .$$

The constants $C_{\mathcal{T},1}$ and $C_{\mathcal{T},2}$ only depend on the approximation order k, the space dimension d as well as on $\sigma_{\mathcal{T}}$ and thus on the regularity and stability parameters of the polytopal mesh $\mathcal{K}_h$.

Next, we prove the continuity of $\mathfrak{I}_{h,H}^k$, i.e. the estimate

$$\|\mathfrak{I}_{h,H}^k v\|_{H^1(K)} \leq c \|v\|_{H^2(K)} \quad \text{for } v \in H^2(K) .$$

Let $v \in H^2(K)$ be fixed. The interpolation $\mathfrak{I}_{h,H}^k v$ satisfies the boundary value problem

$$\begin{aligned} -\Delta u &= 0 \quad \text{in } K , \\ u &= g_v \quad \text{on } \partial K , \end{aligned}$$

where $g_v = \mathfrak{I}^k_{h,H} v\big|_{\partial K}$ is a piecewise polynomial of degree k on the boundary ∂K. We write $u = u_0 + u_g$ with $u_g = \mathfrak{I}_{\mathcal{T}} v$ and obtain the Galerkin formulation

$$\text{Find } u_0 \in H^1_0(K): \quad (\nabla u_0, \nabla w)_{L_2(K)} = -(\nabla u_g, \nabla w)_{L_2(K)} \quad \forall w \in H^1_0(K)\,,$$

which has a unique solution. Testing with $w = u_0$ and applying the Cauchy–Schwarz inequality yield

$$|u_0|^2_{H^1(K)} \le |(\nabla u_g, \nabla u_0)_{L_2(K)}| \le |u_g|_{H^1(K)} |u_0|_{H^1(K)}\,,$$

and consequently

$$|u_0|_{H^1(K)} \le \|u_g\|_{H^1(K)} = \|\mathfrak{I}_{\mathcal{T}} v\|_{H^1(K)}\,.$$

Because of the piecewise smoothness of the boundary of K and since it can be embedded into a square of side length h_K, the Poincaré–Friedrichs inequality reads

$$\|w\|_{L_2(K)} \le h_K\, |w|_{H^1(K)} \quad \text{for } w \in H^1_0(K)\,,$$

see e.g. [38]. By the use of the given estimates and $h_K = 1$, we obtain

$$\begin{aligned}
\|\mathfrak{I}^k_{h,H} v\|_{H^1(K)} &\le \|u_0\|_{H^1(K)} + \|u_g\|_{H^1(K)} \\
&= \left(\|u_0\|^2_{L_2(K)} + |u_0|^2_{H^1(K)}\right)^{1/2} + \|\mathfrak{I}_{\mathcal{T}} v\|_{H^1(K)} \\
&\le \sqrt{2}\, |u_0|_{H^1(K)} + \|\mathfrak{I}_{\mathcal{T}} v\|_{H^1(K)} \\
&\le \left(\sqrt{2} + 1\right) \|\mathfrak{I}_{\mathcal{T}} v\|_{H^1(K)} \\
&\le c\, \|v\|_{H^2(K)}\,.
\end{aligned}$$

Finally, we apply the continuity of $\mathfrak{I}^k_{h,H}$ as well as the property (2.19) of $\mathfrak{I}^k_{h,B}$ with $h_K = 1$ and we get

$$\begin{aligned}
\|\mathfrak{I}^k_h v\|_{H^1(K)} &\le \|\mathfrak{I}^k_{h,H} v\|_{H^1(K)} + \|\mathfrak{I}^k_{h,B} v\|_{H^1(K)} \\
&\le \|\mathfrak{I}^k_{h,H} v\|_{H^1(K)} + \|v - \mathfrak{I}^k_{h,H} v\|_{H^1(K)} \\
&\le \|v\|_{H^1(K)} + 2\|\mathfrak{I}^k_{h,H} v\|_{H^1(K)} \\
&\le c\, \|v\|_{H^2(K)}
\end{aligned}$$

that concludes the proof. □

Remark 2.25 The stability of the mesh $\mathcal{K}_h$ was only needed to ensure the shape-regularity of the auxiliary mesh, such that classical interpolation results on triangular and tetrahedral meshes can be exploited. Thus, the stability of $\mathcal{K}_h$ can be relaxed as long as the interpolation estimates (2.21) on the auxiliary mesh are guaranteed. This yields the following variants:

1. In 2D, it is sufficient to assume the regularity of $\mathcal{K}_h$. According to Corollary 2.9, the auxiliary mesh $\mathcal{T}_h(K)$ thus satisfy a maximum angle condition. Under these assumptions the classical Lagrange interpolation operator fulfils the desired estimates (2.21), see [110, 114].
2. In 3D, we may assume the regularity and weak stability of $\mathcal{K}_h$, which ensures a maximum angle condition for the tetrahedral auxiliary mesh $\mathcal{T}_0(K)$, see Lemma 2.18. For $k = 1$, this is sufficient to prove the interpolation estimates (2.21) on $\mathcal{T}_0(K)$, but it is still an open question whether these estimates hold for $k > 1$, see [115].

According to the previous remark, the lemma stays valid even if the edges ($d = 2$) and faces ($d = 3$) of the polytopal mesh degenerate in their size. Thus, the edge length h_E may decreases faster than the element diameter h_K such that the uniform estimate $h_K \leq c_{\mathcal{K}} h_E$ is violated in two- and three-dimensions.

The condition $h_K = 1$ in Lemma 2.24 is not satisfied in general. Thus, we introduce a scaling for the elements $K \in \mathcal{K}_h$ such that

$$\widehat{K} \ni \widehat{\mathbf{x}} \mapsto \mathbf{x} = \mathfrak{F}_K(\widehat{\mathbf{x}}) = h_K \widehat{\mathbf{x}} \in K \ . \tag{2.22}$$

Consequently, $h_{\widehat{K}} = 1$ and we set $\widehat{v} = v \circ \mathfrak{F}_K$. Simple calculations show that for $v \in H^\ell(K)$, $\ell \in \mathbb{N}_0$ it is $\widehat{v} \in H^\ell(\widehat{K})$ and

$$|\widehat{v}|_{H^\ell(\widehat{K})} = h_K^{\ell-d/2} |v|_{H^\ell(K)} \ . \tag{2.23}$$

Additionally, it holds

$$(u, v)_{L_2(K)} = h_K^d (\widehat{u}, \widehat{v})_{L_2(\widehat{K})} \quad \text{and} \quad (\nabla u, \nabla v)_{L_2(K)} = h_K^{d-2} (\widehat{\nabla}\widehat{u}, \widehat{\nabla}\widehat{v})_{L_2(\widehat{K})}$$

for $u, v \in H^1(K)$, where $\widehat{\nabla}$ denotes the gradient with respect to $\widehat{x}$. According to the definition of the weighted inner product, see (2.17), we obtain

$$(u, v)_{hH^1(K)} = h_K^d (\widehat{u}, \widehat{v})_{hH^1(\widehat{K})} \ . \tag{2.24}$$

Lemma 2.26 *The restrictions of the interpolation operators $\mathfrak{I}_{h,H}^k$ and $\mathfrak{I}_h^k$ onto an element $K \in \mathcal{K}_h$ satisfy for $v \in H^2(K)$*

$$\widehat{\mathfrak{I}_{h,H}^k v} = \widehat{\mathfrak{I}_{h,H}^k}\, \widehat{v} \qquad \text{and} \qquad \widehat{\mathfrak{I}_h^k v} = \widehat{\mathfrak{I}_h^k}\, \widehat{v} \, ,$$

where $\widehat{\mathfrak{I}_h^k} = \widehat{\mathfrak{I}_{h,B}^k} + \widehat{\mathfrak{I}_{h,H}^k}$ and $\widehat{\mathfrak{I}_{h,H}^k}$ as well as $\widehat{\mathfrak{I}_{h,B}^k}$ are the interpolation operators with respect to the scaled element $\widehat{K}$.

Proof Due to the construction of $\mathfrak{I}_{h,H}^k$ by pointwise evaluations on the boundary ∂K and the harmonic extension, it is obvious that $\widehat{\mathfrak{I}_{h,H}^k v} = \mathfrak{I}_{h,H}^k \widehat{v}$. Therefore, we only have to show $\widehat{\mathfrak{I}_{h,B}^k v} = \widehat{\mathfrak{I}_{h,B}^k} \widehat{v}$ with $\widehat{\mathfrak{I}_{h,B}^k} : H^2(\widehat{K}) \to \operatorname{span}\{\psi_{\widehat{K},i}\}$. Here, we explicitly refer to the element basis functions $\psi_{K,i}$ and $\psi_{\widehat{K},i}$, cf. (2.16), in order to emphasize the dependence on K and $\widehat{K}$, respectively. Furthermore, it is sufficient to prove

$$\widehat{\mathfrak{I}_{h,B}^k v} \in \operatorname{span}\{\psi_{\widehat{K},i}\}$$

and

$$(\widehat{\mathfrak{I}_{h,B}^k v}, \varphi)_{H^1(\widehat{K})} = (\widehat{\mathfrak{I}_{h,B}^k} \widehat{v}, \varphi)_{H^1(\widehat{K})} \quad \text{for } \varphi \in \operatorname{span}\{\psi_{\widehat{K},i}\},$$

since for $\varphi = \widehat{\mathfrak{I}_{h,B}^k v} - \widehat{\mathfrak{I}_{h,B}^k}\widehat{v}$, we obtain

$$\|\widehat{\mathfrak{I}_{h,B}^k v} - \widehat{\mathfrak{I}_{h,B}^k}\widehat{v}\|_{H^1(\widehat{K})} = 0 \quad \text{and thus} \quad \widehat{\mathfrak{I}_{h,B}^k v} = \widehat{\mathfrak{I}_{h,B}^k}\widehat{v}.$$

Here, we have skipped the range $i = 1, \ldots, n(k)$ for shorter notation. In the definition of the element basis functions $\psi_{K,i}$, see (2.10), we have made no specific choice of the polynomials $p_{K,i}$. In the following, let the polynomials for the functions $\psi_{\widehat{K},i}$ over $\widehat{K}$ be chosen in dependence of $\psi_{K,i}$ as

$$p_{\widehat{K},i} = h_K^2 \widehat{p}_{K,i}\,.$$

In consequence, we obtain for the scaled element function $\widehat{\psi}_{K,i} = \psi_{K,i} \circ \mathfrak{F}_K$ that

$$-\widehat{\Delta}\widehat{\psi}_{K,i} = h_K^2 \widehat{p}_{K,i} = p_{\widehat{K},i} = -\widehat{\Delta}\psi_{\widehat{K},i} \quad \text{in } \widehat{K}$$

and $\widehat{\psi}_{K,i} = \psi_{\widehat{K},i}$ on ∂K, where $\widehat{\Delta}$ denotes the Laplace operator with respect to $\widehat{x}$. Due to the unique solvability of the Dirichlet problem for the Laplace equation, we get $\psi_{\widehat{K},i} = \widehat{\psi}_{K,i}$ and thus

$$\widehat{\mathfrak{I}_{h,B}^k v} = \sum_{i=1}^{n(k)} v_{K,i}\widehat{\psi}_{K,i} \in \operatorname{span}\{\psi_{\widehat{K},i}\}\,.$$

Next, let $\varphi_{\widehat{K}} \in \operatorname{span}\{\psi_{\widehat{K},i}\}$ and set $\varphi_K = \varphi_{\widehat{K}} \circ \mathfrak{F}_K^{-1} \in \operatorname{span}\{\psi_{K,i}\}$. By the definition of $\mathfrak{I}_{h,B}^k$, we have

$$\left(\mathfrak{I}_{h,B}^k v, \varphi_K\right)_{hH^1(K)} = \left(v - \mathfrak{I}_{h,H}^k v, \varphi_K\right)_{hH^1(K)}\,.$$

Applying (2.24) to both sides of the equation yields

$$
\begin{aligned}
(\widehat{\mathfrak{I}^k_{h,B} v}, \varphi_{\widehat{K}})_{H^1(\widehat{K})} &= (\widehat{v - \mathfrak{I}^k_{h,H} v}, \varphi_{\widehat{K}})_{H^1(\widehat{K})} \\
&= (\widehat{v} - \widehat{\mathfrak{I}^k_{h,H} v}, \varphi_{\widehat{K}})_{H^1(\widehat{K})} \\
&= (\widehat{\mathfrak{I}^k_{h,B}} \widehat{v}, \varphi_{\widehat{K}})_{H^1(\widehat{K})} ,
\end{aligned}
$$

where the last equality comes from the definition of $\widehat{\mathfrak{I}^k_{h,B}}$ and the fact that the inner products $(\cdot,\cdot)_{hH^1(\widehat{K})}$ and $(\cdot,\cdot)_{H^1(\widehat{K})}$ coincide on the scaled element. Since $\varphi_{\widehat{K}}$ is chosen arbitrarily, this equality concludes the proof. □

Theorem 2.27 *Let $\mathscr{K}_h$ be a regular and stable polytopal mesh of the bounded polytopal domain $\Omega \subset \mathbb{R}^d$, $d = 2, 3$. The interpolation operators $\mathfrak{I}^k_{h,H}$ and $\mathfrak{I}^k_h$ satisfy*

$$
\|v - \mathfrak{I}^k_{h,H} v\|_{H^\ell(\Omega)} \le c\, h^{k+1-\ell}\, |v|_{H^{k+1}(\Omega)} \quad \text{for } v \in H^{k+1}_\Delta(\mathscr{K}_h) ,
$$

and

$$
\|v - \mathfrak{I}^k_h v\|_{H^\ell(\Omega)} \le c\, h^{k+1-\ell}\, |v|_{H^{k+1}(\Omega)} \quad \text{for } v \in H^{k+1}(\Omega) ,
$$

respectively, where $h = \max\{h_K : K \in \mathscr{K}_h\}$, $\ell = 0, 1$ and the constant c only depends on the mesh parameters, the dimension d and on k.

Proof First, we consider the second estimate and the case $\ell = 1$. Let us start to examine the error over one element $K \in \mathscr{K}_h$. We scale this element in such a way that its diameter becomes one, see (2.22). With the help of (2.23) and Lemma 2.26, we obtain

$$
\begin{aligned}
\|v - \mathfrak{I}^k_h v\|^2_{H^1(K)} &= \|v - \mathfrak{I}^k_h v\|^2_{L_2(K)} + |v - \mathfrak{I}^k_h v|^2_{H^1(K)} \\
&\le c h_K^d \|\widehat{v} - \widehat{\mathfrak{I}^k_h} \widehat{v}\|^2_{L_2(\widehat{K})} + c h_K^{d-2}\, |\widehat{v} - \widehat{\mathfrak{I}^k_h} \widehat{v}|^2_{H^1(\widehat{K})} \\
&\le c h_K^{d-2}\, \|\widehat{v} - \widehat{\mathfrak{I}^k_h} \widehat{v}\|^2_{H^1(\widehat{K})}
\end{aligned}
$$

since $h_K \le 1$. Let $\widehat{p} \in \mathscr{P}^k(\widehat{K})$ be the polynomial of the Bramble–Hilbert Lemma for star-shaped domains, which closely approximates $\widehat{v}$, see [40]. It satisfies

$$
|\widehat{v} - \widehat{p}|_{H^\ell(\widehat{K})} \le C\, h^{k+1-\ell}_{\widehat{K}}\, |\widehat{v}|_{H^{k+1}(\widehat{K})} \quad \text{for } \ell = 0, 1, \ldots, k+1 \tag{2.25}
$$

with a constant C that only depends on $\sigma_{\mathcal{K}}$, d and k. Due to the scaling $h_{\widehat{K}} = 1$ and by the application of Lemmata 2.23 and 2.24, we obtain

$$\begin{aligned} \|\widehat{v} - \widehat{\mathfrak{I}}_h^k \widehat{v}\|_{H^1(\widehat{K})} &\leq \|\widehat{v} - \widehat{p}\|_{H^1(\widehat{K})} + \|\widehat{\mathfrak{I}}_h^k (\widehat{v} - \widehat{p})\|_{H^1(\widehat{K})} \\ &\leq (1+c)\, \|\widehat{v} - \widehat{p}\|_{H^2(\widehat{K})} \\ &\leq (1+c) C\, |\widehat{v}|_{H^{k+1}(\widehat{K})} , \end{aligned} \tag{2.26}$$

where we have used (2.25) in the last step. Comparing the previous estimates and transforming back to the element K yields

$$\|v - \mathfrak{I}_h^k v\|_{H^1(K)}^2 \leq c h_K^{2k}\, |v|_{H^{k+1}(K)}^2 .$$

Finally, we have to sum up this inequality over all elements of the mesh and apply the square root to it. This gives

$$\|v - \mathfrak{I}_h^k v\|_{H^1(\Omega)} \leq c \left(\sum_{K \in \mathcal{K}_h} h_K^{2k}\, |v|_{H^{k+1}(K)}^2 \right)^{1/2} \leq c\, h^k\, |v|_{H^{k+1}(K)} ,$$

and finishes the proof for $\ell = 1$. The case $\ell = 0$ follows by

$$\|v - \mathfrak{I}_h^k v\|_{L_2(K)} = h_K^{d/2} \|\widehat{v} - \widehat{\mathfrak{I}_h^k \widehat{v}}\|_{L_2(\widehat{K})} \leq h_K^{d/2} \|\widehat{v} - \widehat{\mathfrak{I}_h^k \widehat{v}}\|_{H^1(\widehat{K})} ,$$

and the same arguments as above.

The error estimate for $\mathfrak{I}_{h,H}^k$ follows in the same way. The case $k = 1$ is already proven since $\mathfrak{I}_{h,H}^1 = \mathfrak{I}_h^1$, thus let $k \geq 2$. The main difference is in (2.25), where we have to ensure that $\widehat{p}$ is harmonic. In the formulation of the Bramble–Hilbert Lemma in [40], $\widehat{p}$ is chosen as Taylor polynomial of $\widehat{v}$ averaged over the inscribed circle or ball of K given by the regularity of the mesh, cf. Definitions 2.1 and 2.11. Furthermore, the commutativity is proven for the operator of the weak derivative and the operator for the averaged Taylor polynomial for $k \geq 2$. Thus, since $\widehat{v} \in H^2(\widehat{K})$ and $\widehat{\Delta \widehat{v}} = 0$ in the weak sense, we obtain that the averaged Taylor polynomial $\widehat{p}$ is harmonic. □

Remark 2.28 The stability of the mesh $\mathcal{K}_h$ in the previous theorem is used only in order to apply Lemma 2.24. This assumption can be weakened in certain cases, see Remark 2.25. The statement of Theorem 2.27 still holds for $d = 2$ if solely the regularity of the mesh is assumed, and for $d = 3$ with $k = 1$ if the mesh is regular and weakly stable.

2.5 Galerkin Formulation and Convergence Estimates

In the previous sections we have discussed the discretization of the Sobolev space $H^1(\Omega)$ and investigated approximation properties. Thus, we come back to the model problem (2.1) and formulate the finite element method with the use of the introduced arbitrary order basis functions. Therefore, we consider in the following a bounded polygonal or polyhedral domain $\Omega \subset \mathbb{R}^d$, $d = 2, 3$ which is meshed by a regular polytopal mesh $\mathcal{K}_h$. In the three-dimensional case $d = 3$, we restrict ourselves to polyhedral elements with triangular faces as discussed in Sect. 2.3.4 and postpone the general case for later considerations.

In the case of inhomogeneous Dirichlet data g_D, we extend this boundary data into the interior of the domain. The extension is denoted by g_D again, and we assume that it can be chosen such that $g_D \in V_h^k$. Let

$$V_{h,D}^k = V_h^k \cap H_D^1(\Omega) \quad \text{with} \quad H_D^1(\Omega) = \{v \in H^1(\Omega) : v|_{\Gamma_D} = 0\} .$$

The Galerkin formulation for the model problem (2.1) reads:

$$\begin{aligned} &\text{Find } u \in g_D + H_D^1(\Omega) : \\ &b(u, v) = (f, v)_{L_2(\Omega)} + (g_N, v)_{L_2(\Gamma_N)} \quad \forall v \in H_D^1(\Omega) , \end{aligned} \tag{2.27}$$

and the corresponding discrete Galerkin formulation:

$$\begin{aligned} &\text{Find } u_h \in g_D + V_{h,D}^k : \\ &b(u_h, v_h) = (f, v_h)_{L_2(\Omega)} + (g_N, v_h)_{L_2(\Gamma_N)} \quad \forall v_h \in V_{h,D}^k , \end{aligned} \tag{2.28}$$

where

$$b(u_h, v_h) = (a\nabla u_h, \nabla v_h)_{L_2(\Omega)}$$

is the well known bilinear form for the diffusion problem. Due to the boundedness of the diffusion coefficient, the bilinear form $b(\cdot, \cdot)$ is bounded and elliptic on $H_D^1(\Omega)$. Because of the conforming approximation space $V_D^k \subset H_D^1(\Omega)$, the Galerkin as well as the discrete Galerkin formulation above admit a unique solution according to the Lax–Milgram Lemma. Céa's Lemma yields

$$\|u - u_h\|_{H^1(\Omega)} \leq c \min_{v_h \in g_D + V_{h,D}^k} \|u - v_h\|_{H^1(\Omega)} .$$

This quasi-best approximation gives rise to error estimates for the finite element formulation. The minimum on the right hand side can be estimated from above by setting $v_h = \mathfrak{I}_h^k u$. Thus, the interpolation estimates derived in Sect. 2.4 turn over to the finite element approximation. By the use of the interpolation properties given in Theorem 2.27, we obtain the next result. Since the mesh assumptions are mainly needed in order to apply the interpolation error estimates, these assumptions can be relaxed in the following theorems in certain situations, see Remark 2.28.

Theorem 2.29 *Let $\mathcal{K}_h$ be a regular and stable polytopal mesh of the bounded domain $\Omega \subset \mathbb{R}^d$. The solution $u_h \in V_h^k$ of the Galerkin formulation from above satisfies*

$$\|u - u_h\|_{H^1(\Omega)} \le c\, h^k\, |u|_{H^{k+1}(\Omega)} \quad \textit{for } u \in H^{k+1}(\Omega)\,,$$

where $h = \max\{h_K : K \in \mathcal{K}_h\}$ and the constant c only depends on the mesh parameters, the dimension d and on k.

If we assume more regularity for the model problem, the Aubin–Nitsche trick together with Theorem 2.29 can be used to prove an error estimate in the L_2-norm, see, e.g., [40].

Theorem 2.30 *Let $\mathcal{K}_h$ be a regular and stable polytopal mesh of the bounded domain $\Omega \subset \mathbb{R}^d$ and let there be, for any $g \in L_2(\Omega)$, a unique solution of*

$$\textit{Find } w \in H^1_D(\Omega): \quad b(v, w) = (g, v)_{L_2(\Omega)} \quad \forall v \in H^1_D(\Omega)\,,$$

with $w \in H^2(\Omega)$ such that

$$|w|_{H^2(\Omega)} \le C\, \|g\|_{L_2(\Omega)}\,.$$

The solution $u_h \in V_h^k$ of the Galerkin formulation from above satisfies

$$\|u - u_h\|_{L_2(\Omega)} \le c\, h^{k+1}\, |u|_{H^{k+1}(\Omega)} \quad \textit{for } u \in H^{k+1}(\Omega)\,,$$

where the constant c only depends on the mesh parameters, the dimension d and on k.

Proof Since $u - u_h \in H^1_D(\Omega) \subset L_2(\Omega)$, there is a unique $w \in H^2(\Omega)$ such that

$$b(v, w) = (u - u_h, v)_{L_2(\Omega)} \quad \text{for } v \in H^1_D(\Omega)$$

with

$$|w|_{H^2(\Omega)} \le C\, \|u - u_h\|_{L_2(\Omega)}\,. \tag{2.29}$$

The Galerkin orthogonality

$$b(u - u_h, v_h) = 0 \quad \text{for } v_h \in V^k_{h,D} \subset H^1_D(\Omega)$$

and the continuity of the bilinear form yield for $\mathfrak{I}_h^1 w \in V_{h,D}^k$

$$\begin{aligned}\|u - u_h\|_{L_2(\Omega)}^2 &= (u - u_h, u - u_h)_{L_2(\Omega)} = b(u - u_h, w) \\ &= b(u - u_h, w - \mathfrak{I}_h^1 w) \leq c\,\|u - u_h\|_{H^1(\Omega)} \|w - \mathfrak{I}_h^1 w\|_{H^1(\Omega)}\,.\end{aligned}$$

The first term on the right hand side is estimated using Theorem 2.29 and the second by the use of Theorem 2.27. This yields

$$\|u - u_h\|_{L_2(\Omega)}^2 \leq ch^{k+1}\,|u|_{H^{k+1}(\Omega)} |w|_{H^2(\Omega)}\,.$$

Applying (2.29) and dividing by $\|u - u_h\|_{L_2(\Omega)}$ gives the desired estimate. □

If the boundary value problem (2.1) has vanishing right hand side, i.e. $f = 0$, and thus the solution satisfies $u \in H_\Delta^1(\mathscr{K}_h)$, we can seek the approximation u_h directly in the subspace $V_{h,H}^k = \operatorname{span} \Psi_{h,H}^k \subset V_h^k$. Consequently, we obtain a reduced Galerkin formulation. The same arguments as above yield optimal rates of convergence, when the interpolation operator $\mathfrak{I}_{h,H}^k$ is used instead of $\mathfrak{I}_h^k$.

Theorem 2.31 *Under the same assumptions as in Theorems 2.29 and 2.30, the solution $u_h \in V_{h,H}^k$ of the reduced Galerkin formulation with $f = 0$ satisfies*

$$\|u - u_h\|_{H^\ell(\Omega)} \leq c\,h^{k+1-\ell}\,|u|_{H^{k+1}(\Omega)} \quad \textit{for } u \in H_\Delta^{k+1}(\mathscr{K}_h)\,,$$

where $\ell = 0, 1$ and the constant c only depends on the mesh parameters, the dimension d and on k as well as on ℓ.

Remark 2.32 The stability of the mesh $\mathscr{K}_h$ in the previous theorems can be weakened, cf. Remark 2.25. The statements still hold for $d = 2$ if solely the regularity of the mesh is assumed, and for $d = 3$ with $k = 1$ if the mesh is regular and weakly stable.

In the realization of the discrete Galerkin formulation (2.28), we have to address the evaluation of the bilinear form applied to ansatz functions. Since the diffusion coefficient is assumed to be constant on each element such that $a(\cdot) = a_K \in \mathbb{R}$ on K, for $K \in \mathscr{K}_h$, we have

$$b(\psi, \varphi) = (a\nabla\psi, \nabla\varphi)_{L_2(\Omega)} = \sum_{K \in \mathscr{K}_h} a_K (\nabla\psi, \nabla\varphi)_{L_2(K)} \quad \text{for } \psi, \varphi \in \Psi_h^k\,.$$

We remember that the basis $\Psi_h^k = \Psi_{h,H}^k \cup \Psi_{h,B}^k$ consists of piecewise harmonic functions and element basis functions which vanish on the element boundaries. According to (2.13), it holds

$$b(\psi, \varphi) = 0 \quad \text{for } \psi \in \Psi_{h,H}^k,\ \varphi \in \Psi_{h,B}^k\,, \tag{2.30}$$

and thus the discrete Galerkin formulation (2.28) decouples. If we split the unknown function into

$$u_h = u_{h,H} + u_{h,B} \quad \text{with} \quad u_{h,H} \in V^k_{h,H} \quad \text{and} \quad u_{h,B} \in V^k_{h,B}\,,$$

and take $g_D \in V^k_{h,H}$, we obtain with $V^k_{h,H,D} = V^k_{h,H} \cap H^1_D(\Omega)$

$$\begin{aligned} &\text{Find } u_{h,H} \in g_D + V^k_{h,H,D}: \\ &b(u_{h,H}, v_h) = (f, v_h)_{L_2(\Omega)} + (g_N, v_h)_{L_2(\Gamma_N)} \quad \forall v_h \in V^k_{h,H,D}\,, \end{aligned} \tag{2.31}$$

and

$$\text{Find } u_{h,B} \in V^k_{h,B}: \quad b(u_{h,B}, v_h) = (f, v_h)_{L_2(\Omega)} \quad \forall v_h \in V^k_{h,B}\,. \tag{2.32}$$

The problem (2.31) turns into a global system of linear equations with a symmetric and positive definite matrix. Since the support of each element basis function lies inside a single element, a closer look at (2.32) shows that the equation further decouples. The element contributions $u_{h,B}\big|_K \in H^1_0(K)$, $K \in \mathscr{K}_h$ are given as solution of

$$(\nabla u_{h,B}\big|_K, \nabla v_h)_{L_2(K)} = (f/a_K, v_h)_{L_2(K)} \quad \forall v_h \in V^k_{h,B}\big|_K$$

for each element $K \in \mathscr{K}_h$. Thus, $u_{h,B}\big|_K$ is locally the orthogonal projection of f/a_K into $V^k_{h,B}\big|_K = \operatorname{span} \Psi^k_{h,B}\big|_K$ the space of element bubble functions with respect to the scaler product $(\nabla\cdot, \nabla\cdot)_{L_2(K)}$. Furthermore, $u_{h,B}$ is separated from the global problem and can be computed via these local projections.

In Theorem 2.31, we already observed that in the case of a vanishing source term, i.e. $f = 0$, it is sufficient to seek the approximation $u_h \in V^k_h$ in the subspace $V^k_{h,H}$. This observation is confirmed by the decoupling of the Galerkin formulation. Because of $u_h = u_{h,H} + u_{h,B}$ with $u_{h,H} \in V^k_{h,H}$, and since the part $u_{h,B} \in V^k_{h,B}$ is uniquely defined by (2.32), we get $u_{h,B} = 0$ for $f = 0$ and thus $u_h = u_{h,H}$.

Furthermore, the property (2.30) and, consequently, the decoupling of the system is very practical from the computational point of view. The global system of linear equations reduces to a system which only involves the degrees of freedom corresponding to nodal and edge basis functions. The unknowns for the element basis functions can be computed independently element-by-element in a preprocessing step. Thus, there is no need for static condensation that is often used in high-order methods to eliminate the element-local degrees of freedom.

The decoupling is also an advantage over the virtual element method in [25]. This method has the same number of unknowns, but the system matrix does not decouple. Thus, a larger system of linear equations has to be solved. Another advantage of the presented strategy in this context is that the approximation u_h can be evaluated, or at least be approximated, in every point inside the domain with the help of the representation formula (4.3), see next section. The virtual element method, however, needs some postprocessing for the evaluation in an arbitrary point.

Remark 2.33 In the construction of the approximation space V_h^k, we have used the same order k over all edges and elements. However, these Trefftz-like basis functions can be used directly with variable order. There is no difficulty which has to be addressed. This flexibility is advantageous in hp-adaptivity, see [63].

Remark 2.34 More details on the computational realization as well as on the local treatment of the implicitly defined basis functions can be found in Chap. 4 and in particular in Sect. 4.5.

In the case of a continuously varying diffusion coefficient in the model problem (2.1), it is possible to approximate the coefficient a by a piecewise constant function a_h. To analyse the impact of this approximation, the first Strang Lemma [162] is used. Replacing the exact material coefficient in the bilinear form $b(\cdot,\cdot)$ by an approximated one can be seen as an approximation $b_h(\cdot,\cdot)$ of the bilinear form. Let the approximation a_h of a sufficiently regular diffusion coefficient satisfy

$$0 < a_{\min} \le a_h(\mathbf{x}) \le a_{\max} \quad \text{for } \mathbf{x} \in \Omega \text{ and } h > 0\,, \tag{2.33}$$

and $a_h(\mathbf{x}) = a_K \in \mathbb{R}$ for $\mathbf{x} \in K$ and $K \in \mathcal{K}_h$. Therefore, the bilinear form $b_h(\cdot,\cdot)$ is uniformly elliptic as well as bounded on V_h^k for $h > 0$, and the variational formulation has a unique solution. The Strang Lemma taken from [58] gives the error estimate

$$\|u - u_h\|_{H^1(\Omega)} \le c \inf_{v_h \in V_h^k} \left\{ \|u - v_h\|_{H^1(\Omega)} + \sup_{w_h \in V_h^k} \frac{|b(v_h, w_h) - b_h(v_h, w_h)|}{\|w_h\|_{H^1(\Omega)}} \right\},$$

for the Galerkin approximation. Obviously, the error in the finite element method is estimated by two terms. One which gives the quasi-best approximation error and one which measures the error coming from the inexact bilinear form. The latter one can be written and estimated in the form

$$\sup_{w_h \in V_h^k} \frac{|((a - a_h)\nabla v_h, \nabla w_h)|}{\|w_h\|_{H^1(\Omega)}} \le \sup_{w_h \in V_h^k} \sum_{K \in \mathcal{K}_h} \|a - a_h\|_{L_\infty(K)} \frac{|(\nabla v_h, \nabla w_h)_K|}{\|w_h\|_{H^1(\Omega)}}\,.$$

If the constant values a_K are chosen as averaged Tayler polynomials of order zero over the inscribed circle and ball of Definitions 2.1 and 2.11, respectively, we have $\|a - a_h\|_{L_\infty(K)} \le c h_K \|a\|_{W^1_\infty(K)}$, see [40], and we obtain after some arguments

$$\sup_{w_h \in V_h^k} \frac{|b(v_h, w_h) - b_h(v_h, w_h)|}{\|w_h\|_{H^1(\Omega)}} \le c\, h\, \|a\|_{W^1_\infty(\Omega)} \|v_h\|_{H^1(\Omega)}\,.$$

Choosing $v_h = \mathfrak{I}_h^k u$ in the Strang estimate and applying Theorem 2.27 as well as Lemma 2.24 for the interpolation operator yields

$$\|u - u_h\|_{H^1(\Omega)} \leq c\, h^k\, |u|_{H^{k+1}(\Omega)} + c\, h\, \|a\|_{W^1_\infty(\Omega)} \|u\|_{H^2(\Omega)} \quad \text{for } u \in H^{k+1}(\Omega)\,.$$

For high-order methods with $k > 1$, the convergence of the finite element error is dominated by the second term, which comes from the piecewise constant approximation of the diffusion coefficient.

In order to achieve the desired convergence rates, it is necessary to approximate the diffusion coefficient more accurately. For a sufficient regular coefficient a, one can use its interpolation $\mathfrak{I}_h^{k-1} a$, for example. For a more detailed discussion and for implementation details, see [146]. The ideas given there can be generalized to $k > 2$ directly.

2.6 Numerical Examples

Finally, the theoretical results are verified by some computational experiments. Theorems 2.29 and 2.30 are illustrated on a model problem. The BEM-based FEM is applied on a sequence of uniformly refined polygonal meshes. In each step of the refinement the boundary-value problem

$$-\Delta u = f \quad \text{in } \Omega = (0,1)^2\,, \qquad u = 0 \quad \text{on } \Gamma \tag{2.34}$$

is solved, where f is chosen such that $u(\mathbf{x}) = \sin(\pi x_1)\,\sin(\pi x_2)$ is the unique solution. The initial mesh and some refinements are shown in Fig. 2.12. The successively refined meshes are obtained by dividing each polygonal element as described in Sect. 2.2.3. The Galerkin error $\|u - u_h\|_{H^\ell(\Omega)}$ is computed for the H^1-norm ($\ell = 1$) and the L_2-norm ($\ell = 0$). In Fig. 2.13, the relative errors are plotted with respect to the mesh size $h = \max\{h_K : K \in \mathcal{K}_h\}$ on a logarithmic scale. The slopes of the curves reflect the theoretical rates of convergence for the approximation orders $k = 1, 2, 3$.

Next, we consider the model problem

$$-\Delta u = 0 \quad \text{in } \Omega = (0,1)^2\,, \qquad u = g_D \quad \text{on } \Gamma\,, \tag{2.35}$$

where g_D is chosen such that $u(\mathbf{x}) = \exp(2\pi(x_1 - 0.3))\cos(2\pi(x_2 - 0.3))$ is the unique solution. According to Theorem 2.31, it is sufficient to seek the approximation u_h in the space $V_{h,H}^k$ containing only piecewise weakly harmonic functions. Therefore, the number of degrees of freedom is reduced in the computations. We solve the reduced Galerkin formulation on a sequence of meshes produced by the Matlab tool PolyMesher, see [167], and compute the Galerkin errors as in the previous experiment. Some of the meshes are visualized in Fig. 2.14 and the relative errors are plotted with respect to the mesh size h in Fig. 2.15. The theoretical orders of convergence are achieved by the computations for $k = 1, 2, 3$.

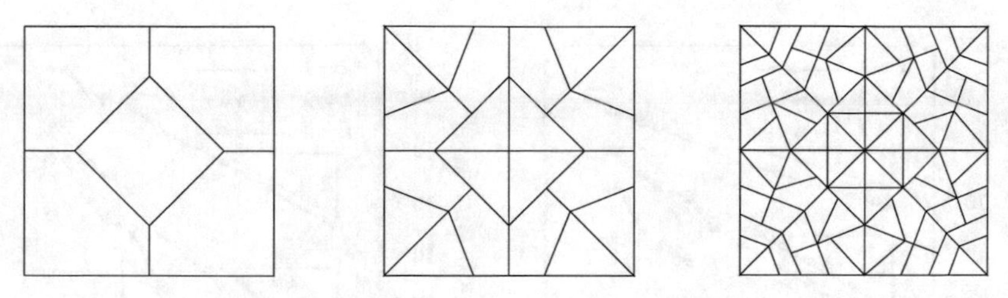

Fig. 2.12 Initial mesh (left), refined mesh after two steps (middle), refined mesh after four steps (right)

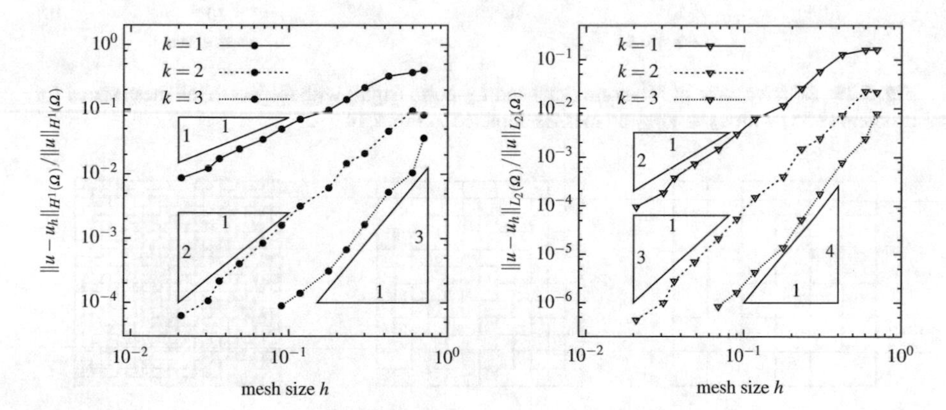

Fig. 2.13 Relative error in H^1-norm (left) and L_2-norm (right) with respect to the mesh size h for problem (2.34) with $u_h \in V_h^k$ on meshes depicted in Fig. 2.12

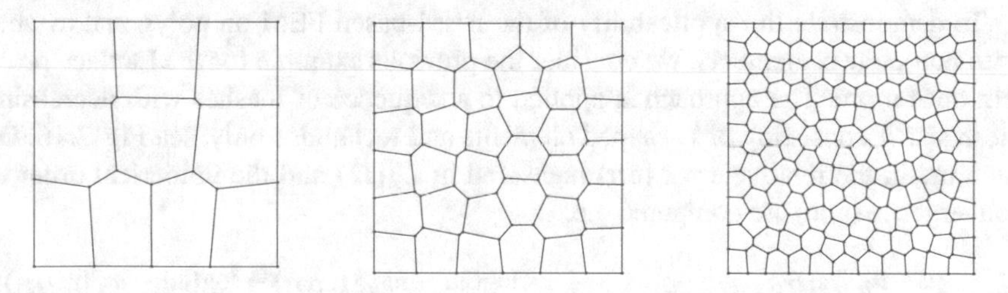

Fig. 2.14 First (left), fourth (middle) and sixth mesh (right) in uniform sequence generated by PolyMesher

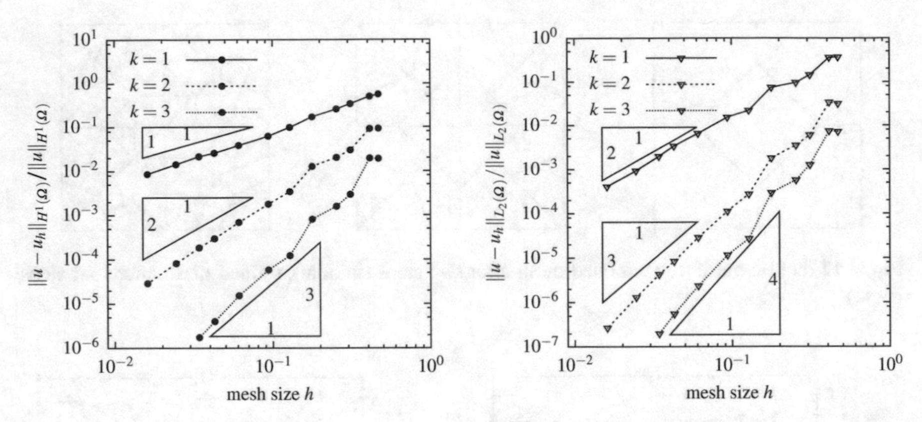

Fig. 2.15 Relative error in H^1-norm (left) and L_2-norm (right) with respect to the mesh size h for problem (2.35) with $u_h \in V^k_{h,H}$ on meshes depicted in Fig. 2.14

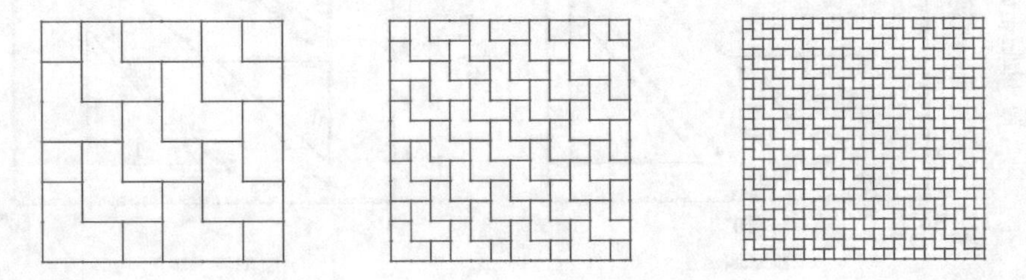

Fig. 2.16 Second (left), third (middle) and fourth mesh (right) in uniform sequence consisting of L-shaped elements and rectangles only

To demonstrate the applicability of the BEM-based FEM on polygonal meshes with non-convex elements, we consider the previous example for the Laplace problem once more. The approach is applied to a sequence of meshes with decreasing mesh size h consisting of L-shaped elements and rectangles only, see Fig. 2.16. On each mesh, the relative error (err) measured in $L_2(\Omega)$ and the numerical order of convergence (noc) are computed, i.e.

$$\text{err} = \frac{\|u - u_h\|_{L_2(\Omega)}}{\|u\|_{L_2(\Omega)}} \quad \text{and} \quad \text{noc} = \frac{\log(\|u - u_{2h}\|_{L_2(\Omega)}) - \log(\|u - u_h\|_{L_2(\Omega)})}{\log 2} .$$

In Table 2.1, the computed values are given together with the degrees of freedom in the trial space $V^k_{h,H,D} = V^k_{h,H} \cap H_D(\Omega)$ for $k = 1, 2, 3$. The results clearly demonstrate the optimal rates of convergence according to Theorem 2.31, where in the finest example for $k = 3$ saturation of accuracy is reached.

Table 2.1 Degrees of freedom (DoF), relative error measured in $L_2(\Omega)$ (err) and numerical order of convergence (noc) for problem (2.35) with $u_h \in V_{h,H}^k$ on meshes depicted in Fig. 2.16

	$k = 1$			$k = 2$			$k = 3$		
h	DoF	err	noc	DoF	err	noc	DoF	err	noc
9.43×10^{-1}	4	3.73×10^{-1}	–	12	3.48×10^{-2}	–	20	7.96×10^{-3}	–
4.71×10^{-1}	25	9.50×10^{-2}	1.97	65	3.50×10^{-3}	3.31	105	4.71×10^{-4}	4.08
2.36×10^{-1}	121	2.57×10^{-2}	1.88	297	3.90×10^{-4}	3.17	473	2.81×10^{-5}	4.07
1.18×10^{-1}	529	6.34×10^{-3}	2.02	1265	4.67×10^{-5}	3.06	2001	1.71×10^{-6}	4.04
5.89×10^{-2}	2209	1.58×10^{-3}	2.01	5217	5.72×10^{-6}	3.03	8225	1.05×10^{-7}	4.03
2.95×10^{-2}	9025	3.94×10^{-4}	2.00	21,185	7.07×10^{-7}	3.02	33,345	6.30×10^{-9}	4.05
1.47×10^{-2}	36,481	9.83×10^{-5}	2.00	85,377	8.30×10^{-8}	3.09	134,273	2.89×10^{-10}	4.44
7.37×10^{-3}	146,689	2.46×10^{-5}	2.00	342,785	1.03×10^{-8}	3.01	538,881	2.77×10^{-10}	–
Theory:			2			3			4

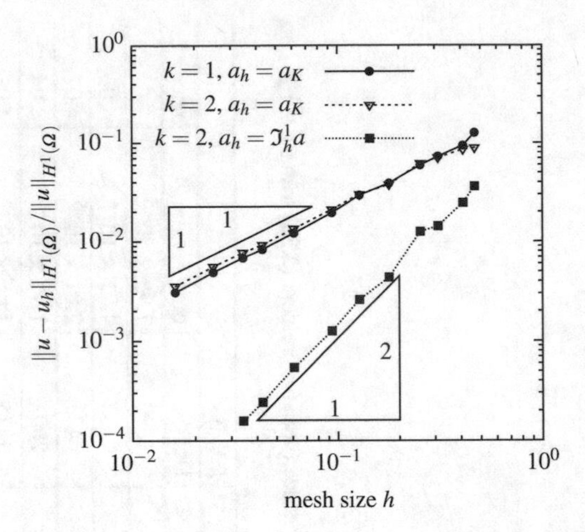

Fig. 2.17 Relative error in H^1-norm with respect to the mesh size h for problem (2.36) with $u_h \in V_h^k$ and different approximations a_h of the diffusion coefficient a on meshes depicted in Fig. 2.14

In order to study the effects of the approximation of the diffusion coefficient we consider the boundary value problem

$$-\operatorname{div}\left(\frac{1}{|\mathbf{x}-\mathbf{x}^*|}\nabla u\right) = 0 \quad \text{in } \Omega = (0,1)^2, \qquad u = g_D \quad \text{on } \Gamma\,, \tag{2.36}$$

where $\mathbf{x}^* = (-0.1, 0.2)^\top$. The Dirichlet boundary data g_D is chosen in such a way that $u(\mathbf{x}) = |\mathbf{x}-\mathbf{x}^*|$ is the exact solution. We apply the approach with the approximation space V_h^k on the uniform sequence generated by the PolyMesher, cf. Fig. 2.14. In the case of a piecewise constant approximation of the diffusion coefficient $a(\mathbf{x}) = |\mathbf{x}-\mathbf{x}^*|^{-1}$, the first order method for $k = 1$ converges with optimal order in the H^1-norm, whereas the second order method for $k = 2$ has a suboptimal convergence rate, see Fig. 2.17. This behaviour has been discussed theoretically in Sect. 2.5, where we observed that the error in the piecewise constant approximation of the diffusion coefficient dominates the convergence process for $k > 1$. Approximating a by $a_h = \mathfrak{I}_h^{k-1} a$, we recover the optimal rates, see Fig. 2.17. For a discussion of the implementation we refer the interested reader to [146].

Finally, a three-dimensional boundary value problem is considered

$$-\operatorname{div}(a\nabla u) = f \quad \text{in } \Omega = (0,1)^3, \qquad u = g_D \quad \text{on } \Gamma\,, \tag{2.37}$$

where $a(\mathbf{x}) = \frac{7}{2} - x_1 - x_2 - x_3$ and f as well as g_D are chosen such that the exact solution is $u(\mathbf{x}) = \cos(\pi x_1)\sin(2\pi x_2)\sin(3\pi \mathbf{x}_3)$. The diffusion coefficient is approximated by a piecewise constant function. The boundary value problem is solved on a uniform sequence of polyhedral meshes, the first one is depicted in

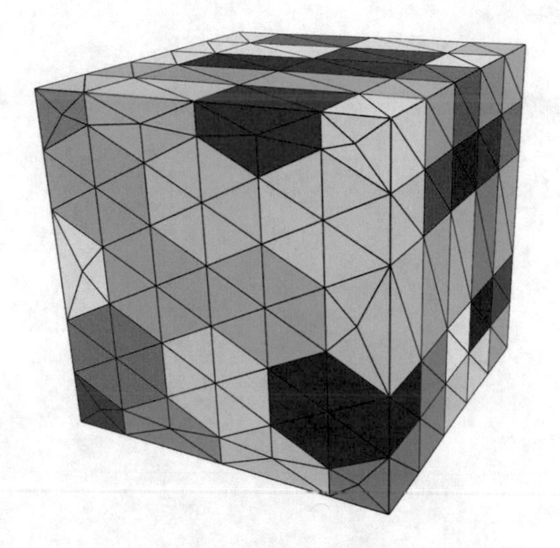

Fig. 2.18 First mesh of unit cube with tessellation of hexahedral bricks and triangular faces

Table 2.2 Degrees of freedom (DoF), relative error measured in the energy norm as well as in the $L_2(\Omega)$-norm with numerical order of convergence (noc) for problem (2.37) with $u_h \in V_h^1$ on meshes with triangulated faces, cf. Fig. 2.18

h	DoF	Energy err	noc	L_2-err	noc
4.23×10^{-1}	152	1.43×10^{0}	–	5.09×10^{-1}	–
2.17×10^{-1}	1176	7.86×10^{-1}	0.90	1.95×10^{-1}	1.44
1.46×10^{-1}	3936	5.40×10^{-1}	0.95	9.61×10^{-2}	1.79
1.10×10^{-1}	9296	4.10×10^{-1}	0.97	5.67×10^{-2}	1.87
8.73×10^{-2}	19,026	3.20×10^{-1}	1.06	3.48×10^{-2}	2.10
7.31×10^{-2}	32,575	2.69×10^{-1}	0.97	2.49×10^{-2}	1.89
6.28×10^{-2}	51,388	2.33×10^{-1}	0.98	1.86×10^{-2}	1.90
5.51×10^{-2}	76,329	2.04×10^{-1}	0.98	1.45×10^{-2}	1.92
4.87×10^{-2}	111,188	1.79×10^{-1}	1.07	1.12×10^{-2}	2.12
Theory:			1		2

Fig. 2.18. The meshes are constructed with the help of hexahedral bricks, where the polygonal faces are triangulated in order to apply the simple generalization for the three-dimensional approximation space in Sect. 2.3.4. The relative errors in the energy norm $\|\cdot\|_b = \sqrt{b(\cdot,\cdot)}$ and the L_2-norm are computed and given in Table 2.2. Furthermore, we give the numerical orders of convergence (noc), cf. (1.7), with respect to these norms. We observe linear convergence in the energy norm and quadratic convergence in the L_2-norm as predicted by the theory.

Chapter 3
Interpolation of Non-smooth Functions and Anisotropic Polytopal Meshes

The solutions of boundary value problems may contain singularities and/or have layers, where the solution changes rapidly. For such non-smooth functions, the application of pointwise interpolation is not well defined and in the presence of layers the use of regular and uniform meshes is not optimal in some sense. For these reasons quasi-interpolation operators for non-smooth functions over polytopal meshes are introduced and analysed in this chapter. In particular, operators of Clément- and Scott–Zhang-type are studied. Furthermore, the notion of anisotropic meshes is introduced. These meshes do not satisfy the classical regularity properties used in the approximation theory and thus they have to be treated in a special way. However, such meshes allow the accurate and efficient approximation of functions featuring anisotropic behaviours near boundary or interior layers.

3.1 Preliminaries

In the theory of classical interpolation it is assumed that the interpolant is at least in the Sobolev space $H^2(\Omega)$ or even smoother, such that point evaluations are well defined. When talking about non-smooth functions, we have those in mind which are only in $H^1(\Omega)$ and do not satisfy any further regularity. Such functions can be solutions of boundary value problems according to existence and uniqueness theory, cf. the Lax–Milgram Lemma given in Theorem 1.6. But, these functions do not fall in the theory of Sect. 2.5 yielding optimal rates of convergence on sequences of uniformly refined meshes. Instead of using pointwise values for the interpolation of non-smooth functions, one has to exploit averages of the function over certain neighbourhoods of the nodes.

© Springer Nature Switzerland AG 2019

S. Weißer, *BEM-based Finite Element Approaches on Polytopal Meshes*, Lecture Notes in Computational Science and Engineering 130,
https://doi.org/10.1007/978-3-030-20961-2_3

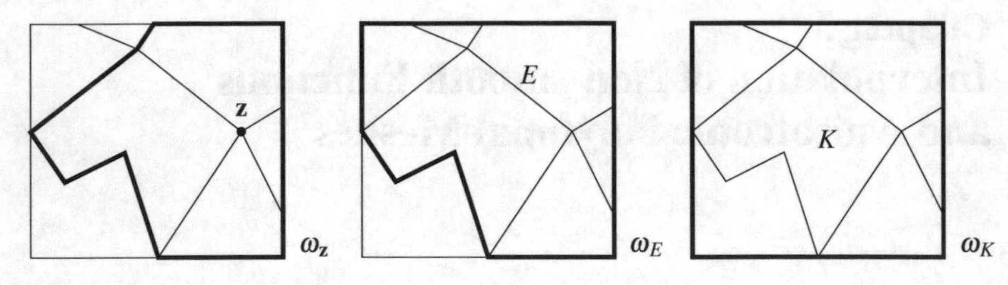

Fig. 3.1 Example of the neighbourhoods of nodes, edges and elements in two space dimensions

Let $\mathscr{K}_h$ be a polytopal mesh of a bounded domain Ω. For each node $\mathbf{z} \in \mathscr{N}_h$ in the mesh we consider its neighbourhood $\omega_{\mathbf{z}}$ defined by

$$\overline{\omega}_{\mathbf{z}} = \bigcup_{\mathbf{z} \in \mathscr{N}(K)} \overline{K}\,, \tag{3.1}$$

where $\mathscr{N}(K)$ denotes the set of all nodes belonging to the element $K \in \mathscr{K}_h$. Furthermore, we introduce the neighbourhoods of edges, faces and elements as

$$\overline{\omega}_E = \bigcup_{\mathbf{z} \in \mathscr{N}(E)} \overline{\omega}_{\mathbf{z}}\,, \qquad \overline{\omega}_F = \bigcup_{\mathbf{z} \in \mathscr{N}(F)} \overline{\omega}_{\mathbf{z}}\,, \qquad \overline{\omega}_K = \bigcup_{\mathbf{z} \in \mathscr{N}(K)} \overline{\omega}_{\mathbf{z}}\,, \tag{3.2}$$

cf. Fig. 3.1 for a visualization in two space dimensions. The neighbourhoods are open sets which are constructed by agglomerating elements next to the corresponding node, edge, face and element, respectively. The diameter of a neighbourhood ω is denoted by h_ω. An important role plays the neighbourhood $\omega_{\mathbf{z}}$. Its diameter $h_{\omega_{\mathbf{z}}}$ is of comparable size to the diameter of $K \subset \omega_{\mathbf{z}}$ as shown in

Lemma 3.1 *Let $\mathscr{K}_h$ be regular and stable mesh of a two- or three-dimensional domain. The following properties hold:*

1. *Each element is covered by a uniformly bounded number of neighbourhoods of elements, i.e. $|\{K' \in \mathscr{K}_h : K \subset \omega_{K'}\}| \leq c, \quad \forall K \in \mathscr{K}_h$.*
2. *For all $\mathbf{z} \in \mathscr{N}_h$ and $K \subset \omega_{\mathbf{z}}$, it is $h_{\omega_{\mathbf{z}}} \leq c h_K$.*

The constants $c > 0$ only depend on $\sigma_{\mathscr{K}}$, $\sigma_{\mathscr{F}}$ and $c_{\mathscr{K}}$.

Proof The first statement is seen easily. Let $K \in \mathscr{K}_h$ be fixed. Due to the above constructions, the number of element neighbourhoods $\omega_{K'}$ with $K \subset \omega_{K'}$ is equal to the number of elements contained in the neighbourhood ω_K. Consequently, the statement follows since

$$\overline{\omega}_K = \bigcup_{\mathbf{z} \in \mathscr{N}(K)} \overline{\omega}_{\mathbf{z}} = \bigcup_{\mathbf{z} \in \mathscr{N}(K)} \; \bigcup_{\mathbf{z} \in \mathscr{N}(K')} \overline{K'}\,,$$

and the number of nodes per element as well as the number of elements containing a node are bounded uniformly, see Lemmata 2.7 and 2.16 as well as Corollaries 2.6 and 2.15 for $d = 2$ and $d = 3$, respectively.

In order to see the second statement, we first recognize that

$$h_{\omega_{\mathbf{z}}} \leq 2 \max_{K' \subset \omega_{\mathbf{z}}} h_{K'} .$$

Let the maximum be reached for some K'. If K and K' share a common edge E, the stability of the mesh, i.e. $h_{K'} \leq c_{\mathscr{K}} h_E \leq c_{\mathscr{K}} h_K$, gives the desired estimate, namely $h_{\omega_{\mathbf{z}}} \leq 2c_{\mathscr{K}} h_K$. If K and K' do not share a common edge, we can construct a sequence of elements $K_i \subset \omega_{\mathbf{z}}$, $i = 1, \ldots, n$ such that $K_1 = K'$, $K_n = K$ and K_i and K_{i+1} share a common edge. Arguing as above yields

$$h_{\omega_{\mathbf{z}}} \leq 2 \left(c_{\mathscr{K}}\right)^{n-1} h_K .$$

Since the number of elements contained in $\omega_{\mathbf{z}}$, and thus in particular n, is uniformly bounded according to Corollaries 2.6 and 2.15, the statement is proven. □

In the forthcoming sections, we treat the two- and three-dimensional cases with $d = 2, 3$ simultaneously. Therefore, if we write F, $\mathscr{F}_h$ and so forth, we mean the faces of the discretization for $d = 3$ and with some abuse of notation the edges for $d = 2$. In this chapter, we restrict ourselves to the first order approximation space V_h^1 with $k = 1$ and we simply write V_h for shorter notation. In the three-dimensional case we may use the simple generalization for the construction of V_h introduced in Sect. 2.3.4 which relies on polyhedral elements with triangular faces. The theory in this chapter is also valid for the case of polyhedral elements with polygonal faces. The detailed description of the approximation space is discussed in the later Sect. 6.2. At this point, however, we give a small outlook in order to present the full theory for quasi-interpolation operators. The generalization of V_h to polyhedral elements with polygonal faces reads

$$V_h = \left\{ v \in H^1(\Omega) : \Delta v\big|_K = 0 \ \forall K \in \mathscr{K}_h \text{ and } v\big|_F \in V_h(F) \ \forall F \in \mathscr{F}_h \right\} , \tag{3.3}$$

where $V_h(F)$ denotes the two-dimensional discretization space over the face F. The nodal basis functions are constructed as in the two-dimensional case but they have to satisfy additionally the Laplace equation in the linear parameter space of each face.

3.2 Trace Inequality and Best Approximation

Before we introduce quasi-interpolation operators and study error estimates, some analytic auxiliary results are reviewed and extended. These include in particular trace inequalities and approximation results for the L_2-projection into the space of constants over patches of elements. If no confusion arises, we write v for both the function and the trace of the function on an edge and face, respectively.

In the two-dimensional setting, Lemma 2.4 guaranties the existence of the isosceles triangles with common angles for non-convex elements in a regular polygonal mesh. This is sufficient to guaranty the following lemma proven in [174].

Lemma 3.2 *Let $\mathscr{K}_h$ be a regular mesh, $v \in H^1(K)$ for $K \in \mathscr{K}_h$ and $E \in \mathscr{E}(K)$. It holds*

$$\|v\|_{L_2(E)} \leq c \left\{ h_E^{-1/2} \|v\|_{L_2(T_E^{\mathrm{iso}})} + h_E^{1/2} |v|_{H^1(T_E^{\mathrm{iso}})} \right\}$$

with the isosceles triangle $T_E^{\mathrm{iso}} \subset K$ from Lemma 2.4, where c only depends on $\alpha_{\mathscr{K}}$, and thus, on the regularity parameter $\sigma_{\mathscr{K}}$.

Under the additional assumption on the stability of the mesh, we can generalize this trace inequality and state a similar result, which is valid for $d = 2, 3$. Remember the convention that F denotes a face or edge depending on the considered dimensions d.

Lemma 3.3 (Trace Inequality) *Let $\mathscr{K}_h$ be a regular and stable mesh, $v \in H^1(K)$ for $K \in \mathscr{K}_h$ and $F \in \mathscr{F}(K)$. It holds*

$$\|v\|_{L_2(F)} \leq c \left(h_F^{-1/2} \|v\|_{L_2(K)} + h_F^{1/2} |v|_{H^1(K)} \right),$$

where c only depends on $\sigma_{\mathscr{K}}$, $\sigma_{\mathscr{F}}$ and $c_{\mathscr{K}}$.

Proof Since $\mathscr{K}_h$ is regular and stable, we have an auxiliary discretization $\mathscr{T}_h(K)$ into tetrahedra such that each face $F \in \mathscr{F}(K)$ is decomposed into triangular facets of these tetrahedra. According to Lemma 2.14 the discretization $\mathscr{T}_h(K)$ is shape regular in the sense of Ciarlet. It is well known, see [2, 40], that there is a constant C only depending on the regularity parameters of the auxiliary discretization such that for $T_{\mathrm{tet}} \in \mathscr{T}_h(K)$ and $v \in H^1(T_{\mathrm{tet}})$ it holds

$$\|v\|^2_{L_2(\partial T_{\mathrm{tet}})} \leq C \left(h_{T_{\mathrm{tet}}}^{-1} \|v\|^2_{L_2(T_{\mathrm{tet}})} + h_{T_{\mathrm{tet}}} \|v\|^2_{H^1(T_{\mathrm{tet}})} \right).$$

Furthermore, it is $h_F / c_{\mathscr{K}} \leq h_{T_{\mathrm{tet}}} \leq c_{\mathscr{K}} h_F$, cf. Sect. 2.2.2, and thus we obtain for the triangle $T \subset \partial T_{\mathrm{tet}} \cap F$ and $v \in H^1(K)$ that

$$\|v\|^2_{L_2(T)} \leq c \left(h_F^{-1} \|v\|^2_{L_2(T_{\mathrm{tet}})} + h_F \|v\|^2_{H^1(T_{\mathrm{tet}})} \right).$$

Summing this inequality for all triangles which lie in F yields the desired result. □

Another important analytical tool is the approximation of functions in Sobolev spaces by polynomials. We already applied such results over polytopal elements in the proof of Theorem 2.27. Since these elements are star-shaped, the well known results from [40, 69] are applicable. Next, we consider the polynomial approximation over the neighbourhoods defined in (3.1) and (3.2) which are not star-shaped in general, and therefore we have to extend the theory.

Lemma 3.4 *Let $\mathcal{K}_h$ be a regular and stable mesh and $k \in \mathbb{N}_0$. There exists for every function $v \in H^{k+1}(\omega)$ and every neighbourhood $\omega \in \{\omega_{\mathbf{z}}, \omega_F, \omega_K\}$ a polynomial $p \in \mathcal{P}^k(\omega)$ such that*

$$|v - p|_{H^\ell(\omega)} \leq C\, h_\omega^{k+1-\ell}\, |v|_{H^{k+1}(\omega)} \quad \textit{for } \ell = 0, \ldots, k+1\,,$$

where C only depends on $\sigma_{\mathcal{K}}$, $\sigma_{\mathcal{F}}$ and $c_{\mathcal{K}}$ as well as on k and the dimension d.

Proof Let $\omega \in \{\omega_{\mathbf{z}}, \omega_F, \omega_K\}$, since $\mathcal{K}_h$ is regular and stable, there is an auxiliary discretization of ω into tetrahedra formed by $\mathcal{T}_h(K)$ of all $K \subset \omega$. This discretization is shape regular in the sense of Ciarlet and the number of tetrahedra is uniformly bounded because there are only finitely many K with $K \subset \omega$ according to Lemma 3.1 and each element is decomposed into a bounded number of tetrahedra according to the Lemmata 2.7 and 2.16. Now, that we have a uniformly bounded number of tetrahedra with uniformly bounded aspect ratios due to the regularity, we can argue as in [8] adapting an iterative procedure already mentioned in [69]. We skip the rest of the proof and refer the interested reader to the cited literature. □

The previous result can be applied to obtain error estimates for the L_2-projection. We only consider the projection into the space of constants. For $v \in H^1(\omega)$ this projection is given by

$$\Pi_\omega v = \frac{1}{|\omega|} \int_\omega v(\mathbf{x})\, \mathrm{d}\mathbf{x}\,.$$

It is known that the Poincaré constant

$$C_P(\omega) = \sup_{v \in H^1(\omega)} \frac{\|v - \Pi_\omega v\|_{L_2(\omega)}}{h_\omega |v|_{H^1(\omega)}} < \infty \tag{3.4}$$

is finite and depends on the shape of ω, see [169]. Exploiting that

$$\|v - \Pi_\omega v\|_{L_2(\omega)} = \min_{q \in \mathbb{R}} \|v - q\|_{L_2(\omega)}\,,$$

we deduce from Lemma 3.4 that the Poincaré constant $C_P(\omega)$ is bounded uniformly for the neighbourhoods $\omega \in \{\omega_{\mathbf{z}}, \omega_F, \omega_K\}$ in a regular and stable mesh.

Lemma 3.5 *Let $\mathcal{K}_h$ be a regular and stable mesh. There exists a uniform constant c, which only depends on $\sigma_{\mathcal{K}}$, $\sigma_{\mathcal{F}}$ and $c_{\mathcal{K}}$, such that for every neighbourhood*

$\omega \in \{\omega_{\mathbf{z}}, \omega_F, \omega_K\}$ with $\mathbf{z} \in \mathcal{N}_h$, $F \in \mathcal{F}_h$ and $K \in \mathcal{K}_h$, it holds

$$\|v - \Pi_\omega v\|_{L_2(\omega)} \leq c h_\omega |v|_{H^1(\omega)} \quad \text{for } v \in H^1(\omega)\,.$$

In the following, we give an alternative proof for the two-dimensional case ($d = 2$) with $\omega = \omega_{\mathbf{z}}$. For convex ω, the authors of [136] showed $C_P(\omega) < 1/\pi$. In our situation, however, $\omega_{\mathbf{z}}$ is a patch of non-convex elements which is itself non-convex in general. We proceed as in [180]. The main tool in the forthcoming proof is Proposition 2.10 (Decomposition) of [169]. As preliminary of this proposition, an admissible decomposition $\{\omega_i\}_{i=1}^n$ of ω with pairwise disjoint domains ω_i and

$$\overline{\omega} = \bigcup_{i=1}^{n} \overline{\omega}_i$$

is needed. Admissible means in this context, that there exist triangles $\{T_i\}_{i=1}^n$ such that $T_i \subset \omega_i$ and for every pair i, j of different indices, there is a sequence $i = k_0, \ldots, k_\ell = j$ of indices such that for every m the triangles $T_{k_{m-1}}$ and T_{k_m} share a complete side. Under these assumptions, the Poincaré constant of ω is bounded by

$$C_P(\omega) \leq \max_{1\leq i\leq n} \left\{ 8(n-1)\left(1 - \min_{1\leq j\leq n} \frac{|\omega_j|}{|\omega|}\right)\left(C_P^2(\omega_i) + 2C_P(\omega_i)\right)\frac{|\omega|\, h_{\omega_i}^2}{|T_i|\, h_\omega^2}\right\}^{1/2}. \tag{3.5}$$

Proof (Lemma 3.5, Alternative for $d = 2$ with $\omega = \omega_{\mathbf{z}}$) Before we prove the estimate, we note that $C_P(K) < c$ for an element K which satisfies the regularity and stability assumptions of Definitions 2.1 and 2.2. This follows by remembering the construction of the auxiliary triangulation $\mathcal{T}_h(K)$. K can be interpreted as patch of triangles corresponding to the point $\mathbf{z}_K$. Thus, we choose $\omega_i = T_i$, $i = 1, \ldots, n$ with $\{T_i\}_{i=1}^n = T_h(K)$ for the admissible decomposition of K. The integer n corresponds to the number of nodes in K and thus it is uniformly bounded according to Lemma 2.7. Furthermore, it is $C_P(\omega_i) < 1/\pi$, $|K| < h_K^2$ and $h_{\omega_i}^2/|T_i| = h_{T_i}^2/|T_i| \leq c$, because of the shape-regularity of the auxiliary triangulation proven in Lemma 2.3. Consequently, the application of Proposition 2.10 (Decomposition) from [169] yields $C_P(K) < c$.

Now, we address the estimate for general $\omega_{\mathbf{z}}$ in the lemma. Therefore, we apply once more Proposition 2.10 of [169]. For this reason, we construct a decomposition $\{\omega_i\}_{i=1}^n$ and show that it is admissible by giving explicitly a set of triangles $\{T_i\}_{i=1}^n$ which satisfy the above mentioned properties. Furthermore, the terms in (3.5) are estimated.

To simplify the construction, we first assume that the patch consists of only one element, i.e. $\omega_{\mathbf{z}} = K \in \mathcal{K}_h$, and let $E_1, E_2 \in \mathcal{E}(K)$ with $\mathbf{z} = \overline{E}_1 \cap \overline{E}_2$. We decompose $\omega_{\mathbf{z}}$, or equivalently K, into ω_1 and ω_2 such that $n = 2$. The decomposition is done by splitting K along the polygonal chain through the points $\mathbf{z}$, $\mathbf{z}_K$ and $\mathbf{z}'$, where $\mathbf{z}' \in \mathcal{N}(K)$ is chosen such that the angle $\beta = \angle \mathbf{z}\mathbf{z}_K\mathbf{z}'$ is

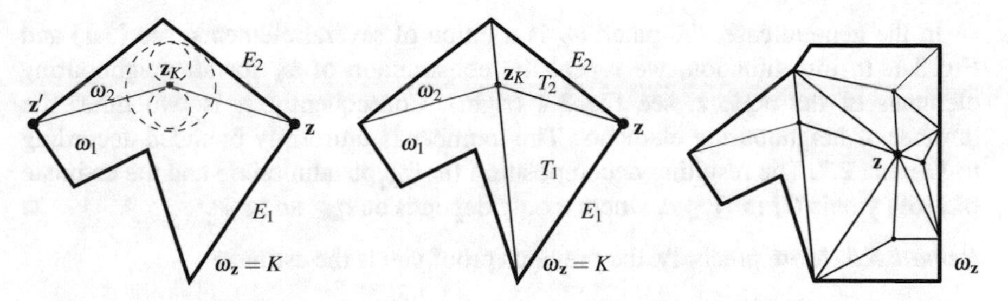

Fig. 3.2 Construction of admissible decomposition for K and $\omega_\mathbf{z}$ from Fig. 3.1

maximized, see Fig. 3.2 left. It is $\beta \in (\pi/2, \pi]$, since K is star-shaped with respect to a circle centered at $\mathbf{z}_K$. The triangles $\{T_i\}_{i=1}^n$ are chosen from the auxiliary triangulation in Lemma 2.3 as $T_i = T_{E_i} \in \mathscr{T}_h(K)$, cf. Fig. 3.2 middle. Obviously, $\{\omega_i\}_{i=1}^n$ is an admissible decomposition. Next, we estimate the terms in (3.5) and show that $C_P(\omega_i) \leq c$. The element K is star-shaped with respect to a circle of radius ρ_K and we have split this circle into two circular sectors during the construction of ω_i, $i = 1, 2$. A small calculation shows that ω_i is also star-shaped with respect to a circle of radius

$$\rho_{\omega_i} = \frac{\rho_K \sin(\beta/2)}{1 + \sin(\beta/2)},$$

which lies inside the mentioned circular sector and consequently satisfies the relation $\rho_K/(1+\sqrt{2}) < \rho_{\omega_i} \leq \rho_K/2$, see Fig. 3.2 (left). Thus, the aspect ratio of ω_i is uniformly bounded, since

$$\frac{h_{\omega_i}}{\rho_{\omega_i}} \leq \frac{(1+\sqrt{2})h_K}{\rho_K} \leq (1+\sqrt{2})\sigma_{\mathscr{K}} .$$

Furthermore, we observe that $h_{\omega_i} \leq h_K \leq \sigma_{\mathscr{K}} \rho_K \leq \sigma_{\mathscr{K}} |\overline{\mathbf{z}\mathbf{z}_K}|$ and accordingly $h_{\omega_i} \leq \sigma_{\mathscr{K}} |\overline{\mathbf{z}'\mathbf{z}_K}|$. Consequently, ω_i, $i = 1, 2$ is a regular element in the sense of Definition 2.1 and thus, we have already proven that $C_P(\omega_i) \leq c$. Additionally, we obtain by (2.2) and by the regularity of the mesh that

$$\frac{h_{\omega_i}^2}{|T_i|} \leq \frac{2h_{\omega_i}^2}{h_{E_i}\rho_K} \leq \frac{2h_K^2}{h_{E_i}\rho_K} \leq 2c_{\mathscr{K}}\sigma_{\mathscr{K}} .$$

This yields together with $|\omega_\mathbf{z}| \leq h_{\omega_\mathbf{z}}^2$ and Proposition 2.10 (Decomposition) of [169] that

$$C_P(\omega_\mathbf{z}) \leq \left(16(n-1)\left(c^2+2c\right)c_{\mathscr{K}}\sigma_{\mathscr{K}}\right)^{1/2} ,$$

and thus, a uniform bound in the case of $\omega_\mathbf{z} = K$ and $n = 2$.

In the general case, the patch $\omega_{\mathbf{z}}$ is a union of several elements, see (3.1) and Fig. 3.1. In this situation, we repeat the construction of ω_i for all neighbouring elements of the node $\mathbf{z}$, see Fig. 3.2 (right). Consequently, n is two times the number of neighbouring elements. This number is uniformly bounded according to Lemma 2.7. The resulting decomposition $\{\omega_i\}_{i=1}^n$ is admissible and the estimate of [169] yields $C_P(\omega_{\mathbf{z}}) \leq c$, where c only depends on $\sigma_{\mathcal{K}}$ and $c_{\mathcal{K}}$. □

Remark 3.6 More precisely, the previous proof yields the estimate

$$C_P(\omega_{\mathbf{z}}) \leq \left(16(n-1)c_{\mathcal{K}}\sigma_{\mathcal{K}} \max_{1\leq i\leq n}\left\{C_P^2(\omega_i) + 2C_P(\omega_i)\right\}\right)^{1/2},$$

where n is two times the number of elements in $\omega_{\mathbf{z}}$ that is usually a small number. Consequently, $C_P(\omega_{\mathbf{z}})$ is controlled by $C_P(\omega_i)$, $i = 1, \ldots, n$ which only depend on the chunkiness parameter $h_{\omega_i}/\rho_{\omega_i}$ according to [40].

3.3 Quasi-Interpolation of Non-smooth Functions

In the case of smooth functions like in $H^2(\Omega)$, it is possible to use nodal interpolation. Such interpolation operators have been constructed and studied in Sect. 2.4, and they yield optimal approximation error estimates. The goal of this section, however, is to define interpolation for general functions in $H^1(\Omega)$. Consequently, quasi-interpolation operators are applied, which utilizes certain neighbourhoods of the nodes. Classical results on simplicial meshes go back to Clément [59] and to Scott and Zhang [154]. They use L_2-projections instead of point evaluations in order to specify the expansion coefficients in the given basis.

For $v \in H^1(\Omega)$, we are interested in quasi-interpolation operators of the form

$$\mathfrak{I}v = \sum_{\mathbf{z}\in\mathcal{N}_*} (\Pi_{\omega(\psi_{\mathbf{z}})}v)\psi_{\mathbf{z}} \in V_h\,, \tag{3.6}$$

where the set of nodes $\mathcal{N}_*$ and the neighbourhoods $\omega(\psi_{\mathbf{z}})$, which depend on the first order basis functions, have to be specified. The Clément and Scott–Zhang interpolation operators differ in the choice of $\mathcal{N}_*$ and $\omega(\psi_{\mathbf{z}})$. Furthermore, it is desirable that homogeneous Dirichlet data is preserved such that $\mathfrak{I}v \in H_D^1(\Omega)$ for $v \in H_D^1(\Omega)$.

3.3.1 Clément-Type Interpolation

The Clément interpolation operator $\mathfrak{I}_C$ is defined as usual by (3.6), where we choose $\mathcal{N}_* = \mathcal{N}_h \setminus \mathcal{N}_{h,D}$ as all nodes which do not lie on the Dirichlet boundary, and

$\omega(\psi_{\mathbf{z}}) = \omega_{\mathbf{z}}$ as neighbourhood of the nodes. Thus, the interpolation is given as a linear combination of the basis functions $\psi_{\mathbf{z}}$ associated to the nodes in the interior of Ω and the Neumann boundary Γ_N. The expansion coefficients are chosen as average over the neighbourhood of the corresponding nodes. For $v \in H^1_D(\Omega)$, it is $\mathfrak{I}_C v \in H^1_D(\Omega)$ by construction.

Theorem 3.7 *Let $\mathscr{K}_h$ be a regular and stable mesh and let $F \in \mathscr{F}_h$ and $K \in \mathscr{K}_h$. The Clément interpolation operator satisfies for $v \in H^1_D(\Omega)$ the interpolation error estimates*

$$\|v - \mathfrak{I}_C v\|_{L_2(K)} \leq c h_K |v|_{H^1(\omega_K)} \quad \text{and} \quad \|v - \mathfrak{I}_C v\|_{L_2(F)} \leq c h_F^{1/2} |v|_{H^1(\omega_F)} ,$$

where the constants c only depend on $\sigma_{\mathscr{K}}$, $\sigma_{\mathscr{F}}$ and $c_{\mathscr{K}}$.

Proof The proof follows the arguments of [170, 174] with several modifications for the treatment of polytopal meshes. We start with the first estimate. For $K \in \mathscr{K}_h$, we have the partition of unity property, i.e. $\sum_{\mathbf{z}\in\mathscr{N}(K)} \psi_{\mathbf{z}} = 1$ on $\overline{K}$ and $\|\psi_{\mathbf{z}}\|_{L_\infty(\overline{K})} = 1$ for $\mathbf{z} \in \mathscr{N}(K)$. We distinguish two cases, let all nodes $\mathbf{z} \in \mathscr{N}(K)$ of the element K be located in the interior of Ω or in the interior of the boundary Γ_N, i.e. $\mathbf{z} \in \mathscr{N}_h \setminus \mathscr{N}_{h,D}$. Applying the best approximation result Lemma 3.5, we obtain

$$\begin{aligned}
\|v - \mathfrak{I}_C v\|_{L_2(K)} &\leq \sum_{\mathbf{z}\in\mathscr{N}(K)} \|\psi_{\mathbf{z}}(v - \Pi_{\omega_{\mathbf{z}}} v)\|_{L_2(K)} \\
&\leq \sum_{\mathbf{z}\in\mathscr{N}(K)} \|v - \Pi_{\omega_{\mathbf{z}}} v\|_{L_2(\omega_{\mathbf{z}})} \\
&\leq \sum_{\mathbf{z}\in\mathscr{N}(K)} c h_{\omega_{\mathbf{z}}} |v|_{H^1(\omega_{\mathbf{z}})} \\
&\leq c h_K |v|_{H^1(\omega_K)} .
\end{aligned}$$

In the last step we used that the number of nodes in $\mathscr{N}(K)$ is uniformly bounded, see Lemmata 2.7, 2.16, and 3.1, which gives $h_{\omega_{\mathbf{z}}} \leq c h_K$. In the case that at least one node of the element K is on the Dirichlet boundary Γ_D, i.e. $\mathbf{z} \in \mathscr{N}_{h,D}$, we write

$$\begin{aligned}
v - \mathfrak{I}_C v &= \sum_{\mathbf{z}\in\mathscr{N}(K)} \psi_{\mathbf{z}} v - \sum_{\mathbf{z}\in\mathscr{N}(K)\setminus\mathscr{N}_{h,D}} \psi_{\mathbf{z}} \Pi_{\omega_{\mathbf{z}}} v \\
&= \sum_{\mathbf{z}\in\mathscr{N}(K)} \psi_{\mathbf{z}}(v - \Pi_{\omega_{\mathbf{z}}} v) + \sum_{\mathbf{z}\in\mathscr{N}(K)\cap\mathscr{N}_{h,D}} \psi_{\mathbf{z}} \Pi_{\omega_{\mathbf{z}}} v ,
\end{aligned}$$

and obtain

$$\|v - \mathfrak{I}_C v\|_{L_2(K)} \leq \sum_{\mathbf{z}\in\mathscr{N}(K)} \|\psi_{\mathbf{z}}(v - \Pi_{\omega_{\mathbf{z}}} v)\|_{L_2(K)} + \sum_{\mathbf{z}\in\mathscr{N}(K)\cap\mathscr{N}_{h,D}} \|\psi_{\mathbf{z}} \Pi_{\omega_{\mathbf{z}}} v\|_{L_2(K)} .$$

The first sum has already been treated and the term in the second sum can be estimated by

$$\|\psi_{\mathbf{z}} \Pi_{\omega_{\mathbf{z}}} v\|_{L_2(K)} \leq |\Pi_{\omega_{\mathbf{z}}} v| \, \|\psi_{\mathbf{z}}\|_{L_\infty(K)} \, |K|^{1/2} \leq h_K^{d/2} |\Pi_{\omega_{\mathbf{z}}} v| \, .$$

Because of $\mathbf{z} \in \mathcal{N}_{h,D}$, there is an element $K' \subset \omega_{\mathbf{z}}$ and a face $F' \in \mathcal{F}(K')$ in the Dirichlet boundary, such that $\mathbf{z} \in \mathcal{N}(F')$ and $F' \in \mathcal{F}_{h,D}$. Therefore, v vanishes on F' and we obtain with the trace inequality, see Lemma 3.3,

$$\begin{aligned}
|\Pi_{\omega_{\mathbf{z}}} v| &= |F'|^{-1/2} \, \|v - \Pi_{\omega_{\mathbf{z}}} v\|_{L_2(F')} \\
&\leq c|F'|^{-1/2} h_{F'}^{1/2} \left\{ h_{F'}^{-1} \|v - \Pi_{\omega_{\mathbf{z}}} v\|_{L_2(\omega_{\mathbf{z}})} + |v|_{H^1(\omega_{\mathbf{z}})} \right\} \\
&\leq c h_{K'}^{1-d/2} \left\{ h_{\omega_{\mathbf{z}}}^{-1} \|v - \Pi_{\omega_{\mathbf{z}}} v\|_{L_2(\omega_{\mathbf{z}})} + |v|_{H^1(\omega_{\mathbf{z}})} \right\} ,
\end{aligned}$$

where we exploit $h_{K'}^{d-1} \leq c|F'|$, see Remark 2.13, $h_{F'} \leq h_{K'}$, and $h_{\omega_{\mathbf{z}}} \leq c h_{K'} \leq c h_{F'}$ according to Lemma 3.1 and the stability of the mesh. The best approximation, see Lemma 3.5, and the observations $h_K \leq h_{\omega_{\mathbf{z}}} \leq c h_{K'}$ as well as $1 - d/2 \leq 0$ gives

$$|\Pi_{\omega_{\mathbf{z}}} v| \leq c h_K^{1-d/2} |v|_{H^1(\omega_{\mathbf{z}})} \, . \tag{3.7}$$

Putting all estimates together proves the first statement of the theorem.

To prove the second estimate of the theorem, we proceed in a similar manner. Let $F \in \mathcal{F}_h$ be an edge ($d = 2$) or a face ($d = 3$). We have $\sum_{\mathbf{z} \in \mathcal{N}(F)} \psi_{\mathbf{z}} = 1$ on $\overline{F}$ and $\|\psi_{\mathbf{z}}\|_{L_\infty(F)} = 1$ for $\mathbf{z} \in \mathcal{N}(F)$. First, let $F \in \mathcal{F}_h$ be such that all its nodes $\mathbf{z} \in \mathcal{N}(F)$ are located in the interior of Ω or in the interior of the boundary Γ_N, i.e. $\mathbf{z} \in \mathcal{N}_h \setminus \mathcal{N}_{h,D}$. Applying the trace inequality, see Lemma 3.3, with an element $K' \in \mathcal{K}_h$ that satisfies $K' \subset \omega_{\mathbf{z}}$ and $F \in \mathcal{F}(K')$, as well as the best approximation, see Lemma 3.5, we obtain as above

$$\begin{aligned}
\|v - \mathfrak{I}_h v\|_{L_2(F)} &\leq \sum_{\mathbf{z} \in \mathcal{N}(F)} \|v - \Pi_{\omega_{\mathbf{z}}} v\|_{L_2(F)} \\
&\leq \sum_{\mathbf{z} \in \mathcal{N}(F)} c h_F^{1/2} \left\{ h_F^{-1} \|v - \Pi_{\omega_{\mathbf{z}}} v\|_{L_2(\omega_{\mathbf{z}})} + |v|_{H^1(\omega_{\mathbf{z}})} \right\} \\
&\leq \sum_{\mathbf{z} \in \mathcal{N}(F)} c h_F^{1/2} |v|_{H^1(\omega_{\mathbf{z}})} \\
&\leq c h_F^{1/2} |v|_{H^1(\omega_F)} \, .
\end{aligned}$$

If at least one node of F is on Γ_D, i.e. $\mathbf{z} \in \mathcal{N}_{h,D}$, we have

$$\|v - \mathfrak{I}_h v\|_{L_2(F)} \leq \sum_{\mathbf{z}\in\mathcal{N}(F)} \|\psi_\mathbf{z}(v - \Pi_{\omega_\mathbf{z}} v)\|_{L_2(F)} + \sum_{\mathbf{z}\in\mathcal{N}(F)\cap\mathcal{N}_{h,D}} \|\psi_\mathbf{z}\Pi_{\omega_\mathbf{z}} v\|_{L_2(F)} \,.$$

The first sum is treated as before, so let us have a look at the second sum. For $\mathbf{z} \in \mathcal{N}(F) \cap \mathcal{N}_{h,D}$ and some element $K' \in \mathcal{K}_h$ with $F \in \mathcal{F}(K')$, we have according to (3.7)

$$\|\psi_\mathbf{z}\Pi_{\omega_\mathbf{z}} v\|_{L_2(F)} \leq |F|^{1/2}\,|\Pi_{\omega_\mathbf{z}} v| \leq c\,|F|^{1/2}\,h_{K'}^{1-d/2}|v|_{H^1(\omega_\mathbf{z})} \leq c\,h_F^{1/2}|v|_{H^1(\omega_\mathbf{z})} \,,$$

where in the last estimate we have used $|F| \leq h_F^{d-1}$ and $h_{K'}^{1-d/2} \leq h_F^{1-d/2}$. Putting all estimates together and exploiting that the number of nodes per edge ($d = 2$) and face ($d = 3$) is uniformly bounded, see Lemma 2.7 and Definition 2.10, yields the second statement of the theorem and concludes the proof. □

3.3.2 Scott–Zhang-Type Interpolation

The Scott–Zhang interpolation operator $\mathfrak{I}_{SZ} : H^1(\Omega) \to V_h$ is defined as usual by (3.6), where we choose $\mathcal{N}_* = \mathcal{N}_h$ and $\omega(\psi_\mathbf{z}) = F_\mathbf{z}$, where $F_\mathbf{z} \in \mathcal{F}_h$ is an edge ($d = 2$) or face ($d = 3$) with $\mathbf{z} \in \overline{F}_\mathbf{z}$ and

$$F_\mathbf{z} \subset \Gamma_D \text{ if } \mathbf{z} \in \overline{\Gamma}_D \qquad \text{and} \qquad F_\mathbf{z} \subset \Omega \cup \Gamma_N \text{ if } \mathbf{z} \in \Omega \cup \Gamma_N \,.$$

Thus, the interpolation is given as a linear combination of all basis functions $\psi_\mathbf{z}$. The expansion coefficients are chosen as average over edges and faces. By construction, it is $\mathfrak{I}_{SZ} v \in H_D^1(\Omega)$ for $v \in H_D^1(\Omega)$, such that homogeneous Dirichlet data is preserved. We have the following local stability result, which can be utilized to derive interpolation error estimates.

Lemma 3.8 *Let $\mathcal{K}_h$ be a regular and stable mesh and $K \in \mathcal{K}_h$. The Scott–Zhang interpolation operator satisfies for $v \in H^1(\Omega)$ the local stability*

$$\|\mathfrak{I}_{SZ} v\|_{L_2(K)} \leq c\left(\|v\|_{L_2(\omega_K)} + h_K|v|_{H^1(\omega_K)}\right) \,,$$

where the constant c only depends on $\sigma_\mathcal{K}$, $\sigma_\mathcal{F}$ and $c_\mathcal{K}$.

Proof The only non-vanishing basis functions $\psi_\mathbf{z}$ over K in the expansion of $\mathfrak{I}_{SZ} v$ are those with $\mathbf{z} \in \mathcal{N}(K)$. Due to the stability of the L_2-projection $\Pi_{F_\mathbf{z}}$ we have $\|\Pi_{F_\mathbf{z}} v\|_{L_2(F_\mathbf{z})} \leq \|v\|_{L_2(F_\mathbf{z})}$. Furthermore, there exists $K_\mathbf{z} \in \mathcal{K}_h$ with $F_\mathbf{z} \subset \partial K_\mathbf{z}$ such that $K_{\mathbf{z}} \subset \omega_K$. Therefore, we obtain with the trace inequality, see Lemma 3.3,

$$\begin{aligned}|\Pi_{F_{\mathbf{z}}} v| &= |F_{\mathbf{z}}|^{-1/2} \, \|\Pi_{F_{\mathbf{z}}} v\|_{L_2(F_{\mathbf{z}})} \\ &\leq c\,|F_{\mathbf{z}}|^{-1/2} h_{F_{\mathbf{z}}}^{-1/2} \left(\|v\|_{L_2(K_{\mathbf{z}})} + h_{F_{\mathbf{z}}} |v|_{H^1(K_{\mathbf{z}})}\right) \\ &\leq c\,|K_{\mathbf{z}}|^{-1/2} \left(\|v\|_{L_2(K_{\mathbf{z}})} + h_{K_{\mathbf{z}}} |v|_{H^1(K_{\mathbf{z}})}\right),\end{aligned}$$

since $|K_{\mathbf{z}}| \leq h_{K_{\mathbf{z}}}^d \leq c_{\mathscr{K}}^d h_{F_{\mathbf{z}}}^{d-1} h_{F_{\mathbf{z}}} \leq c|F_{\mathbf{z}}| h_{F_{\mathbf{z}}}$ and $h_{F_{\mathbf{z}}} \leq h_{K_{\mathbf{z}}}$ due to the regularity and stability of the mesh. Utilizing this estimate and $\|\psi_{\mathbf{z}}\|_{L_\infty(K)} = 1$ yields

$$\begin{aligned}\|\mathfrak{I}_{SZ} v\|_{L_2(K)} &\leq \sum_{\mathbf{z}\in\mathscr{N}(K)} \|(\Pi_{F_{\mathbf{z}}} v)\, \psi_{\mathbf{z}}\|_{L_2(K)} \\ &\leq \sum_{\mathbf{z}\in\mathscr{N}(K)} |(\Pi_{F_{\mathbf{z}}} v)|\, \|\psi_{\mathbf{z}}\|_{L_\infty(K)}\, |K|^{1/2} \\ &\leq c \sum_{\mathbf{z}\in\mathscr{N}(K)} \left(\frac{|K|}{|K_{\mathbf{z}}|}\right)^{1/2} \left(\|v\|_{L_2(K_{\mathbf{z}})} + h_{K_{\mathbf{z}}} |v|_{H^1(K_{\mathbf{z}})}\right).\end{aligned}$$

Furthermore, it is $K, K_{\mathbf{z}} \subset \omega_{\mathbf{z}}$ and thus $h_{K_{\mathbf{z}}} \leq h_{\omega_{\mathbf{z}}}$. Lemma 3.1 yields $h_{\omega_{\mathbf{z}}} \leq ch_K$ and consequently $h_{K_{\mathbf{z}}} \leq ch_K$. Additionally, we can bound $|K|/|K_{\mathbf{z}}|$ uniformly, because of $|K| \leq h_K^d \leq ch_{K_{\mathbf{z}}}^d \leq c\sigma_{\mathscr{K}}^d \rho_{K_{\mathbf{z}}}^d \leq c|K_{\mathbf{z}}|$, since the d-dimensional ball of radius $\rho_{K_{\mathbf{z}}}$ is inscribed in $K_{\mathbf{z}}$. Exploiting that $K_{\mathbf{z}} \subset \omega_K$ and that the number of nodes per element is uniformly bounded, see Lemmata 2.7 and 2.16, finishes the proof. □

Theorem 3.9 *Let $\mathscr{K}_h$ be a regular and stable mesh and $K \in \mathscr{K}_h$. The Scott–Zhang interpolation operator satisfies for $v \in H^1(\Omega)$ the interpolation error estimate*

$$\|v - \mathfrak{I}_{SZ} v\|_{L_2(K)} \leq ch_K |v|_{H^1(\omega_K)},$$

where the constant c only depends on $\sigma_{\mathscr{K}}$, $\sigma_{\mathscr{F}}$ and $c_{\mathscr{K}}$.

Proof For $p = \Pi_{\omega_K} v \in \mathbb{R}$ it is obviously $p = \mathfrak{I}_{SZ} p$ and $\nabla p = 0$. The estimate in the theorem follows by Lemma 3.8 and the application of Lemma 3.4, since

$$\begin{aligned}\|v - \mathfrak{I}_{SZ} v\|_{L_2(K)} &\leq \|v - p\|_{L_2(K)} + \|\mathfrak{I}_{SZ}(v - p)\|_{L_2(K)} \\ &\leq c\left(\|v - p\|_{L_2(\omega_K)} + h_K |v|_{H^1(\omega_K)}\right) \\ &\leq ch_K |v|_{H^1(\omega_K)}.\end{aligned}$$

□

3.4 Anisotropic Polytopal Meshes

When dealing with highly anisotropic solutions of boundary value problems, it is widely recognized that anisotropic mesh refinements have significant potential for improving the efficiency of the solution process. Pioneering works for the analysis of finite element methods on anisotropic meshes have been performed by Apel [10] as well as by Formaggia and Perotto [78, 79]. The meshes usually consist of triangular and quadrilateral elements in two-dimension as well as on tetrahedral and hexahedral elements in three-dimension. First results on a posteriori error estimates for driving adaptive mesh refinement with anisotropic elements have been derived by Kunert [119] for triangular and tetrahedral meshes. For the mesh generation and adaptation different concepts are available which rely on metric-based strategies, see, e.g., [108, 125], or on splitting of elements, see [152] and the references therein. The anisotropic splitting of classical elements, however, results in certain restrictions why several authors combine this approach with additional strategies like edge swapping, node removal and local node movement. These restrictions come from the limited element shapes and the necessity to remove or handle hanging nodes in the discretization. For three-dimensional elements the situation is even more difficult. In contrast, anisotropic polytopal elements promise a high potential in the accurate resolution of sharp layers in the solutions of boundary value problems due to their enormous flexibility. An appropriate framework is developed in this section.

3.4.1 Characterisation of Anisotropy and Regularity

Let $K \subset \mathbb{R}^d$, $d = 2, 3$ be a bounded polytopal element. Furthermore, we assume that K is not degenerated, i.e. $|K| = \text{meas}_d(K) > 0$. We define the center or mean of K as

$$\bar{\mathbf{x}}_K = \frac{1}{|K|}\int_K \mathbf{x}\,\mathrm{d}\mathbf{x}$$

and the covariance matrix of K as

$$M_{\text{Cov}}(K) = \frac{1}{|K|}\int_K (\mathbf{x}-\bar{\mathbf{x}}_K)(\mathbf{x}-\bar{\mathbf{x}}_K)^\top\,\mathrm{d}\mathbf{x} \in \mathbb{R}^{d\times d}\,.$$

This matrix has already been used in Sect. 2.2.3 for the bisection of elements in the discussion of mesh refinement. Obviously, M_{Cov} is real valued, symmetric and positive definite since K is not degenerated. Therefore, it admits an eigenvalue decomposition

$$M_{\text{Cov}}(K) = U_K \Lambda_K U_K^\top$$

with

$$U_K^\top = U_K^{-1} \quad \text{and} \quad \Lambda_K = \operatorname{diag}(\lambda_{K,1}, \ldots, \lambda_{K,d}) \,.$$

Without loss of generality, let the eigenvalues satisfy $\lambda_{K,1} \geq \ldots \geq \lambda_{K,d} > 0$ and we assume that the corresponding eigenvectors $\mathbf{u}_{K,1}, \ldots, \mathbf{u}_{K,d}$, collected in U, form a basis of $\mathbb{R}^d$ with the same orientation for all considered elements $K \in \mathcal{K}_h$.

The eigenvectors of $M_{\mathrm{Cov}}(K)$ give the characteristic directions of K. This fact is, e.g., also used in the principal component analysis (PCA). The eigenvalue $\lambda_{K,j}$ is the variance of the underlying data in the direction of the corresponding eigenvector $\mathbf{u}_{K,j}$. Thus, the square root of the eigenvalues give the standard deviations in a statistical setting. Consequently, if

$$M_{\mathrm{Cov}}(K) = cI$$

for $c > 0$, there are no dominant directions in the element K. We characterise the anisotropy with the help of the quotient $\lambda_{K,1}/\lambda_{K,d} \geq 1$ and call an element

$$\text{isotropic, if} \quad \frac{\lambda_{K,1}}{\lambda_{K,d}} \approx 1 \,,$$

$$\text{and anisotropic, if} \quad \frac{\lambda_{K,1}}{\lambda_{K,d}} \gg 1 \,.$$

For $d = 3$, we might even characterise whether the element is anisotropic in one or more directions by comparing the different combinations of eigenvalues.

Exploiting the spectral information of the polytopal elements, we next introduce a linear transformation of an anisotropic element K onto a kind of reference element $\widehat{K}$. For each $\mathbf{x} \in K$, we define the mapping by

$$\mathbf{x} \mapsto \widehat{\mathbf{x}} = \mathfrak{F}_K(\mathbf{x}) = A_K\mathbf{x} \quad \text{with} \quad A_K = \alpha_K \Lambda_K^{-1/2} U_K^\top \,, \tag{3.8}$$

and $\alpha_K > 0$, which will be chosen later. $\widehat{K} = \mathfrak{F}_K(K)$ is called reference configuration later on.

Lemma 3.10 *Under the above transformation, it holds*

1. $|\widehat{K}| = |K|\,|\det(A_K)| = \alpha_K^d |K| / \sqrt{\prod_{j=1}^d \lambda_{K,j}}$,
2. $\bar{\mathbf{x}}_{\widehat{K}} = \mathfrak{F}_K(\bar{\mathbf{x}}_K)$,
3. $M_{\mathrm{Cov}}(\widehat{K}) = \alpha_K^2 I$.

Proof First, we recognize that

$$\det(A_K) = \alpha_K^d \det(\Lambda_K^{-1/2} U_K^\top) = \alpha_K^d / \sqrt{\det(\Lambda_K)} = \alpha_K^d / \sqrt{\textstyle\prod_{j=1}^d \lambda_{K,j}} \,.$$

Consequently, we obtain by the transformation

$$|\widehat{K}| = \int_{\widehat{K}} \mathrm{d}\widehat{\mathbf{x}} = |K|\,|\det(A_K)| = \alpha_K^d |K| / \sqrt{\det(M_{\mathrm{Cov}}(K))}\,,$$

that proves the first statement. For the center, we have

$$\bar{\mathbf{x}}_{\widehat{K}} = \frac{1}{|\widehat{K}|}\int_{\widehat{K}} \widehat{\mathbf{x}}\,\mathrm{d}\widehat{\mathbf{x}} = \frac{|\det(A_K)|}{|\widehat{K}|} A_K \int_K \mathbf{x}\,\mathrm{d}\mathbf{x} = A_K \bar{\mathbf{x}}_K = \mathfrak{F}_K(\bar{\mathbf{x}}_K)\,.$$

The covariance matrix has the form

$$\begin{aligned} M_{\mathrm{Cov}}(\widehat{K}) &= \frac{1}{|\widehat{K}|}\int_{\widehat{K}} (\widehat{\mathbf{x}} - \bar{\mathbf{x}}_{\widehat{K}})(\widehat{\mathbf{x}} - \bar{\mathbf{x}}_{\widehat{K}})^\top \mathrm{d}\widehat{\mathbf{x}} \\ &= \frac{|\det(A_K)|}{|\widehat{K}|}\int_K A_K(\mathbf{x} - \bar{\mathbf{x}}_K)\,(A_K(\mathbf{x} - \bar{\mathbf{x}}_K))^\top \mathrm{d}\mathbf{x} \\ &= A_K M_{\mathrm{Cov}}(K) A_K^\top \\ &= \alpha_K^2 (\Lambda_K^{-1/2} U_K^\top)(U_K \Lambda_K U_K^\top)(\Lambda_K^{-1/2} U_K^\top)^\top \\ &= \alpha_K^2\, I\,, \end{aligned}$$

that finishes the proof. □

According to the previous lemma, the reference configuration $\widehat{K}$ is isotropic, since $\lambda_{\widehat{K},1}/\lambda_{\widehat{K},d} = 1$, and thus, it has no dominant direction. We can still choose the parameter α_K in the mapping. We might use $\alpha_K = 1$ such that the variance of the element in every direction is equal to one. On the other hand, we can use the parameter α_K in order to normalise the volume of $\widehat{K}$ such that $|\widehat{K}| = 1$. This is achieved by

$$\alpha_K = \left(\frac{\sqrt{\det(M_{\mathrm{Cov}}(K))}}{|K|} \right)^{1/d} = \left(\frac{\sqrt{\prod_{j=1}^d \lambda_{K,j}}}{|K|} \right)^{1/d}, \tag{3.9}$$

see Lemma 3.10, and will be used in the following.

Example 3.11 The transformation (3.8) for α_K according to (3.9) is demonstrated for an anisotropic element $K \subset \mathbb{R}^2$, i.e. $d = 2$. The element K is depicted in Fig. 3.3 (left). The eigenvalues of $M_{\mathrm{Cov}}(K)$ are

$$\lambda_{K,1} \approx 111.46 \quad \text{and} \quad \lambda_{K,2} \approx 1.18\,,$$

and thus

$$\frac{\lambda_{K,1}}{\lambda_{K,2}} \approx 94.40 \gg 1\,.$$

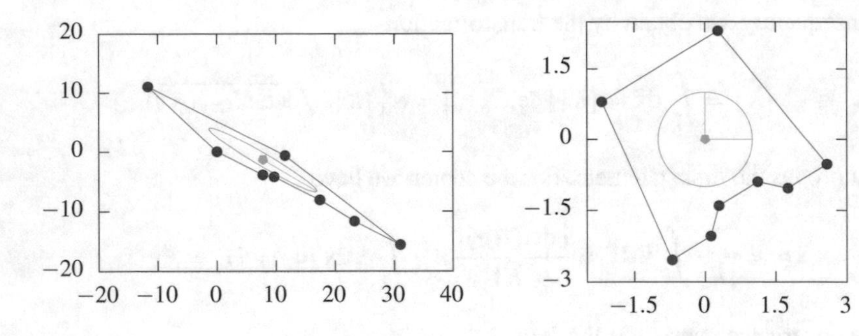

Fig. 3.3 Demonstration of transformation (3.8): original anisotropic element (left) and transformed element centered at the origin (right)

In Fig. 3.3, we additionally visualize the eigenvectors of $M_{\text{Cov}}(K)$ scaled by the square root of their corresponding eigenvalue and centered at the mean of the element. The ellipse is the one given uniquely by the scaled vectors. In the right picture of Fig. 3.3, the transformed element $\widehat{K} = \mathfrak{F}_K(K)$ is given with the scaled eigenvectors of its covariant matrix $M_{\text{Cov}}(\widehat{K})$. The computation verifies $|\widehat{K}| = 1$, and we have

$$M_{\text{Cov}}(\widehat{K}) \approx \begin{pmatrix} 8.59 \cdot 10^{-2} & -3.93 \cdot 10^{-17} \\ -3.93 \cdot 10^{-17} & 8.59 \cdot 10^{-2} \end{pmatrix}.$$

In view of the quasi-interpolation and interpolation operators and their approximation properties, the meshes have to guaranty certain requirements. In the previous analysis of such operators, we made use of isotropic polytopal elements in regular and stable meshes $\mathscr{K}_h$. The corresponding definitions of Sect. 2.2 are summarized in the following remark.

Remark 3.12 Let $\mathscr{K}_h$ be a polytopal mesh. $\mathscr{K}_h$ is called a regular and stable (isotropic) mesh, if all elements $K \in \mathscr{K}_h$ satisfy:

1. K is a star-shaped polygon/polyhedron with respect to a circle/ball of radius ρ_K and midpoint $\mathbf{z}_K$.
2. The aspect ratio is uniformly bounded from above by $\sigma_{\mathscr{K}}$, i.e. $h_K/\rho_K < \sigma_{\mathscr{K}}$.
3. For the element K and all its edges $E \in \mathscr{E}(K)$ it holds $h_K \le c_{\mathscr{K}} h_E$.
4. In the case $d = 3$, all polygonal faces $F \in \mathscr{F}(K)$ of the polyhedral element K are star-shaped with respect to a circle of radius ρ_F and midpoint $\mathbf{z}_F$ and their aspect ratio is uniformly bounded, i.e. $h_F/\rho_F < \sigma_{\mathscr{F}}$.

Obviously, these assumptions are not satisfied in the case of anisotropic meshes. The aspect ratio of the element depicted in Fig. 3.3 (left) is very large and one of its edges degenerates compared with the element diameter. In the definition of regular and stable anisotropic meshes, we make use of the previously introduced reference configuration.

Definition 3.13 (Regular and Stable Anisotropic Mesh) Let $\mathscr{K}_h$ be a polytopal mesh. $\mathscr{K}_h$ is called a regular and stable anisotropic mesh, if:

1. The reference configuration $\widehat{K}$ for all $K \in \mathscr{K}_h$ obtained by (3.8) is a regular and stable polytopal element according to Sect. 2.2, see Remark 3.12.
2. Neighbouring elements behave similarly in their anisotropy. More precisely, for two neighbouring elements K_1 and K_2, i.e. $\overline{K}_1 \cap \overline{K}_2 \neq \varnothing$, with covariance matrices

$$M_{\mathrm{Cov}}(K_1) = U_{K_1} \Lambda_{K_1} U_{K_1}^\top \quad \text{and} \quad M_{\mathrm{Cov}}(K_2) = U_{K_2} \Lambda_{K_2} U_{K_2}^\top$$

as defined above, we can write

$$\Lambda_{K_2} = (I + \Delta^{K_1,K_2})\Lambda_{K_1} \quad \text{and} \quad U_{K_2} = R^{K_1,K_2} U_{K_1}$$

with

$$\Delta^{K_1,K_2} = \operatorname{diag}\left(\delta_j^{K_1,K_2} : j = 1, \ldots, d\right),$$

and a rotation matrix $R^{K_1,K_2} \in \mathbb{R}^{d\times d}$ such that for $j = 1, \ldots, d$

$$0 \le |\delta_j^{K_1,K_2}| < c_\delta < 1 \quad \text{and} \quad 0 \le \|R^{K_1,K_2} - I\|_2 \left(\frac{\lambda_{K_1,1}}{\lambda_{K_1,d}}\right)^{1/2} < c_R$$

uniformly for all neighbouring elements, where $\|\cdot\|_2$ denotes the spectral norm.

In the rest of the chapter, the generic constant c may also depend on c_δ and c_R.

Remark 3.14 For $d = 2$, the rotation matrix has the form

$$R^{K_1,K_2} = \begin{pmatrix} \cos\phi^{K_1,K_2} & -\sin\phi^{K_1,K_2} \\ \sin\phi^{K_1,K_2} & \cos\phi^{K_1,K_2} \end{pmatrix},$$

with an angle ϕ^{K_1,K_2}. For the spectral norm $\|R^{K_1,K_2} - I\|_2$, we recognize that

$$(R^{K_1,K_2} - I)^\top (R^{K_1,K_2} - I) = \left(\sin^2\phi^{K_1,K_2} + (1 - \cos\phi^{K_1,K_2})^2\right) I,$$

and consequently

$$\begin{aligned} \|R^{K_1,K_2} - I\|_2 &= \left(\sin^2\phi^{K_1,K_2} + (1 - \cos\phi^{K_1,K_2})^2\right)^{1/2} \\ &= 2\left|\sin\left(\frac{\phi^{K_1,K_2}}{2}\right) - \sin(0)\right| \\ &\le |\phi^{K_1,K_2}|, \end{aligned}$$

according to the mean value theorem. The assumption on the spectral norm in Definition 3.13 can thus be replaced by

$$|\phi^{K_1,K_2}| \left(\frac{\lambda_{K_1,1}}{\lambda_{K_1,2}}\right)^{1/2} < c_\phi \,.$$

This implies that neighbouring highly anisotropic elements have to be aligned in almost the same directions, whereas isotropic or moderately anisotropic elements might vary in their characteristic directions locally.

Let us study the reference configuration $\widehat{K} \subset \mathbb{R}^d$, $d = 2, 3$ of $K \in \mathscr{K}_h$, which is regular and stable. Due to the scaling with α_K, it is $|\widehat{K}| = 1$ and we obtain

$$1 = |\widehat{K}| \le h_{\widehat{K}}^d \le \sigma_{\mathscr{K}}^d \rho_{\widehat{K}}^d = \sigma_{\mathscr{K}}^d \; \nu\pi\rho_{\widehat{K}}^d \,/\, (\nu\pi) \le \sigma_{\mathscr{K}}^d \; |\widehat{K}| \,/\, (\nu\pi) = \sigma_{\mathscr{K}}^d \,/\, (\nu\pi)\,,$$

where $\nu = 1$ for $d = 2$ and $\nu = 4/3$ for $d = 3$, since the circle/ball is inscribed the element $\widehat{K}$. Consequently, we obtain

$$1 \le h_{\widehat{K}} \le \frac{\sigma_{\mathscr{K}}}{(\nu\pi)^{1/d}} \,. \tag{3.10}$$

Furthermore, for $d = 3$, let $\widehat{F}$ be a face of $\widehat{K}$ and denote by $\widehat{E}$ one of its edges, i.e., $\widehat{E} \in \mathscr{E}(\widehat{F})$. Due to the regularity and stability, we find

$$|\widehat{F}| \ge \pi\rho_{\widehat{F}}^2 \ge \pi h_{\widehat{F}}^2/\sigma_{\mathscr{F}}^2 \ge \pi h_{\widehat{E}}^2/\sigma_{\mathscr{F}}^2 \ge \pi h_{\widehat{K}}^2/(c_{\mathscr{K}}\sigma_{\mathscr{F}}^2)\,,$$

and thus for $d = 2, 3$

$$h_{\widehat{K}}^{d-1} \le c|\widehat{F}| \,. \tag{3.11}$$

A regular and stable anisotropic element can be mapped according to the previous definition onto a regular and stable polytopal element in the usual sense. In the definition of quasi-interpolation operators, we deal, however, with patches of elements instead of single elements. Thus, we have to study the mapping of such patches. These include in particular the patches $\omega_{\mathbf{z}}$, ω_F and ω_K defined in Sect. 3.1.

Lemma 3.15 *Let $\mathscr{K}_h$ be a regular and stable anisotropic mesh, $\omega_{\mathbf{z}}$ be the patch of elements corresponding to the node $\mathbf{z} \in \mathscr{N}_h$, and $K_1, K_2 \in \mathscr{K}_h$ with $K_1, K_2 \subset \omega_{\mathbf{z}}$. The mapped element $\mathfrak{F}_{K_1}(K_2)$ is regular and stable in the sense of Sect. 2.2, see Remark 3.12, with slightly perturbed regularity and stability parameters $\widetilde{\sigma}_{\mathscr{K}}$ and $\widetilde{c}_{\mathscr{K}}$ depending only on the regularity and stability of $\mathscr{K}_h$. Consequently, the mapped patch $\mathfrak{F}_K(\omega_{\mathbf{z}})$ consists of regular and stable polytopal elements for all $K \in \mathscr{K}_h$ with $K \subset \omega_{\mathbf{z}}$.*

Proof We verify Remark 3.12 for the mapped element $\widetilde{K}_2 = \mathfrak{F}_{K_1}(K_2)$.

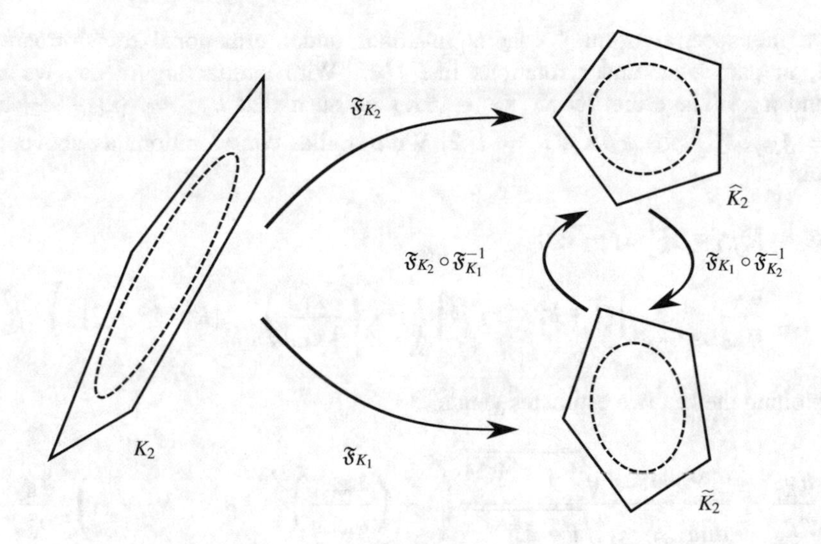

Fig. 3.4 Anisotropic element K_2 with mapped regular and stable element $\widehat{K}_2$ (reference configuration) and perturbed mapped element $\widetilde{K}_2 = \mathfrak{F}_{K_1}(K_2)$

First, we address 1. of Remark 3.12. $\widehat{K}_2 = \mathfrak{F}_{K_2}(K_2)$ is regular and thus, star-shaped with respect to a circle/ball $\widehat{B}$. If we transform $\widehat{K}_2$ into $\widetilde{K}_2$ with the mapping $\mathfrak{F}_{K_1} \circ \mathfrak{F}_{K_2}^{-1}$, see Fig. 3.4, the circle/ball $\widehat{B}$ is transformed into an ellipse/ellipsoid $\widetilde{B} = \mathfrak{F}_{K_1} \circ \mathfrak{F}_{K_2}^{-1}(\widehat{B})$. Since the transformations are linear, the element $\widetilde{K}_2$ is star-shaped with respect to the ellipse/ellipsoid $\widetilde{B}$ and in particular with respect to the circle/ball inscribed $\widetilde{B}$.

Next, we address 2. of Remark 3.12 and we bound the aspect ratio. The radius $\rho_{\widetilde{K}_2}$ of the inscribed circle/ball as above is equal to the smallest semi-axis of the ellipse/ellipsoid $\widetilde{B}$. Let $\widetilde{\mathbf{x}}_1$ and $\widetilde{\mathbf{x}}_2$ be the intersection of $\widetilde{B}$ and the inscribed circle/ball. Thus, we obtain

$$
\begin{aligned}
2\rho_{\widehat{K}_2} &= |\mathfrak{F}_{K_2} \circ \mathfrak{F}_{K_1}^{-1}(\widetilde{\mathbf{x}}_1 - \widetilde{\mathbf{x}}_2)| \\
&= \left| \alpha_{K_2} \Lambda_{K_2}^{-1/2} U_{K_2}^\top \frac{1}{\alpha_{K_1}} U_{K_1} \Lambda_{K_1}^{1/2} (\widetilde{\mathbf{x}}_1 - \widetilde{\mathbf{x}}_2) \right| \\
&= \frac{\alpha_{K_2}}{\alpha_{K_1}} \left| \Lambda_{K_1}^{-1/2} (I + \Delta^{K_1,K_2})^{-1/2} U_{K_1}^\top \left(R^{K_1,K_2}\right)^\top U_{K_1} \Lambda_{K_1}^{1/2} (\widetilde{\mathbf{x}}_1 - \widetilde{\mathbf{x}}_2) \right| \\
&= \frac{\alpha_{K_2}}{\alpha_{K_1}} \left| (I + \Delta^{K_1,K_2})^{-1/2} \left(\Lambda_{K_1}^{-1/2} U_{K_1}^\top \left(R^{K_1,K_2} - I\right)^\top U_{K_1} \Lambda_{K_1}^{1/2} + I \right) (\widetilde{\mathbf{x}}_1 - \widetilde{\mathbf{x}}_2) \right| \\
&\le \frac{\alpha_{K_2}}{\alpha_{K_1}} \left\| (I + \Delta^{K_1,K_2})^{-1/2} \right\|_2 \left(\|\Lambda_{K_1}^{-1/2}\|_2 \|R^{K_1,K_2} - I\|_2 \|\Lambda_{K_1}^{1/2}\|_2 + 1 \right) 2\rho_{\widetilde{K}_2} \\
&= \frac{\alpha_{K_2}}{\alpha_{K_1}} \max_{j=1,\dots,d} \left\{ (1 + \delta_j^{K_1,K_2})^{-1/2} \right\} \left(1 + \left(\frac{\lambda_{K_1,1}}{\lambda_{K_1,d}} \right)^{1/2} \|R^{K_1,K_2} - I\|_2 \right) 2\rho_{\widetilde{K}_2} ,
\end{aligned}
$$

since the spectral norm $\|\cdot\|_2$ is invariant under orthogonal transformations and, in particular, under rotations like U_{K_1}. With similar arguments, we can bound $h_{\widetilde{K}_2}$. Therefore, let $\widetilde{\mathbf{x}}_1, \widetilde{\mathbf{x}}_2 \in \partial\widetilde{K}_2$ be such that $h_{\widetilde{K}_2} = |\widetilde{\mathbf{x}}_1 - \widetilde{\mathbf{x}}_2|$ and $\widehat{\mathbf{x}}_i = \mathfrak{F}_{K_2} \circ \mathfrak{F}_{K_1}^{-1}(\widetilde{\mathbf{x}}_i) \in \partial\widehat{K}_2$, $i = 1, 2$. With similar considerations as above, we obtain

$$\begin{aligned} h_{\widetilde{K}_2} &= |\mathfrak{F}_{K_1} \circ \mathfrak{F}_{K_2}^{-1}(\widehat{\mathbf{x}}_1 - \widehat{\mathbf{x}}_2)| \\ &\le \frac{\alpha_{K_1}}{\alpha_{K_2}} \max_{j=1,\dots,d} \left\{(1+\delta_j^{K_1,K_2})^{1/2}\right\} \left(1 + \left(\frac{\lambda_{K_1,1}}{\lambda_{K_1,d}}\right)^{1/2} \|R^{K_1,K_2} - I\|_2\right) h_{\widehat{K}_2} \,. \end{aligned}$$

Exploiting the last two estimates yields

$$\begin{aligned} \frac{h_{\widetilde{K}_2}}{\rho_{\widetilde{K}_2}} &\le \frac{\max_{j=1,\dots,d}\sqrt{1+\delta_j^{K_1,K_2}}}{\min_{j=1,\dots,d}\sqrt{1+\delta_j^{K_1,K_2}}} \left(1 + \left(\frac{\lambda_{K_1,1}}{\lambda_{K_1,d}}\right)^{1/2} \|R^{K_1,K_2} - I\|_2\right)^2 \frac{h_{\widehat{K}_2}}{\rho_{\widehat{K}_2}} \\ &\le \sqrt{\frac{1+c_\delta}{1-c_\delta}}\,(1+c_R)^2\,\frac{h_{\widehat{K}_2}}{\rho_{\widehat{K}_2}} \le \sqrt{\frac{1+c_\delta}{1-c_\delta}}\,(1+c_R)^2\,\sigma_{\mathcal{K}} = \widetilde{\sigma}_{\mathcal{K}} \,. \end{aligned}$$

Obviously, the aspect ratio is uniformly bounded from above by a perturbed regularity parameter $\widetilde{\sigma}_{\mathcal{K}}$.

Finally we address 3. of Remark 3.12. Let $\widetilde{E}$ be an edge of $\widetilde{K}_2$ with endpoints $\widetilde{\mathbf{x}}_1$ and $\widetilde{\mathbf{x}}_2$. Furthermore, let $\widehat{E}$ be the corresponding edge of $\widehat{K}_2$ with endpoints $\widehat{\mathbf{x}}_1$ and $\widehat{\mathbf{x}}_2$. In the penultimate equation we estimated $h_{\widetilde{K}_2}$ by a term times $h_{\widehat{K}_2}$. Due to the stability it is $h_{\widehat{K}_2} \le c_{\mathcal{K}} h_{\widehat{E}}$ and, as in the estimate of $\rho_{\widehat{K}_2}$ above, we find that

$$\begin{aligned} h_{\widehat{E}} &= |\widehat{\mathbf{x}}_1 - \widehat{\mathbf{x}}_2| = |\mathfrak{F}_{K_2} \circ \mathfrak{F}_{K_1}^{-1}(\widetilde{\mathbf{x}}_1 - \widetilde{\mathbf{x}}_2)| \\ &\le \frac{\alpha_{K_2}}{\alpha_{K_1}} \max_{j=1,\dots,d} \left\{(1+\delta_j^{K_1,K_2})^{-1/2}\right\} \left(1 + \left(\frac{\lambda_{K_1,1}}{\lambda_{K_1,d}}\right)^{1/2} \|R^{K_1,K_2} - I\|_2\right) h_{\widetilde{E}} \,. \end{aligned}$$

Summarizing, we obtain

$$h_{\widetilde{K}_2} \le \sqrt{\frac{1+c_\delta}{1-c_\delta}}\,(1+c_R)^2\,c_{\mathcal{K}}\,h_{\widetilde{E}} = \widetilde{c}_{\mathcal{K}} h_{\widetilde{E}} \,.$$

□

Remark 3.16 According to the previous proof, the perturbed regularity and stability parameters are given by

$$\widetilde{\sigma}_{\mathcal{K}} = \sqrt{\frac{1+c_\delta}{1-c_\delta}}\,(1+c_R)^2\,\sigma_{\mathcal{K}} \quad\text{and}\quad \widetilde{c}_{\mathcal{K}} = \sqrt{\frac{1+c_\delta}{1-c_\delta}}\,(1+c_R)^2\,c_{\mathcal{K}} \,.$$

Proposition 3.17 *Let $K \in \mathcal{K}_h$ be a polytopal element of a regular and stable anisotropic mesh $\mathcal{K}_h$ and $F \in \mathcal{F}_h$ one of its edges ($d = 2$) or faces ($d = 3$). The mapped patches $\mathfrak{F}_K(\omega_K)$ and $\mathfrak{F}_K(\omega_F)$ consist of regular and stable polytopal elements.*

Proof The mapped patches $\mathfrak{F}_K(\omega_\mathbf{z})$, $\mathbf{z} \in \mathcal{N}(K)$ consist of regular and stable polytopal elements according to Lemma 3.15. Since ω_K and ω_F are given as union of the neighbourhoods $\omega_\mathbf{z}$, see (3.2), the statement of the proposition follows. □

Proposition 3.18 *Each node $\mathbf{z} \in \mathcal{N}_h$ of a regular and stable anisotropic mesh $\mathcal{K}_h$ belongs to a uniformly bounded number of elements and, vice versa, each element $K \in \mathcal{K}_h$ has a uniformly bounded number of nodes on its boundary.*

Proof Let $\omega_\mathbf{z}$ be the neighbourhood of the node $\mathbf{z}$. According to Lemma 3.15, the mapped neighbourhood $\widetilde{\omega}_\mathbf{z}$ consists of regular and stable polytopal elements, which admit a shape-regular decomposition into simplices (triangles or tetrahedra). The mapped node $\widetilde{\mathbf{z}}$ therefore belongs to a uniformly bounded number of simplices and thus to finitely many polytopal elements, cf. Sect. 2.2. Since $\widetilde{\omega}_\mathbf{z}$ is obtained by a linear transformation, we follow that $\mathbf{z}$ belongs to a uniformly bounded number of anisotropic elements. With the same argument we see that $\widetilde{K}$ and thus K has a uniformly bounded number of nodes on its boundary. □

Remark 3.19 In the publications of Apel and Kunert (see e.g. [10, 119]), it is assumed that neighbouring triangles/tetrahedra behave similarly. More precisely, they assume:

- The number of tetrahedra containing a node $\mathbf{z}$ is bounded uniformly.
- The dimension of adjacent tetrahedra must not change rapidly, i.e.

$$h_{i,T} \sim h_{i,T'} \quad \forall T, T' \text{ with } T \cap T' \neq \varnothing,\ i = 1, 2, 3,$$

  where $h_{1,T} \geq h_{2,T} \geq h_{3,T}$ are the heights of the tetrahedron T over its faces.

The first point is always satisfied in our setting according to the previous proposition. The second point corresponds to our assumption that Λ_{K_1} and Λ_{K_2} differ moderately for neighbouring elements K_1 and K_2, see Definition 3.13. The assumption on U_{K_1} and U_{K_2} in the definition ensure that the heights are aligned in the same directions, this is also hidden in the assumption of Apel and Kunert.

The regularity of the mapped patches has several consequences, which are exploited in later proofs.

Lemma 3.20 *Let K_1, K_2 be polytopal elements of a regular and stable anisotropic mesh $\mathcal{K}_h$, $\omega_\mathbf{z}$ and ω_{K_1} be the neighbourhoods of the node $\mathbf{z} \in \mathcal{N}_h$ and the element K_1, respectively. Furthermore, let $K_1, K_2 \subset \omega_\mathbf{z}$. We have for the mapped patch $\widetilde{\omega} \in \{\mathfrak{F}_{K_1}(\omega_\mathbf{z}), \mathfrak{F}_{K_1}(\omega_{K_1})\}$ and the neighbouring elements, that*

$$h_{\widetilde{\omega}} \leq c \qquad \text{and} \qquad \frac{|K_2|}{|K_1|} \leq c\,,$$

where the constants only depend on the regularity and stability parameters of the mesh.

Proof According to Lemma 3.15 and Proposition 3.17 the patch $\widetilde{\omega}$ consists of regular and stable polytopal elements. Obviously, it is $h_{\widetilde{\omega}} \leq C \max\{h_{\widetilde{K}} : \widetilde{K} \subset \widetilde{\omega}\}$, where the constant takes the value $C = 2$ for $\widetilde{\omega} = \mathfrak{F}_{K_1}(\omega_{\mathbf{z}})$ and $C = 3$ for $\widetilde{\omega} = \mathfrak{F}_{K_1}(\omega_{K_1})$, respectively. Let us assume without loss of generality that the maximum is reached for $\widetilde{K}$ which shares a common edge $\widetilde{E}$ with $\widetilde{K}_1$. Otherwise consider a sequence of polytopal elements in $\widetilde{\omega}$, cf. Lemma 3.1. Due to the regularity and stability of the elements, it is

$$h_{\widetilde{\omega}} \leq 3h_{\widetilde{K}} \leq 3c_{\mathscr{K}} h_{\widetilde{E}} \leq 3c_{\mathscr{K}} h_{\widetilde{K}_1} \leq \frac{3c_{\mathscr{K}}\sigma_{\mathscr{K}}}{(\nu\pi)^{1/d}}$$

according to (3.10), since $\widetilde{K}_1 = \mathfrak{F}_{K_1}(K_1) = \widehat{K}_1$.

In order to prove the second estimate, we observe that $|K_1| = |\widehat{K}_1|/|\det(A_{K_1})|$, see Lemma 3.10. The same variable transform yields $|K_2| = |\widetilde{K}_2|/|\det(A_{K_1})|$, where $\widetilde{K}_2 = \mathfrak{F}_{K_1}(K_2)$. Thus, we obtain

$$\frac{|K_2|}{|K_1|} = \frac{|\widetilde{K}_2|}{|\widehat{K}_1|} = |\widetilde{K}_2| \leq |\widetilde{\omega}_{\mathbf{z}}| \leq h_{\widetilde{\omega}_{\mathbf{z}}}^d \leq c$$

and finish the proof. □

3.4.2 Approximation Space

The approximation space V_h is defined in such a way that the functions $v_h \in V_h$ are harmonic on each element, cf. (3.3). This property originates from the definition of basis functions ψ in Sect. 2.3 as local solutions of Laplace and Poisson problems over the physical elements $K \in \mathscr{K}_h$. In classical finite element methods, however, the basis functions are usually introduced over a reference element. In order to obtain the approximation space over a general physical element these basis functions from the reference element are mapped to the physical one. This strategy has not been addressed so far for polytopal elements due to the lack of an appropriate reference element. But, in the previous section we introduced a reference configuration $\widehat{K}$ for an element K. Thus, we can define basis functions $\widehat{\psi}$ on $\widehat{K}$ as in Sect. 2.3 which are in the lowest order case harmonic and map them onto the physical element K such that $\psi^{\text{ref}} = \widehat{\psi} \circ \mathfrak{F}_K$. In general, these functions are not harmonic anymore on the physical elements, i.e. $\Delta\psi^{\text{ref}} \neq 0$ in K. More precisely, we obtain by the transformation (3.8)

$$\operatorname{div}\left(M_{\text{Cov}}(K)\nabla\psi^{\text{ref}}\right) = 0 \quad \text{in } K\ .$$

Hence, ψ^{ref} is defined to fulfil an anisotropic diffusion equation on K. This is consistent in the sense, that if K is already a reference configuration, i.e. $K = \widehat{K}$, then it is $\Delta\psi^{\text{ref}} = 0$ because of $M_{\text{Cov}}(K) = \alpha_K^2 I$, cf. Lemma 3.10. Thus, nodal basis functions $\psi_{\mathbf{z}}^{\text{ref}}$ constructed this way coincide with the nodal basis function $\psi_{\mathbf{z}}$ defined by (2.6) in Sect. 2.3.1.

The approximation space constructed as described above is denoted by V_h^{ref} since the reference configuration is exploited. For the sake of simplicity we restrict ourselves here to $k = 1$ as well as to the two-dimensional case and to the three-dimensional case with solely triangular faces of the polyhedra. Then, we can also write

$$V_h^{\text{ref}} = \left\{ v \in H^1(\Omega) : \operatorname{div}\left(M_{\text{Cov}}(K)\nabla v\right)\big|_K = 0 \text{ and } v|_{\partial K} \in \mathscr{P}_{\text{pw}}^1(\partial K)\ \forall K \in \mathscr{K}_h \right\}.$$

The spaces V_h and V_h^{ref} share two important properties which are used in the forthcoming proofs, namely

$$\mathscr{P}^1(K) \subset V_h\big|_K, \quad \mathscr{P}^1(K) \subset V_h^{\text{ref}}\big|_K \quad \text{and} \quad 0 \le \psi_{\mathbf{z}}, \psi_{\mathbf{z}}^{\text{ref}} \le 1, \tag{3.12}$$

where $\psi_{\mathbf{z}}$ and $\psi_{\mathbf{z}}^{\text{ref}}$ denote the corresponding nodal basis functions of V_h and V_h^{ref}, respectively.

3.4.3 Anisotropic Trace Inequality and Best Approximation

In this section we transfer some of the results of Sect. 3.2 to the regime of anisotropic meshes. Here, the mapping (3.8) is employed to transform a regular and stable anisotropic element K onto its reference configuration $\widehat{K}$, which is regular and stable in the sense of Sect. 2.2, see also Remark 3.12.

Lemma 3.21 (Anisotropic Trace Inequality) *Let $K \in \mathscr{K}_h$ be a polytopal element of a regular and stable anisotropic mesh $\mathscr{K}_h$ with edge ($d = 2$) or face ($d = 3$) $F \in \mathscr{F}_h$, $F \subset \partial K$. It holds*

$$\|v\|_{L_2(F)}^2 \le c\,\frac{|F|}{|K|}\left(\|v\|_{L_2(K)}^2 + \|A_K^{-\top}\nabla v\|_{L_2(K)}^2\right),$$

where the constant c only depends on the regularity and stability parameters of the mesh.

Proof In order to prove the estimate, we make use of the transformation (3.8) to the reference configuration $\widehat{K}$ with $\widehat{v} = v\circ\mathfrak{F}_K^{-1}$, a trace inequality on $\widehat{K}$, see Lemma 3.3, as well as of (3.10), (3.11) and $h_{\widehat{K}}^{-d} \le |\widehat{K}|^{-1} = 1$. These tools yield

$$\begin{aligned}
\|v\|_{L_2(F)}^2 &= \frac{|F|}{|\widehat{F}|}\|\widehat{v}\|_{L_2(\widehat{F})}^2 \\
&\le c\frac{|F|}{|\widehat{F}|}\left(h_{\widehat{K}}^{-1}\|\widehat{v}\|_{L_2(\widehat{K})}^2 + h_{\widehat{K}}|\widehat{v}|_{H^1(\widehat{K})}^2\right) \\
&\le c|F|h_{\widehat{K}}^{-d}\left(\|\widehat{v}\|_{L_2(\widehat{K})}^2 + h_{\widehat{K}}^2|\widehat{v}|_{H^1(\widehat{K})}^2\right) \\
&\le c|F|\left(\|\widehat{v}\|_{L_2(\widehat{K})}^2 + \|\widehat{\nabla}\widehat{v}\|_{L_2(\widehat{K})}^2\right) \\
&= c\frac{|F|}{|K|}\left(\|v\|_{L_2(K)}^2 + \|A_K^{-\top}\nabla v\|_{L_2(K)}^2\right).
\end{aligned}$$

□

Remark 3.22 If we plug in the definition of $A = \alpha_K \Lambda_K^{-1/2} U_K^\top$, we have the anisotropic trace inequality

$$\|v\|_{L_2(F)}^2 \le c\frac{|F|}{|K|}\left(\|v\|_{L_2(K)}^2 + \|\alpha_K^{-1}\Lambda_K^{1/2}U_K^\top\nabla v\|_{L_2(K)}^2\right).$$

Obviously, the derivatives of v in the characteristic directions $\mathbf{u}_{K,j}$ are scaled by the characteristic lengths $\lambda_j^{1/2}$, $j = 1, \ldots, d$ of the element K. This seems to be appropriate for functions with anisotropic behaviour which are aligned with the mesh.

For later comparisons with other methods, we bound the term $|F|/|K|$ in case of $F \subset \partial K$. Let $\mathbf{z}_{\widehat{K}}$ be the midpoint of the circle/ball in Definitions 2.1 and 2.11, respectively, of the regular and stable reference configuration $\widehat{K}$. Obviously, it is $|K| \ge |P|$ for the d-dimensional pyramid P with base side F and apex point $\mathfrak{F}_K^{-1}(z_{\widehat{K}})$, since $P \subset K$ due to the linearity of $\mathfrak{F}_K$. Denote by $h_{P,F}$ the hight of this pyramid, then it is $|P| = \frac{1}{3}|F|h_{P,F}$ and we obtain

$$\frac{|F|}{|K|} \le ch_{P,F}^{-1}. \tag{3.13}$$

In the derivation of approximation estimates, the Poincaré constant also plays a crucial role on anisotropic meshes. This constant is given in (3.4).

Lemma 3.23 *Let $\mathcal{K}_h$ be a regular and stable anisotropic mesh, $\omega_{\mathbf{z}}$ and ω_K be neighbourhoods as described in Sect. 3.1, and $K \in \mathcal{K}_h$ with $K \subset \omega_{\mathbf{z}}$. The Poincaré constants $C_P(\widetilde{\omega}_{\mathbf{z}})$ and $C_P(\widetilde{\omega}_K)$ for the mapped patches $\widetilde{\omega}_{\mathbf{z}} = \mathfrak{F}_K(\omega_{\mathbf{z}})$ as well as $\widetilde{\omega}_K = \mathfrak{F}_K(\omega_K)$, can be bounded uniformly depending only on the regularity and stability parameters of the mesh.*

Proof According to Lemma 3.15 and Proposition 3.17, the patches $\widetilde{\omega}_{\mathbf{z}}$ and $\widetilde{\omega}_K$ consist of regular and stable polytopal elements. Thus, we utilize Lemma 3.5 on the mapped patches and the statement follows. □

Next, we derive a best approximation result on patches of anisotropic elements.

Lemma 3.24 *Let $\mathscr{K}_h$ be a regular and stable anisotropic mesh with node $\mathbf{z} \in \mathscr{N}_h$ and element $K \in \mathscr{K}_h$. Furthermore, let $\omega_{\mathbf{z}}$ and ω_K be the neighbourhood of $\mathbf{z}$ and K, respectively, and we assume $K \subset \omega_{\mathbf{z}}$. For $\omega \in \{\omega_{\mathbf{z}}, \omega_K\}$ it holds*

$$\|v - \Pi_\omega v\|_{L_2(\omega)} \le c\, \|A_K^{-\top} \nabla v\|_{L_2(\omega)}\,,$$

and furthermore

$$\|v - \Pi_\omega v\|_{L_2(\omega)} \le c \left(\sum_{K' \in \mathscr{K}_h : K' \subset \omega} \|A_{K'}^{-1} \nabla v\|_{L_2(K')}^2 \right)^{1/2},$$

where the constant c only depends on the regularity and stability parameters of the mesh.

Proof We make use of the mapping (3.8) and indicate the objects on the mapped geometry with a tilde, e.g., $\widetilde{\omega} = \mathfrak{F}_K(\omega)$. Furthermore, we exploited that the mapped L_2-projection coincides with the L_2-projection on the mapped patch, consequently $\widetilde{\Pi_\omega v} = \Pi_{\widetilde{\omega}} \widetilde{v}$. This yields together with Lemma 3.23

$$\begin{aligned} \|v - \Pi_\omega v\|_{L_2(\omega)} &= |K|^{1/2}\, \|\widetilde{v} - \Pi_{\widetilde{\omega}} \widetilde{v}\|_{L_2(\widetilde{\omega})} \\ &\le c h_{\widetilde{\omega}} |K|^{1/2}\, |\widetilde{v}|_{H^1(\widetilde{\omega})} \\ &= c h_{\widetilde{\omega}} |K|^{1/2}\, \|\widetilde{\nabla} \widetilde{v}\|_{L_2(\widetilde{\omega})} \\ &= c h_{\widetilde{\omega}}\, \|A_K^{-\top} \nabla v\|_{L_2(\omega)}\,. \end{aligned}$$

The term $h_{\widetilde{\omega}}$ is uniformly bounded according to Lemma 3.20, and thus the first estimate is proven.

In order to prove the second estimate, we employ the first one and write

$$\|v - \Pi_\omega v\|_{L_2(\omega)}^2 \le c\, \|A_K^{-\top} \nabla v\|_{L_2(\omega)}^2 = c \sum_{K' \in \mathscr{K}_h : K' \subset \omega} \|A_K^{-\top} \nabla v\|_{L_2(K')}^2\,.$$

Therefore, it remains to estimate $\|A_K^{-\top} \nabla v\|_{L_2(K')}$ by $\|A_{K'}^{-\top} \nabla v\|_{L_2(K')}$ for any element $K' \subset \omega$. We make use of the mesh regularity and stability, see Definition 3.13, and proceed similar as in the proof of Lemma 3.15.

$$\begin{aligned}
\|A_K^{-\top}\nabla v\|_{L_2(K')} &= \frac{\alpha_{K'}}{\alpha_K}\|\alpha_{K'}^{-1}((I+\Delta^{K',K})A_{K'})^{1/2}(R^{K',K}U_{K'})^\top\nabla v\|_{L_2(K')}\\
&= \frac{\alpha_{K'}}{\alpha_K}\|\alpha_{K'}^{-1}(I+\Delta^{K',K})^{1/2}\Lambda_{K'}^{1/2}U_{K'}^\top(R^{K',K})^\top\nabla v\|_{L_2(K')}\\
&= \frac{\alpha_{K'}}{\alpha_K}\|\alpha_{K'}^{-1}(I+\Delta^{K',K})^{1/2}\Lambda_{K'}^{1/2}U_{K'}^\top(R^{K',K})^\top U_{K'}\Lambda_{K'}^{-1/2}\Lambda_{K'}^{1/2}U_{K'}^\top\nabla v\|_{L_2(K')}\\
&\le \frac{\alpha_{K'}}{\alpha_K}\|(I+\Delta^{K',K})^{1/2}\Lambda_{K'}^{1/2}U_{K'}^\top(R^{K',K})^\top U_{K'}\Lambda_{K'}^{-1/2}\|_2\|A_{K'}^{-\top}\nabla v\|_{L_2(K')}\,,
\end{aligned}$$

where we substituted $A_K^{-\top} = \alpha_{K'}^{-1}\Lambda_{K'}^{1/2}U_{K'}^\top$. Finally, we have to bound the ratio $\alpha_{K'}/\alpha_K$ and the matrix norm. According to the choice (3.9) and Lemma 3.20, it is

$$\begin{aligned}
\left(\frac{\alpha_{K'}}{\alpha_K}\right)^2 &= \frac{|K|\sqrt{\prod_{j=1}^d\lambda_{K',j}}}{|K'|\sqrt{\prod_{j=1}^d\lambda_{K,j}}} = \frac{|K|\sqrt{\prod_{j=1}^d(1+\delta_j^{K,K'})\lambda_{K,j}}}{|K'|\sqrt{\prod_{j=1}^d\lambda_{K,j}}}\\
&\le (1+c_\delta)^{d/2}\frac{|K|}{|K'|}\le c\,,
\end{aligned}$$

and for the matrix norm, we have

$$\begin{aligned}
&\|(I+\Delta^{K',K})^{1/2}\Lambda_{K'}^{1/2}U_{K'}^\top(R^{K',K})^\top U_{K'}\Lambda_{K'}^{-1/2}\|_2\\
&\le \|(I+\Delta^{K',K})^{1/2}\|_2\|\Lambda_{K'}^{1/2}U_{K'}^\top(R^{K',K}-I)^\top U_{K'}\Lambda_{K'}^{-1/2}+I\|_2\\
&\le \sqrt{1+c_\delta}(1+c_R)\,,
\end{aligned}$$

that finishes the proof. □

Remark 3.25 In the previous proof, we have seen in particular that for neighbouring elements $K, K'\subset\omega_K$, it is

$$\|A_K^{-\top}\nabla v\|_{L_2(K')}\le c\,\|A_{K'}^{-\top}\nabla v\|_{L_2(K')}$$

with a constant depending only on the regularity and stability of the mesh.

3.4.4 Quasi-Interpolation of Anisotropic Non-smooth Functions

In this section, we consider the quasi-interpolation operators from Sect. 3.3 on anisotropic polygonal and polyhedral meshes. The analysis relies on the mapping to the reference configuration of regular and stable anisotropic polytopal elements as in [181]. Earlier results for quasi-interpolation operators on anisotropic simplicial meshes can be found in [10, 79, 119], for example. Some comparisons are also drawn in the following.

The general form of the Clément and to Scott–Zhang operator is given in (3.6) for $v \in H^1(\Omega)$, namely

$$\mathfrak{I}v = \sum_{\mathbf{z}\in\mathcal{N}_*} (\Pi_{\omega(\psi_\mathbf{z})} v)\psi_\mathbf{z} \in V_h \ ,$$

where the set of nodes $\mathcal{N}_*$ and the neighbourhoods $\omega(\psi_\mathbf{z})$ are chosen accordingly. We point out, that the results of this section stay valid if we replace the basis functions $\psi_\mathbf{z}$ by $\psi_\mathbf{z}^{\mathrm{ref}}$, which have been discussed in Sect. 3.4.2. In this case the quasi-interpolation operator maps into the approximation space defined with the help of the reference configurations, i.e. $\mathfrak{I} : H^1(\Omega) \to V_h^{\mathrm{ref}}$. In the forthcoming proofs, we only employ the properties (3.12) which are shared by V_h and V_h^{ref}.

3.4.4.1 Clément-Type Interpolation

The Clément interpolation operator $\mathfrak{I}_C$ is defined by (3.6) with $\mathcal{N}_* = \mathcal{N}_h \setminus \mathcal{N}_{h,D}$ and $\omega(\psi_\mathbf{z}) = \omega_\mathbf{z}$, see Sect. 3.3.1 for details. For $v \in H_D^1(\Omega)$, it is $\mathfrak{I}_C v \in H_D^1(\Omega)$ by construction.

Theorem 3.26 *Let $\mathcal{K}_h$ be a regular and stable anisotropic mesh and $K \in \mathcal{K}_h$. The Clément interpolation operator satisfies for $v \in H_D^1(\Omega)$ the interpolation error estimate*

$$\|v - \mathfrak{I}_C v\|_{L_2(K)} \le c\, \|A_K^{-\top}\nabla v\|_{L_2(\omega_K)} \ ,$$

and for an edge/face $F \in \mathcal{F}(K) \setminus \mathcal{F}_{h,D}$

$$\|v - \mathfrak{I}_C v\|_{L_2(F)} \le c\, \frac{|F|^{1/2}}{|K|^{1/2}}\, \|A_K^{-\top}\nabla v\|_{L_2(\omega_F)} \ ,$$

where the constants c only depend on the regularity and stability parameters of the mesh.

Proof We can follow classical arguments as for isotropic meshes, cf. Theorem 3.7. The main ingredients are the observation that the basis functions $\psi_\mathbf{z}$ form a partition of unity on $\overline{K}$, and that they are bounded by one. Furthermore, anisotropic approximation estimates, see Lemma 3.24, the anisotropic trace inequality in Lemmata 3.21 and 3.20 and Remark 3.25 are employed. We only sketch the proof of the second estimate.

The partition of unity property is used, which also holds on each edge/face F, i.e. $\sum_{\mathbf{z}\in\mathcal{N}(F)} \psi_\mathbf{z} = 1$ on $\overline{F}$. We distinguish two cases, first let $\mathcal{N}(F) \cap \mathcal{N}_{h,D} = \emptyset$.

With the help of Lemmata 3.21 and 3.24, we obtain

$$\begin{aligned}\|v - \mathfrak{I}_C v\|_{L_2(F)} &= \sum_{\mathbf{z}\in\mathscr{N}(F)} \|\psi_{\mathbf{z}}(v - \Pi_{\omega_{\mathbf{z}}} v)\|_{L_2(F)} \leq \sum_{\mathbf{z}\in\mathscr{N}(F)} \|v - \Pi_{\omega_{\mathbf{z}}} v\|_{L_2(F)} \\ &\leq c \sum_{\mathbf{z}\in\mathscr{N}(F)} \frac{|F|^{1/2}}{|K|^{1/2}} \left(\|v - \Pi_{\omega_{\mathbf{z}}} v\|^2_{L_2(K)} + \|A_K^{-\top}\nabla v\|^2_{L_2(K)}\right)^{1/2} \\ &\leq c \sum_{\mathbf{z}\in\mathscr{N}(F)} \frac{|F|^{1/2}}{|K|^{1/2}} \|A_K^{-\top}\nabla v\|_{L_2(\omega_{\mathbf{z}})}.\end{aligned}$$

For the second case with $\mathscr{N}(F) \cap \mathscr{N}_{h,D} \neq \varnothing$, we find

$$\|v - \mathfrak{I}_C v\|_{L_2(F)} \leq \sum_{\mathbf{z}\in\mathscr{N}(F)} \|\psi_{\mathbf{z}}(v - \Pi_{\omega_{\mathbf{z}}} v)\|_{L_2(F)} + \sum_{\mathbf{z}\in\mathscr{N}(F)\cap\mathscr{N}_{h,D}} \|\psi_{\mathbf{z}}\Pi_{\omega_{\mathbf{z}}} v\|_{L_2(F)} . \tag{3.14}$$

The first sum has already been estimated, thus we consider the term in the second sum. For $\mathbf{z} \in \mathscr{N}(F) \cap \mathscr{N}_{h,D}$, i.e. $\mathbf{z} \in \overline{\Gamma}_D$, there is an element $K' \subset \omega_{\mathbf{z}}$ and an edge/face $F' \in \mathscr{F}(K')$ such that $\mathbf{z} \in \mathscr{N}(F')$ and $F' \in \mathscr{F}_{h,D}$. Since v vanishes on F', Lemmata 3.21 and 3.24 as well as Remark 3.25 yield

$$|\Pi_{\omega_{\mathbf{z}}} v| = |F'|^{-1/2} \|v - \Pi_{\omega_{\mathbf{z}}} v\|_{L_2(F')} \leq c\, |K'|^{-1/2} \|A_K^{-\top}\nabla v\|_{L_2(\omega_{\mathbf{z}})} .$$

Because $|K'|/|K|$ is uniformly bounded according to Lemma 3.20, we obtain

$$\|\psi_{\mathbf{z}}\Pi_{\omega_{\mathbf{z}}} v\|_{L_2(F)} \leq |\Pi_{\omega_{\mathbf{z}}} v|\, \|\psi_{\mathbf{z}}\|_{L_\infty(F)}\, |F|^{1/2} \leq c\, \frac{|F|^{1/2}}{|K|^{1/2}} \|A_K^{-\top}\nabla v\|_{L_2(\omega_{\mathbf{z}})} .$$

Finally, since the number of nodes per element is uniformly bounded according to Proposition 3.18, this estimate as well as the one derived in the first case applied to (3.14) yield the second interpolation error estimate in the theorem. □

Remark 3.27 In the case of an isotropic polytopal element K with edge/face F it is

$$\lambda_1 \approx \ldots \approx \lambda_d \sim h_K^2 , \quad \text{and thus} \quad \alpha_K \sim 1 .$$

Therefore, we obtain from Theorem 3.26 with $A_K^{-\top} = \alpha_K^{-1} \Lambda_K^{1/2} U_K^\top$ that

$$\|v - \mathfrak{I}_C v\|_{L_2(K)} \leq c h_K \|U_K^\top \nabla v\|_{L_2(\omega_K)} = c h_K\, |v|_{H^1(\omega_K)} ,$$

and

$$\|v - \mathfrak{I}_C v\|_{L_2(F)} \leq c\, \frac{h_K |F|^{1/2}}{|K|^{1/2}} \|U_K^\top \nabla v\|_{L_2(\omega_F)} \leq c h_F^{1/2}\, |v|_{H^1(\omega_F)} ,$$

since $|F| \leq h_F^{d-1}$ as well as $|K| \geq ch_K^d$ and $h_K \leq ch_F$ in consequence of the regularity and stability, cf. Remark 3.12. Obviously, we recover the classical interpolation error estimates for the Clément interpolation operator, cf. Theorem 3.7.

In the following, we rewrite our results in order to compare them with the work of Formaggia and Perotto [79]. It is $A_K^{-\top} = \alpha_K^{-1} \Lambda_K^{1/2} U_K^\top$ with $U_K = (\mathbf{u}_{K,1}, \ldots, \mathbf{u}_{K,d})$. Thus, we observe

$$\|A_K^{-\top} \nabla v\|^2_{L_2(\omega_K)} = \alpha_K^{-2} \sum_{j=1}^{d} \lambda_{K,j} \, \|\mathbf{u}_{K,j} \cdot \nabla v\|^2_{L_2(\omega_K)} ,$$

and since $\mathbf{u}_{K,j} \cdot \nabla v : \mathbb{R}^d \to \mathbb{R}$, we obtain

$$\|\mathbf{u}_{K,j} \cdot \nabla v\|^2_{L_2(\omega_K)} = \sum_{K' \subset \omega_K} \int_{K'} \mathbf{u}_{K,j}^\top \nabla v (\nabla v)^\top \mathbf{u}_{K,j} \, d\mathbf{x} = \mathbf{u}_{K,j}^\top \, G_K(v) \, \mathbf{u}_{K,j}$$

with

$$G_K(v) = \sum_{K' \subset \omega_K} \left(\int_{K'} \frac{\partial v}{\partial x_i} \frac{\partial v}{\partial x_j} \, d\mathbf{x} \right)_{i,j=1}^{d} \in \mathbb{R}^{d \times d} , \quad \mathbf{x} = (x_1, \ldots, x_d)^\top .$$

Therefore, we can deduce from Theorem 3.26 an equivalent formulation.

Proposition 3.28 *Let $\mathscr{K}_h$ be a regular and stable anisotropic mesh and $K \in \mathscr{K}_h$. The Clément interpolation operator satisfies for $v \in H^1_D(\Omega)$ the interpolation error estimate*

$$\|v - \mathfrak{I}_C v\|_{L_2(K)} \leq c \, \alpha_K^{-1} \left(\sum_{j=1}^{d} \lambda_{K,j} \, \mathbf{u}_{K,j}^\top \, G_K(v) \, \mathbf{u}_{K,j} \right)^{1/2} ,$$

and for an edge/face $F \in \mathscr{F}(K) \setminus \mathscr{F}_{h,D}$

$$\|v - \mathfrak{I}_C v\|_{L_2(F)} \leq c \, \alpha_K^{-1} \, \frac{|F|^{1/2}}{|K|^{1/2}} \left(\sum_{j=1}^{d} \lambda_{K,j} \, \mathbf{u}_{K,j}^\top \, G_K(v) \, \mathbf{u}_{K,j} \right)^{1/2} ,$$

where the constant c only depends on the regularity and stability parameters of the mesh.

Now we are ready to compare the interpolation error estimates with the ones derived by Formaggia and Perotto. These authors considered the case of anisotropic triangular meshes in two-dimensions, i.e. $d = 2$. The inequalities in Proposition 3.28 correspond to the derived estimates (2.12) and (2.15) in [79] but they are valid on much more general meshes. When comparing these estimates to the results of Formaggia and Perotto, one has to take care on the powers of the lambdas. The triangular elements in their works are scaled with $\lambda_{i,K}, i = 1, 2$ in the characteristic directions whereas the scaling in this section is $\lambda_{K,i}^{1/2}, i = 1, 2$.

Obviously, the first inequality of the previous proposition corresponds to the derived estimate (2.12) in [79] up to the scaling factor α_K^{-1}. However, for convex elements the assumption

$$\alpha_K \sim 1\,, \qquad \text{i.e.,} \qquad |K| \sim \sqrt{\lambda_{K,1}\lambda_{K,2}}\,,$$

seems to be convenient, since this means that the area $|K|$ of the element is proportional to the area $\pi\sqrt{\lambda_{K,1}}\sqrt{\lambda_{K,2}}$ of the inscribed ellipse, which is given by the scaled characteristic directions of the element.

In order to recognize the relation of the second inequality under these assumptions, we estimate the term $|F|/|K|$ by (3.13) and by applying $h_{P,F} \geq \lambda_{K,2}^{1/2}$. This yields

$$\|v - \mathfrak{I}_C v\|_{L_2(F)} \leq c \left(\frac{1}{\lambda_{K,2}^{1/2}}\right)^{1/2} \left(\lambda_{K,1}\,\mathbf{u}_{K,1}^\top G_K(v)\,\mathbf{u}_{K,1} + \lambda_{K,2}\,\mathbf{u}_{K,2}^\top G_K(v)\,\mathbf{u}_{K,2}\right)^{1/2},$$

and shows the correspondence to [79], since h_K and $\lambda_{1,K}$ are proportional in the referred work.

3.4.4.2 Scott–Zhang-Type Interpolation

The Scott–Zhang interpolation operator $\mathfrak{I}_{SZ} : H^1(\Omega) \to V_h$ is defined by (3.6) with $\mathscr{N}_* = \mathscr{N}$ and $\omega(\psi_\mathbf{z}) = F_\mathbf{z}$, where $F_\mathbf{z} \in \mathscr{F}_h$ is an edge ($d = 2$) or face ($d = 3$) with $\mathbf{z} \in \overline{F}_\mathbf{z}$ and

$$F_\mathbf{z} \subset \Gamma_D \text{ if } \mathbf{z} \in \overline{\Gamma}_D \qquad \text{and} \qquad F_\mathbf{z} \subset \Omega \cup \Gamma_N \text{ if } \mathbf{z} \in \Omega \cup \Gamma_N\,.$$

By construction, it is $\mathfrak{I}_{SZ}v \in H_D^1(\Omega)$ for $v \in H_D^1(\Omega)$, such that homogeneous Dirichlet data is preserved. We have the following local stability result on anisotropic meshes.

Lemma 3.29 *Let $\mathscr{K}_h$ be a regular and stable anisotropic mesh and $K \in \mathscr{K}_h$. The Scott–Zhang interpolation operator satisfies for $v \in H^1(\Omega)$ the local stability*

$$\|\mathfrak{I}_{SZ}v\|_{L_2(K)} \leq c\left(\|v\|_{L_2(\omega_K)} + \|A_K^{-\top}\nabla v\|_{L_2(\omega_K)}\right),$$

where the constant c only depends on the regularity and stability parameters of the mesh.

Proof The proof is analog to the isotropic version in Lemma 3.8. The difference is that the anisotropic trace inequality Lemmata 3.21 and 3.20, Remark 3.25 and Proposition 3.18 are used. For details see [181]. □

Theorem 3.30 *Let $\mathcal{K}_h$ be a regular and stable anisotropic mesh and $K \in \mathcal{K}_h$. The Scott–Zhang interpolation operator satisfies for $v \in H^1(\Omega)$ the interpolation error estimate*

$$\|v - \mathfrak{I}_{SZ}v\|_{L_2(K)} \leq \|A_K^{-\top}\nabla v\|_{L_2(\omega_K)} ,$$

where the constant c only depends on the regularity and stability parameters of the mesh.

Proof For $p = \Pi_{\omega_K} v \in \mathbb{R}$ it is obviously $p = \mathfrak{I}_{SZ}p$ and $\nabla p = 0$. The estimate in the theorem follows by Lemma 3.29 and the application of Lemma 3.24, since

$$\begin{aligned}\|v - \mathfrak{I}_{SZ}v\|_{L_2(K)} &\leq \|v - p\|_{L_2(K)} + \|\mathfrak{I}_{SZ}(v - p)\|_{L_2(K)}\\ &\leq c\left(\|v - p\|_{L_2(\omega_K)} + \|A_K^{-\top}\nabla v\|_{L_2(\omega_K)}\right)\\ &\leq c\,\|A_K^{-\top}\nabla v\|_{L_2(\omega_K)} .\end{aligned}$$

□

3.4.5 Interpolation of Anisotropic Smooth Functions

In the previous section, we considered quasi-interpolation of functions in $H^1(\Omega)$. However, we may also address classical interpolation employing point evaluations in the case that the function to be interpolated is sufficiently regular as in Sect. 2.4. This is possible for functions in $H^2(\Omega)$. In the following, we consider the pointwise interpolation of lowest order with $k = 1$ into the approximation space V_h^{ref} on anisotropic meshes. V_h^{ref} has been discussed in Sect. 3.4.2 and its basis functions $\psi_{\mathbf{z}}^{\mathrm{ref}}$ are constructed such that $\widehat{\psi}_{\mathbf{z}}^{\mathrm{ref}}$ coincide on the reference configuration $\widehat{K}$ with the usual harmonic basis functions from Sect. 2.3. The interpolation operator is given as

$$\mathfrak{I}_h v = \sum_{\mathbf{z}\in\mathcal{N}_h} v(\mathbf{z})\,\psi_{\mathbf{z}}^{\mathrm{ref}} \in V_h^{\mathrm{ref}} \tag{3.15}$$

for $v \in H^2(\Omega)$, on anisotropic meshes. In the analysis, it is sufficient to study the restriction of $\mathfrak{I}_h : H^2(\Omega) \to V_h^{\mathrm{ref}}$ onto a single element $K \in \mathcal{K}_h$ and we denote this restriction by the same symbol

$$\mathfrak{I}_h : H^2(K) \to V_h^{\mathrm{ref}}|_K .$$

Furthermore, we make use of the mapping to and from the reference configuration, cf. (3.8). As earlier, we mark the operators and functions defined over the reference configuration by a hat, as, for instance, $\widehat{v} = v \circ \mathfrak{F}_K^{-1} : \widehat{K} \to K$. We have already used $\nabla v = \alpha_K U_K \Lambda_K^{-1/2} \widehat{\nabla} \widehat{v}$, and by employing some calculus we find

$$\widehat{H}(\widehat{v}) = \alpha_K^{-2} \Lambda_K^{1/2} U_K^\top H(v) U_K \Lambda_K^{1/2}\,, \tag{3.16}$$

where $H(v)$ denotes the Hessian matrix of $v \in H^2(\Omega)$ and $\widehat{H}(\widehat{v})$ the corresponding Hessian on the reference configuration. Additionally, we observe the relation between the interpolation $\mathfrak{I}_h v$ transferred to the reference configuration $\widehat{K}$ and the interpolation $\widehat{\mathfrak{I}}_h \widehat{v}$ defined directly on $\widehat{K}$. Namely, it is

$$\widehat{\mathfrak{I}_h v} = \widehat{\mathfrak{I}}_h \widehat{v}\,, \tag{3.17}$$

since only function evaluations in the nodes are involved and the mapped basis functions coincide with the basis functions defined directly on $\widehat{K}$, see Sect. 3.4.2. Furthermore, the interpolation $\widehat{\mathfrak{I}}_h$ coincides with the pointwise interpolation in Sect. 2.4 since the functions $\psi_{\mathbf{z}}^{\mathrm{ref}}$ are harmonic. Thus, we can apply known results for the interpolation error on the reference configuration.

First, we consider the scaling of the H^1-seminorm when K is mapped to $\widehat{K}$.

Lemma 3.31 *Let $K \in \mathscr{K}_h$ be a polytopal element of a regular and stable anisotropic mesh $\mathscr{K}_h$. For $v \in H^1(K)$, it is*

$$\sqrt{\frac{\prod_{j=2}^{d} \lambda_{K,j}}{\lambda_{K,1}}}\, |\widehat{v}|^2_{H^1(\widehat{K})} \le |v|^2_{H^1(K)} \le \sqrt{\frac{\prod_{j=1}^{d-1} \lambda_{K,j}}{\lambda_{K,d}}}\, |\widehat{v}|^2_{H^1(\widehat{K})}\,.$$

Proof Applying the transformation to the reference configuration yields

$$\begin{aligned} |v|^2_{H^1(K)} &= \|\nabla v\|^2_{L_2(K)} = |K|\, \|\alpha_K U_K \Lambda_K^{-1/2} \widehat{\nabla} \widehat{v}\|^2_{L_2(\widehat{K})} \\ &= |K| \alpha_K^2\, \|\Lambda_K^{-1/2} \widehat{\nabla} \widehat{v}\|^2_{L_2(\widehat{K})} = |K| \alpha_K^2 \sum_{j=1}^{d} \lambda_{K,j}^{-1} \left\| \frac{\partial \widehat{v}}{\partial \widehat{x}_j} \right\|^2_{L_2(\widehat{K})}. \end{aligned}$$

Since $\lambda_{K,1} \ge \ldots \ge \lambda_{K,d}$, we obtain

$$\frac{|K| \alpha_K^2}{\lambda_{K,1}}\, |\widehat{v}|^2_{H^1(\widehat{K})} \le |v|^2_{H^1(K)} \le \frac{|K| \alpha_K^2}{\lambda_{K,d}}\, |\widehat{v}|^2_{H^1(\widehat{K})}\,.$$

Due to the choice (3.9) for α_K, it is $|K| \alpha_K^2 = \sqrt{\prod_{j=1}^{d} \lambda_{K,j}}$, that completes the proof. □

Next, we address the interpolation error. Therefore, we use the convention that $H^0(K) = L_2(K)$.

Theorem 3.32 *Let $K \in \mathscr{K}_h$ be a polytopal element of a regular and stable anisotropic mesh $\mathscr{K}_h$. For $v \in H^2(\Omega)$, it is*

$$|v - \mathfrak{I}_h v|^2_{H^\ell(K)} \leq c\alpha_K^{-4}\, S_\ell(K) \sum_{i,j=1}^{d} \lambda_{K,i}\lambda_{K,j} L_K(\mathbf{u}_{K,i}, \mathbf{u}_{K,j}; v)$$

with

$$S_\ell(K) = \begin{cases} 1, & \text{for } \ell = 0, \\ \dfrac{1}{|K|}\sqrt{\dfrac{\prod_{j=1}^{d-1}\lambda_{K,j}}{\lambda_{K,d}}}, & \text{for } \ell = 1, \end{cases}$$

where

$$L_K(\mathbf{u}_{K,i}, \mathbf{u}_{K,j}; v) = \int_K \left(\mathbf{u}_{K,i}^\top H(v)\mathbf{u}_{K,j}\right)^2 \mathrm{d}\mathbf{x} \quad \text{for } i, j = 1, \ldots, d\,.$$

and the constant c only depends on the regularity and stability parameters of the mesh.

Proof Property (3.17) together with the scaling to the reference configuration and Lemma 3.31 as well as (3.10) yield for $\ell = 0, 1$

$$\begin{aligned} |v - \mathfrak{I}_h v|^2_{H^\ell(K)} &\leq |K|\, S_\ell(K)\, |\widehat{v} - \widehat{\mathfrak{I}}_h \widehat{v}|^2_{H^\ell(\widehat{K})} \\ &\leq c h_{\widehat{K}}^{2(2-\ell)} |K|\, S_\ell(K)\, |\widehat{v}|^2_{H^2(\widehat{K})} \\ &\leq c|K|\, S_\ell(K)\, |\widehat{v}|^2_{H^2(\widehat{K})}\,, \end{aligned}$$

where the interpolation estimate in Theorem 2.27 has been applied on $\widehat{K}$. Next, we transform the H^2-semi-norm back to the element K. Employing the mapping and the relation (3.16) gives

$$\begin{aligned} |\widehat{v}|^2_{H^2(\widehat{K})} &= \int_{\widehat{K}} \|\widehat{H}(\widehat{v})\|_F^2 \,\mathrm{d}\widehat{\mathbf{x}} \\ &= \frac{\alpha_K^{-4}}{|K|} \int_K \|\Lambda_K^{1/2} U_K^\top H(v) U_K \Lambda_K^{1/2}\|_F^2 \,\mathrm{d}\mathbf{x}\,, \end{aligned}$$

where $\|\cdot\|_F$ denotes the Frobenius norm of a matrix. A small exercise yields

$$\|\Lambda_K^{1/2} U_K^\top H(v) U_K \Lambda_K^{1/2}\|_F^2 = \sum_{i,j=1}^{d} \lambda_{K,i}\lambda_{K,j} \left(\mathbf{u}_{K,i}^\top H(v)\mathbf{u}_{K,j}\right)^2 ,$$

and consequently

$$|\widehat{v}|_{H^2(\widehat{K})}^2 = \frac{\alpha_K^{-4}}{|K|} \sum_{i,j=1}^{d} \lambda_{K,i}\lambda_{K,j} L_K(\mathbf{u}_{K,i}, \mathbf{u}_{K,j}; v) .$$

Combining the derived results yields the desired estimates. □

For the comparison with the work of Formaggia and Perotto developed in two-dimensions, we remember that their lambdas behave like $\lambda_{i,K} \sim \sqrt{\lambda_{K,i}}$, $i = 1, 2$. Employing the assumption $\alpha_K \sim 1$ raised in the comparison of Sect. 3.3.1, we find

$$\frac{\sqrt{\lambda_{K,1}/\lambda_{K,2}}}{|K|} \sim \frac{1}{\lambda_{K,2}} .$$

Therefore, we recognize that the estimates in Theorem 3.32 match the results of Lemma 2 in [79], but on much more general meshes.

3.4.6 Numerical Assessment of Anisotropic Meshes

In the introduction of Sect. 3.4, we already mentioned that polygonal and polyhedral meshes are much more flexible in meshing than classical finite element shapes. This is in particular true for the generation of anisotropic meshes. In this section we give a first numerical assessment on polytopal anisotropic mesh refinement. We propose a bisection approach that does not rely on any initially prescribed direction and which is applicable in two- and three-dimensions. Classical bisection approaches for triangular and tetrahedral meshes do not share this versatility and they have to be combined with additional strategies like edge swapping, node removal and local node movement, see [152].

Starting from the local interpolation error estimate in Theorem 3.26, we obtain the global version

$$\|v - \mathfrak{I}_C v\|_{L_2(\Omega)} \leq c \left(\sum_{K \in \mathscr{K}_h} \|A_K^{-\top} \nabla v\|_{L_2(K)}^2 \right)^{1/2}$$

by exploiting Remark 3.25 and Proposition 3.18. As in the derivation of Proposition 3.28, we easily see that

$$\eta = \sqrt{\sum_{K\in\mathscr{K}_h} \eta_K^2} \qquad \text{with} \qquad \eta_K^2 = \alpha_K^{-2} \sum_{j=1}^{d} \lambda_{K,j}\, \mathbf{u}_{K,j}^\top\, G_K^*(v)\, \mathbf{u}_{K,j}$$

and

$$G_K^*(v) = \left(\int_K \frac{\partial v}{\partial x_i} \frac{\partial v}{\partial x_j} \mathbf{dx} \right)_{i,j=1}^{d} \in \mathbb{R}^{d\times d}, \quad \mathbf{x} = (x_1, \ldots, x_d)^\top$$

is a good error measure and the local values η_K may serve as error indicators over the polytopal elements. This estimate also remains meaningful on isotropic polytopal meshes, cf. Remark 3.27. In the case that $v \in H^1(\Omega)$ and its derivatives are known, we can thus apply the following adaptive mesh refinement algorithm:

1. Let $\mathscr{K}_0$ be a given initial mesh and $\ell = 0$.
2. Compute the error indicators η_K and η with the knowledge of the exact function v and its derivatives.
3. Mark all elements K for refinement which satisfy $\eta_K > 0.95\eta/\sqrt{|\mathscr{K}_\ell|}$, where $|\mathscr{K}_\ell|$ is the number of elements in the current mesh.
4. Refine the marked elements as described below in order to obtain a refined mesh $\mathscr{K}_{\ell+1}$.
5. Go to 2.

In step 3, we have chosen a equidistribution strategy which marks all elements for refinement whose error indicator is larger than the mean value. The factor 0.95 has been chosen for stabilizing reasons in the computations when the error is almost uniformly distributed. For the refinement in step 4, we have a closer look at the first term in the sum of η_K, which reads

$$\lambda_{K,1} \frac{\mathbf{u}_{K,1}^\top\, G_K^*(v)\, \mathbf{u}_{K,1}}{\mathbf{u}_{K,1}^\top\, \mathbf{u}_{K,1}},$$

because of $|\mathbf{u}_{K,1}| = 1$. Since $\lambda_{K,1} \gg \lambda_{K,d}$ for anisotropic elements, the refinement process should try to minimize the quotient such that the whole term does not dominate the error over K. Obviously, we are dealing here with the Rayleigh quotient, which is minimal if $\mathbf{u}_{K,1}$ is the eigenvector to the smallest eigenvalue of $G_K^*(v)$. As consequence, the longest stretching of the polytopal element K should be aligned with the direction of this eigenvector. In order to achieve the correct alignment for the next refined mesh, we may bisect the polytopal element orthogonal to the eigenvector which belongs to the largest eigenvalue of $G_K^*(v)$. Thus, we propose the following refinement strategies:

ISOTROPIC The elements are bisected as introduced in Sect. 2.2.3, i.e., they are split orthogonal to the eigenvector corresponding to the largest eigenvalue of $M_{\mathrm{Cov}}(K)$.

ANISOTROPIC In order to respect the anisotropic nature of v, we split the elements orthogonal to the eigenvector corresponding to the largest eigenvalue of $G_K^*(v)$.

Both refinement strategies do not guaranty the regularity of the meshes since there is no control on the edge lengths due to the naïve bisection. This might be imposed additionally in the realization, but the approach also works well in the forthcoming tests without this extra control.

For the numerical experiments we consider $\Omega = (0,1)^2$ and the function

$$v(x_1, x_2) = \tanh(60x_2) - \tanh(60(x_1 - x_2) - 30) , \tag{3.18}$$

taken from [109], which has two sharp layers: one along the x_1-axis and one along the line given by $x_2 = x_1 - 1/2$. The function as well as the initial mesh is depicted in Fig. 3.5. We apply the BEM-based FEM as usual, although the local BEM solver is not tailored for the anisotropic elements. For the details on the realization see Chap. 4.

Test 1: Mesh Refinement

In the first test we generate several sequences of polygonal meshes starting from an initial grid, see Fig. 3.5 right. These meshes contain naturally hanging nodes and their element shapes are quite general. First, the initial mesh is refined uniformly,

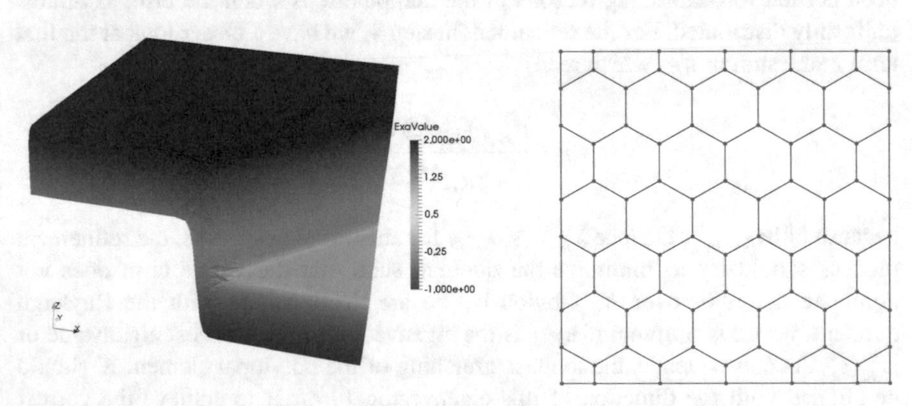

Fig. 3.5 Visualization of function with anisotropic behaviour (left) and initial mesh (right)

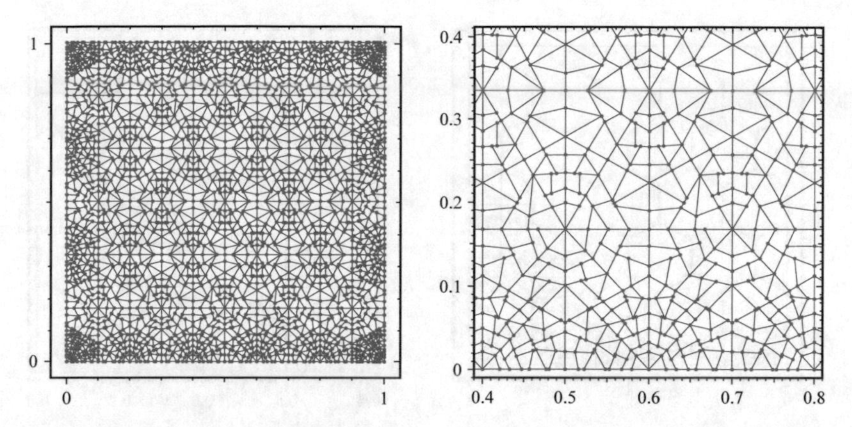

Fig. 3.6 Mesh after six uniform refinement steps using the ISOTROPIC strategy and zoom-in

i.e., all elements of the discretization are bisected in each refinement step. Here, the ISOTROPIC strategy is performed for the bisection. The mesh after six refinements as well as a zoom-in is depicted in Fig. 3.6. The uniform refinement clearly generates a lot of elements in regions where the function (3.18) is flat and where only a few elements would be sufficient for the approximation.

Next, we perform the adaptive refinement algorithm as described above for the different bisection strategies. The generated meshes after 6 refinement steps are visualized in Figs. 3.7 and 3.8 together with a zoom-in of the region where the two layers of the function (3.18) meet. Both strategies detect the layers and adapt the refinement to the underlying function. The adaptive strategies clearly outperform the uniform refinement with respect to the number of nodes which are needed to resolve the layers. Whereas the ISOTROPIC strategy in Fig. 3.7 keeps the aspect ratio of the polygonal elements bounded, the ANISOTROPIC bisection produces highly anisotropic elements, see Fig. 3.8. These anisotropic elements coincide with the layers of the function very well.

Finally, we compare the error measure η for the different strategies. This value is given with respect to the number of degrees of freedom, which coincides with the number of nodes, in a double logarithmic plot in Fig. 3.9. The error measure decreases most rapidly for the ANISOTROPIC strategy and consequently these meshes are most appropriate for the approximation of the function (3.18). The convergence order for η has not been studied analytically, however, we observe faster decrease for the ANISOTROPIC refinement in this test for the considered range. This behaviour might result from a pre-asymptotic regime. A slope of $1/2$ for $d = 2$ corresponds to linear convergence in finite element analysis.

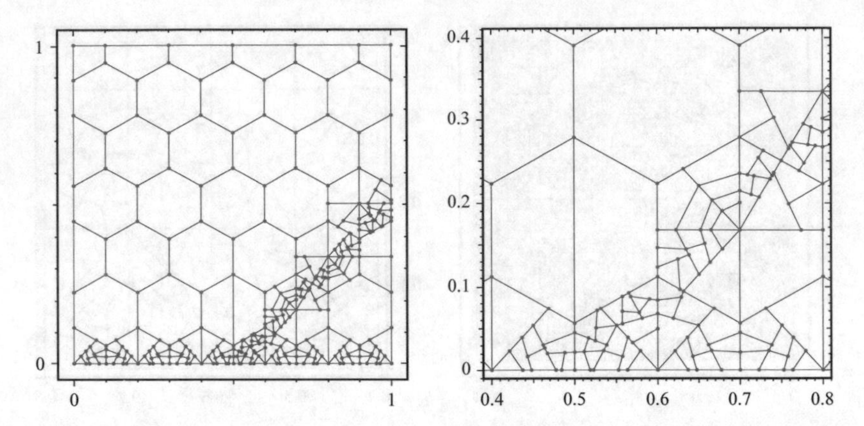

Fig. 3.7 Mesh after six adaptive refinement steps for the ISOTROPIC strategy and zoom-in

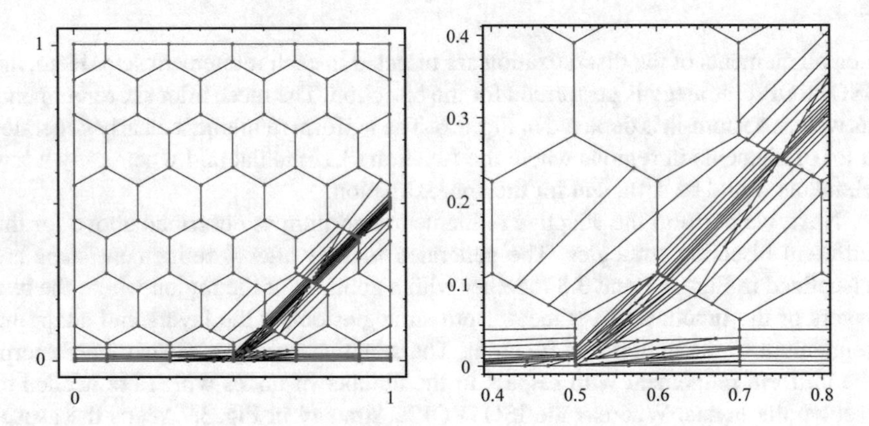

Fig. 3.8 Mesh after six adaptive refinement steps for the ANISOTROPIC strategy and zoom-in

Test 2: Mesh Properties

We analyse the meshes more carefully. For this purpose we pick the 13th mesh of the sequence generated with the ISOTROPIC and the ANISOTROPIC adaptive refinement strategy. In Sect. 3.4.1, we have introduced the ratio $\lambda_{K,1}/\lambda_{K,2}$ for the characterisation of the anisotropy of an element. In Fig. 3.10, we give this ratio with respect to the element ids for the two chosen meshes. For the ISOTROPIC refined mesh the ratio is clearly bounded by 10 and therefore the mesh consists of isotropic elements according to our characterisation. In the ANISOTROPIC refined mesh, however, the ratio varies in a large interval. The mesh consists of several isotropic elements, but there are mainly anisotropic polygons. The ratio of the most anisotropic elements exceeds 10^5 in this example.

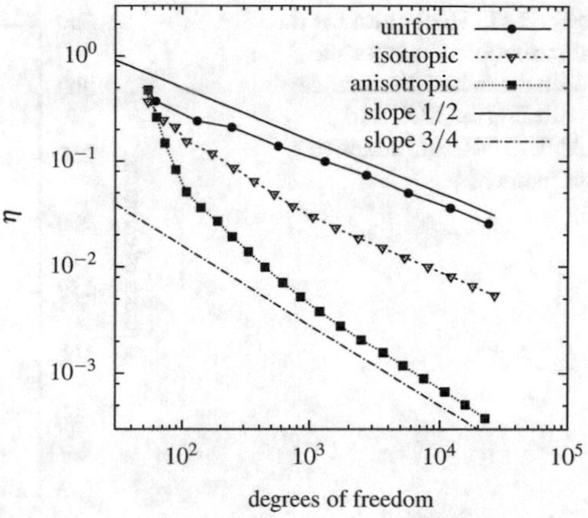

Fig. 3.9 Convergence graph of the anisotropic error measure η with respect to the number of degrees of freedom for the different refinement strategies

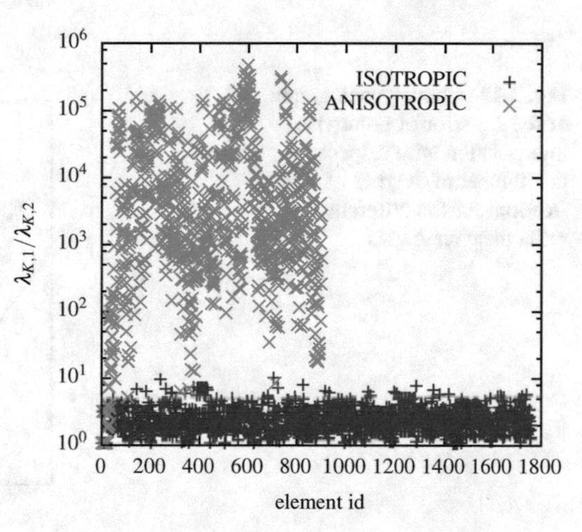

Fig. 3.10 Quotient $\lambda_{K,1}/\lambda_{K,2}$ for all elements in the 13th mesh of the sequence with ISOTROPIC and ANISOTROPIC adaptive refinement

Next we address the scaling parameter α_K in these meshes. In the comparison of the derived estimates with those of Formaggia and Perotto [79], it has been assumed that $\alpha_K \sim 1$. In Fig. 3.11, we present a histogram for the distribution of α_K in the two selected meshes. As expected the values stay bounded for the ISOTROPIC refined mesh. Furthermore, α_K stays in the same range for the ANISOTROPIC refinement. In our example, all values lie in the interval $(0.28, 0.32)$ although we are dealing with elements of quite different aspect ratios, cf. Fig. 3.10.

Test 3: Interpolation Error

In the final test we apply the pointwise interpolation into the space V_h to the function (3.18) over the meshes generated in this section. The convergence of the interpolation is studied numerically for the different sequences. We consider the interpolation error in the L_2-norm. In Fig. 3.12, we give $\|v - \mathfrak{I}_h v\|_{L_2(\Omega)}$ with respect to the number of degrees of freedom in a double logarithmic plot, where $\mathfrak{I}_h : H^2(\Omega) \to V_h$ is defined as in Sect. 2.4. Since $v \in H^2(\Omega)$ in this experiment,

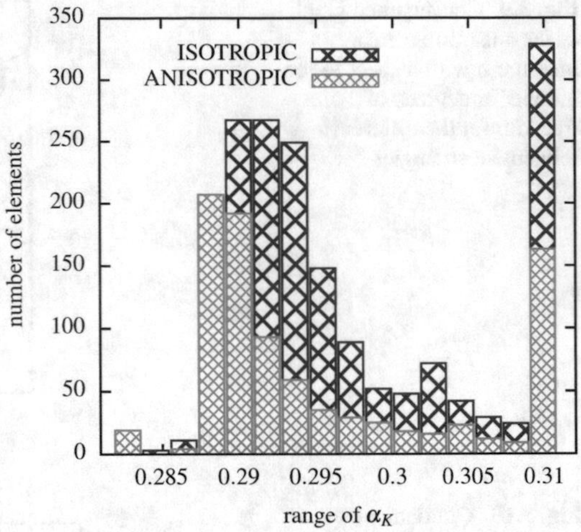

Fig. 3.11 Histogram for the distribution of α_K for the 13th mesh in the sequence with ISOTROPIC and ANISOTROPIC adaptive refinement

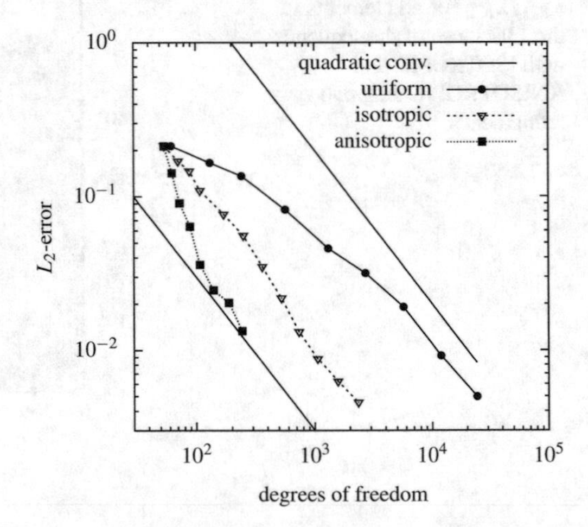

Fig. 3.12 Convergence graph of the L_2-error of pointwise interpolation with respect to the number of degrees of freedom for the different refinement strategies

we expect quadratic convergence with respect to the mesh size on the sequence of uniformly refined meshes. This convergence rate corresponds to a slope of one in the double logarithmic plot in two-dimensions. In Fig. 3.12, we observe that the uniform refinement reaches indeed quadratic convergence after a pre-asymptotic regime. The optimal rate of convergence is achieved as soon as the layers are resolved in the mesh. On the adaptively generated meshes, however, the interpolation error converges with optimal rates from the beginning. We can even recognize in Fig. 3.12 that the ANISOTROPIC refined meshes outperform the others. The layers are captured within a few refinement steps. Therefore, the error reduces faster than for the ISOTROPIC refined meshes before it reaches the optimal convergence rate.

Let us compare the seventh meshes in the sequences which are obtained after six refinements and which are visualized in Figs. 3.6, 3.7, and 3.8. For the uniformly refined mesh we have 2709 nodes and it is $\|v - \mathfrak{I}_h v\|_{L_2(\Omega)} \approx 3.17 \times 10^{-2}$. The adaptively refined mesh using ISOTROPIC bisection contains only 363 nodes but yields a comparable error $\|v - \mathfrak{I}_h v\|_{L_2(\Omega)} \approx 3.49 \times 10^{-2}$. The most accurate approximation is achieved on the ANISOTROPIC refined mesh with $\|v - \mathfrak{I}_h v\|_{L_2(\Omega)} \approx 2.04 \times 10^{-2}$ and only 189 nodes. A comparable interpolation error to the other refinement strategies is obtained on the fifth mesh of the sequence of ANISOTROPIC refined meshes. This mesh consists of 108 nodes only.

Chapter 4
Boundary Integral Equations and Their Approximations

The local problems in the definition of basis functions for the BEM-based FEM are treated by means of boundary integral equations. This chapter gives a short introduction into this topic with a special emphasis on its application in the BEM-based FEM. Therefore, the boundary integral operators for the Laplace problem are reviewed in two- and three-dimensions and corresponding boundary integral equations are derived. Their discretization is realized by a Galerkin boundary element method, which is used in the numerical examples and tests throughout the book. However, we also give an alternative approach for the discretization of boundary integral equations that relies on the Nyström method. The application of these approaches as local solvers for the BEM-based FEM is discussed in details and some comparisons highlighting advantageous and disadvantageous of these two solvers are given.

4.1 Preliminaries

Boundary element methods (BEM) are alternative approaches to finite element methods for the approximation of boundary value problems. They play an important role in modern numerical computations in the applied and engineering sciences. These methods rely on equivalent boundary integral equations of the corresponding boundary value problems, which are known in many cases. The key ingredient is the knowledge of a fundamental solution of the differential operator. Although the existence of such functions can be guarantied for a wide class of partial differential equations, see [100], the explicit construction is a more difficult task. However, the fundamental solution is known for important operators with constant coefficients such as for the Laplace and Helmholtz operators as well as for the system of elasticity and for Stokes equations, for instance. These include the most important applications of the boundary element methods. The advantage of the BEM over the

© Springer Nature Switzerland AG 2019

S. Weißer, *BEM-based Finite Element Approaches on Polytopal Meshes*,
Lecture Notes in Computational Science and Engineering 130,
https://doi.org/10.1007/978-3-030-20961-2_4

FEM is that the d-dimensional problems are reduced to $d-1$-dimensional ones on the boundary of the underlying domain. Furthermore, due to the formulation on the boundary, the BEM is naturally applicable for unbounded exterior domains, which are of particular interest in scattering problems, for example. When discretizing a boundary integral equation, we generally speak about a boundary element method. But, if it is referred to BEM in this book, we usually mean a Galerkin approach for the approximation of the boundary integral equation as described in Sect. 4.3. The Galerkin methods perfectly fit to the variational formulation of these integral equations. Their theoretical study is complete and provides a powerful tool for the analysis. In the engineering community, collocation methods are often preferred because of their easier practical implementation. However, the stability and convergence theory for these methods is only available for two-dimensional problems. Alternatively, a Nyström discretization of the boundary integral equation can be chosen, where the integrals of the operators are replaced by appropriate quadrature formulas. This strategy is discussed in Sect. 4.4. For more details on the theory of integral and in particular boundary integral equations we refer to the literature [13, 61, 105, 107, 118, 127, 128]. Galerkin boundary element methods are studied and discussed in [151, 158, 159] for elliptic differential operators. The collocation and Nyström approaches can be found beside others in [13, 118] and we especially mention [117, 133] for the Nyström discretizations.

In the following presentation, we restrict ourselves to the pure Laplace problem

$$-\Delta u = 0 \quad \text{in } K\,, \qquad u = g \quad \text{on } \partial K \tag{4.1}$$

with Dirichlet boundary conditions on a bounded polytopal domain K in two- and three-dimensions. Note that K will be a polytopal element and g a piecewise polynomial function in our application later on. This problem setting is sufficient for the approximation in the BEM-based FEM as seen in Sect. 4.5. The approach is also applicable to the before mentioned differential operators and in particular to convection-diffusion-reaction problems. Furthermore, other types of boundary conditions can be incorporated when needed, for instance, in Neumann or mixed boundary value problems. Some of the possible modifications are discussed in Chap. 6.

4.2 Boundary Integral Formulations

Let $K \subset \mathbb{R}^d$, $d = 2, 3$ be a bounded open domain with polygonal or polyhedral boundary, and we consider the boundary value problem (4.1) with some given function $g \in H^{1/2}(\partial K)$. For the following theory of boundary integral formulations, we need the usual trace operator γ_0^K. For sufficiently smooth functions, it is given as restriction of the function to the boundary. For Lipschitz domains, and thus in particular for polytopal domains, the trace is a bounded linear operator with $\gamma_0^K : H^s(K) \to H^{s-1/2}(\partial K)$ for $1/2 < s \leq 1$ and it has a continuous right inverse.

Here, the superscript indicates that the trace is taken with respect to the domain K. Let $v \in H^1(K)$ with Δv in the dual of $H^1(K)$. Due to Green's first identity [128], there exists a unique function $\gamma_1^K v \in H^{-1/2}(\partial K)$ such that

$$\int_K \nabla v(\mathbf{y}) \cdot \nabla w(\mathbf{y})\, \mathrm{d}\mathbf{y} = \int_{\partial K} \gamma_1^K v(\mathbf{y}) \gamma_0^K w(\mathbf{y})\, \mathrm{d}s_{\mathbf{y}} - \int_K w(\mathbf{y}) \Delta v(\mathbf{y})\, \mathrm{d}\mathbf{y} \tag{4.2}$$

for $w \in H^1(K)$. We call $\gamma_1^K v$ the conormal derivative of v. If v is sufficiently smooth, e.g. $v \in H^2(K)$, we have

$$(\gamma_1^K v)(\mathbf{x}) = \mathbf{n}_K(\mathbf{x}) \cdot (\gamma_0^K \nabla v)(\mathbf{x}) \quad \text{for } \mathbf{x} \in \partial K\,,$$

where $\mathbf{n}_K(\mathbf{x})$ denotes the outer normal vector of the domain K at $\mathbf{x}$. The trace and the conormal derivative are also called Dirichlet and Neumann trace for the Laplace equation. Additionally, we make use of the fundamental solution of the Laplacian. This singular function is given as

$$U^*(\mathbf{x}, \mathbf{y}) = \begin{cases} -\dfrac{1}{2\pi} \ln |\mathbf{x} - \mathbf{y}| & \text{for } \mathbf{x}, \mathbf{y} \in \mathbb{R}^2\,, \\[2ex] \dfrac{1}{4\pi\, |\mathbf{x} - \mathbf{y}|} & \text{for } \mathbf{x}, \mathbf{y} \in \mathbb{R}^3\,. \end{cases}$$

The fundamental solution satisfies the equation

$$-\Delta_{\mathbf{y}} U^*(\mathbf{x}, \mathbf{y}) = \delta_0(\mathbf{y} - \mathbf{x})$$

in the distributional sense, where δ_0 is the Dirac delta distribution. If we substitute $v(\mathbf{y}) = U^*(\mathbf{x}, \mathbf{y})$ in Green's second identity

$$\int_K (v(\mathbf{y})\Delta u(\mathbf{y}) - u(\mathbf{y})\Delta v(\mathbf{y}))\, \mathrm{d}\mathbf{y} = \int_{\partial K} \left(\gamma_0^K v(\mathbf{y}) \gamma_1^K u(\mathbf{y}) - \gamma_0^K u(\mathbf{y}) \gamma_1^K v(\mathbf{y})\right) \mathrm{d}s_{\mathbf{y}}\,,$$

see [128], we obtain a representation formula for the solution u in every point $\mathbf{x} \in K$. It reads

$$u(\mathbf{x}) = \int_{\partial K} U^*(\mathbf{x}, \mathbf{y}) \gamma_1^K u(\mathbf{y})\, \mathrm{d}s_{\mathbf{y}} - \int_{\partial K} \gamma_{1,\mathbf{y}}^K U^*(\mathbf{x}, \mathbf{y}) \gamma_0^K u(\mathbf{y})\, \mathrm{d}s_{\mathbf{y}}\,, \tag{4.3}$$

where $\gamma_{1,\mathbf{y}}^K$ denotes the conormal derivative operator with respect to the variable $\mathbf{y}$. By differentiation of (4.3), we obtain formulas for the derivatives of u. Consequently, if the data $\gamma_0^K u$ and $\gamma_1^K u$ is known, it is possible to evaluate the function u and its derivatives everywhere in the domain K. Furthermore, it is possible to compute the Neumann data if the Dirichlet data is known as in (4.1). We apply the trace and the conormal derivative operator to the representation formula and obtain a system of equations

$$\begin{pmatrix} \gamma_0^K u \\ \gamma_1^K u \end{pmatrix} = \begin{pmatrix} (1-\varsigma)\mathbf{I} - \mathbf{K}_K & \mathbf{V}_K \\ \mathbf{D}_K & \varsigma\mathbf{I} + \mathbf{K}_K' \end{pmatrix} \begin{pmatrix} \gamma_0^K u \\ \gamma_1^K u \end{pmatrix}, \tag{4.4}$$

where

$$\varsigma(\mathbf{x}) = \lim_{\varepsilon\to 0} \frac{1}{2(d-1)\pi} \frac{1}{\varepsilon^{d-1}} \int_{\mathbf{y}\in K:|\mathbf{y}-\mathbf{x}|=\varepsilon} \mathrm{d}s_{\mathbf{y}} \quad \text{for } \mathbf{x} \in \partial K . \tag{4.5}$$

The system (4.4) contains the standard boundary integral operators which are well studied, see, e.g., [128, 151, 159]. For $\mathbf{x} \in \partial K$, we have the single-layer potential operator

$$(\mathbf{V}_K\zeta)(\mathbf{x}) = \gamma_0^K \int_{\partial K} U^*(\mathbf{x},\mathbf{y})\zeta(\mathbf{y})\,\mathrm{d}s_{\mathbf{y}} \quad \text{for } \zeta \in H^{-1/2}(\partial K) ,$$

the double-layer potential operator

$$(\mathbf{K}_K\xi)(\mathbf{x}) = \lim_{\varepsilon\to 0} \int_{\mathbf{y}\in\partial K:|\mathbf{y}-\mathbf{x}|\geq\varepsilon} \gamma_{1,\mathbf{y}}^K U^*(\mathbf{x},\mathbf{y})\xi(\mathbf{y})\,\mathrm{d}s_{\mathbf{y}} \quad \text{for } \xi \in H^{1/2}(\partial K) ,$$

and the adjoint double-layer potential operator

$$(\mathbf{K}_K'\zeta)(\mathbf{x}) = \lim_{\varepsilon\to 0} \int_{\mathbf{y}\in\partial K:|\mathbf{y}-\mathbf{x}|\geq\varepsilon} \gamma_{1,\mathbf{x}}^K U^*(\mathbf{x},\mathbf{y})\zeta(\mathbf{y})\,\mathrm{d}s_{\mathbf{y}} \quad \text{for } \zeta \in H^{-1/2}(\partial K) ,$$

as well as the hypersingular integral operator

$$(\mathbf{D}_K\xi)(\mathbf{x}) = -\gamma_1^K \int_{\partial K} \gamma_{1,\mathbf{y}}^K U^*(\mathbf{x},\mathbf{y})\xi(\mathbf{y})\,\mathrm{d}s_{\mathbf{y}} \quad \text{for } \xi \in H^{1/2}(\partial K) .$$

These integral operators

$$\begin{aligned} \mathbf{V}_K &: H^{-1/2+s}(\partial K) \to H^{1/2+s}(\partial K) , \\ \mathbf{K}_K &: H^{1/2+s}(\partial K) \to H^{1/2+s}(\partial K) , \\ \mathbf{K}_K' &: H^{-1/2+s}(\partial K) \to H^{-1/2+s}(\partial K) , \\ \mathbf{D}_K &: H^{1/2+s}(\partial K) \to H^{-1/2+s}(\partial K) \end{aligned}$$

are linear and continuous for $s \in [-1/2, 1/2]$, see [61, 128]. The system (4.4) can be utilized to derive the following relations between the boundary integral operators

$$\begin{aligned}
\mathbf{V}_K\mathbf{D}_K &= (\varsigma\mathbf{I}+\mathbf{K})((1-\varsigma)\mathbf{I}-\mathbf{K}_K)\,,\\
\mathbf{D}_K\mathbf{V}_K &= (\varsigma\mathbf{I}+\mathbf{K}_K')((1-\varsigma)\mathbf{I}-\mathbf{K}_K')\,,\\
\mathbf{V}_K\mathbf{K}_K' &= \mathbf{K}_K\mathbf{V}_K\,,\\
\mathbf{K}_K'\mathbf{D}_K &= \mathbf{D}_K\mathbf{K}_K\,.
\end{aligned}$$

Remark 4.1 The function $u = 1$ obviously satisfies the Laplace equation and it is $\gamma_0^K u = 1$ and $\gamma_1^K u = 0$. Consequently, we obtain from the first equation in (4.4) that

$$\varsigma(\mathbf{x}) = -\int_{\partial K} \gamma_{1,\mathbf{y}}^K U^*(\mathbf{x},\mathbf{y})\,\mathrm{d}s_{\mathbf{y}} \quad \text{for } \mathbf{x}\in\partial K\,. \tag{4.6}$$

If the boundary ∂K is smooth in a neighbourhood of the point $\mathbf{x} \in \partial K$, i.e., it can be represented locally by a differentiable parametrization, then (4.5) yields

$$\varsigma(\mathbf{x}) = \frac{1}{2}\,.$$

Thus, we have $\varsigma = 1/2$ almost everywhere on ∂K for a polytopal domain K. On the other hand, if $\mathbf{x} \in \partial K$ is on an edge in 3D or it is a vertex, then ς is related to the interior angle of K at the point $\mathbf{x}$. In the two-dimensional case $K \subset \mathbb{R}^2$, it can be shown that

$$\varsigma(\mathbf{x}) = \frac{\alpha}{2\pi}$$

for a corner point $\mathbf{x}$ of a polygonal domain, where $\alpha \in (0, 2\pi)$ denotes the interior angle of the polygon at $\mathbf{x}$, see, e.g., [118].

4.2.1 Direct Approach for Dirichlet Problem

For $K \subset \mathbb{R}^2$ with $h_K < 1$ and $K \subset \mathbb{R}^3$, the single-layer potential operator induces a bilinear form $(\mathbf{V}_K\,\cdot,\cdot)_{L_2(\partial K)}$, which is $H^{-1/2}(\partial K)$-elliptic and continuous on $H^{-1/2}(\partial K)$, see [128, 159]. Here, the L_2-inner product has to be interpreted as duality pairing. According to the Lax–Milgram Lemma the single-layer potential operator is invertible. Therefore, the first equation of system (4.4) yields a relation between the Dirichlet and the Neumann trace, namely

$$\gamma_1^K u = \mathbf{S}_K\gamma_0^K u \quad \text{with} \quad \mathbf{S}_K = \mathbf{V}_K^{-1}\left(\tfrac{1}{2}\mathbf{I}+\mathbf{K}_K\right)\,. \tag{4.7}$$

The operator

$$\mathbf{S}_K : H^{1/2}(\partial K) \to H^{-1/2}(\partial K)$$

is called Steklov–Poincaré operator and (4.7) is its non-symmetric representation. This operator is linear and continuous due to its definition. With the help of the second equation in the system (4.4), we find the symmetric representation

$$\mathbf{S}_K = \mathbf{D}_K + \left(\tfrac{1}{2}\mathbf{I} + \mathbf{K}_K'\right)\mathbf{V}_K^{-1}\left(\tfrac{1}{2}\mathbf{I} + \mathbf{K}_K\right). \tag{4.8}$$

The inversion of the single-layer potential operator is not desirable in the evaluation of the Steklov–Poincaré operator. In order to compute the unknown Neumann data $t = \gamma_1^K u \in H^{-1/2}(\partial K)$ from given Dirichlet data $g = \gamma_0^K u \in H^{1/2}(\partial K)$, it is more convenient to use the Galerkin formulation

$$\begin{aligned}&\text{Find } t \in H^{-1/2}(\partial K):\\ &(\mathbf{V}_K t, \zeta)_{L_2(\partial K)} = \left(\left(\tfrac{1}{2}\mathbf{I} + \mathbf{K}_K\right) g, \zeta\right)_{L_2(\partial K)} \quad \forall \zeta \in H^{-1/2}(\partial K).\end{aligned} \tag{4.9}$$

This formulation admits a unique solution according to the Lax–Milgram Lemma and is consequently equivalent to the evaluation of $\mathbf{S}_K$. Thus, in order to solve the Dirichlet problem for the Laplace equation (4.1), we may choose the representation formula (4.3) for u and compute its Neumann trace with the help of the Galerkin formulation (4.9). The solution obtained this way satisfies $u \in H^1(K)$, see [128, 159]. This is a direct approach since the Dirichlet and Neumann traces of the unknown solution are either known or computed and used in the representation formula.

4.2.2 Indirect Approach for Dirichlet Problem

Alternatively, one may follow an indirect approach. Instead of computing traces of the unknown function, the solution is sought as a potential of an unknown density. It is known, see, e.g., [128, 151, 159], that the double-layer potential

$$u(\mathbf{x}) = \int_{\partial K} \gamma_{1,\mathbf{y}}^K U^*(\mathbf{x}, \mathbf{y}) \xi(\mathbf{y}) \, \mathrm{d}s_{\mathbf{y}} \quad \text{for } \mathbf{x} \in K \tag{4.10}$$

with arbitrary density $\xi \in H^{1/2}(\partial K)$ satisfies the Laplace equation. Thus, the density ξ has to be determined such that the Dirichlet boundary condition in (4.1) is satisfied. Applying the trace operator to (4.10) yields the following boundary integral equation of second kind

$$(1 - \varsigma(\mathbf{x}))\xi(\mathbf{x}) - (\mathbf{K}_K \xi)(\mathbf{x}) = -g(\mathbf{x}) \quad \text{for } \mathbf{x} \in \partial K. \tag{4.11}$$

It admits a unique solution $\xi \in H^{1/2}(\partial K)$ which is formally given as a Neumann series

$$\xi(\mathbf{x}) = -\sum_{\ell=0}^{\infty} (\varsigma\mathbf{I} + \mathbf{K}_K)^{\ell} g(\mathbf{x}) \quad \text{for } \mathbf{x} \in \partial K .$$

Furthermore, the series is convergent since $\varsigma\mathbf{I} + \mathbf{K}_K$ is a contraction in $H^{1/2}(\partial K)$, see [160].

4.2.3 Direct Approach for Neumann Problem

Although this chapter focuses on the Dirichlet problem for the Laplace equation, we briefly consider the Neumann problem:

$$-\Delta u = 0 \quad \text{in } K , \qquad \gamma_1^K u = t \quad \text{on } \partial K ,$$

where $t \in H^{-1/2}(\partial K)$ satisfies the solvability condition

$$\int_{\partial K} t \, \mathrm{d}s_{\mathbf{x}} = 0 \tag{4.12}$$

such that there exists a unique solution

$$u \in H_*^1(K) = \{v \in H^1(K) : (v, 1)_{L_2(\partial K)} = 0\} .$$

We follow a direct approach and derive a boundary integral equation for the unknown Dirichlet data $g = \gamma_0^K u \in H^{1/2}(\partial K)$. Afterwards, the representation formula (4.3) gives the solution of the boundary value problem.

In order to find a connection between the Dirichlet and Neumann traces we consider this time the second equation in (4.4), which yields

$$\mathbf{D}_K \gamma_0^K u = \left(\tfrac{1}{2}\mathbf{I} - \mathbf{K}_K'\right) \gamma_1^K u . \tag{4.13}$$

The hypersingular integral operator $\mathbf{D}_K$ is self-adjoint and has a non-trivial kernel on $H^{1/2}(\partial K)$, namely it is $\ker \mathbf{D}_K = \operatorname{span}\{1\}$ for a simply connected domain K. Thus, we define the subspace

$$H_*^{1/2}(\partial K) = \{\xi \in H^{1/2}(\partial K) : (\xi, 1)_{L_2(\partial K)} = 0\}$$

of $H^{1/2}(\partial K)$, containing the functions with vanishing mean value, on which $\mathbf{D}_K$ is bounded and elliptic. $H_*^{1/2}(\partial K)$ can be interpreted as trace space of $H_*^1(K)$.

Consequently, (4.13) has a unique solution $\gamma_0^K u$ in $H_*^{1/2}(\partial K)$ for given data $t = \gamma_1^K u$. With a slight abuse of notation, we denote by $\mathbf{D}_K^{-1}$ the inverse of the hypersingular integral operator on the subspace $H_*^{1/2}(\partial K)$, and thus we can write

$$\gamma_0^K u = \mathbf{P}_K \gamma_1^K u \quad \text{with} \quad \mathbf{P}_K = \mathbf{D}_K^{-1}\left(\tfrac{1}{2}\mathbf{I} - \mathbf{K}_K'\right) \tag{4.14}$$

on $H_*^{1/2}(\partial K)$. The operator $\mathbf{P}_K : H^{-1/2}(\partial K) \to H_*^{1/2}(\partial K)$, which maps the Neumann to the Dirichlet data, is called Poincaré–Steklov operator. (Depending on the literature it is sometimes also called Steklov–Poincaré operator.) Employing the properties of the boundary integral operators, the symmetric representation

$$\mathbf{P}_K = \mathbf{V}_K + \left(\tfrac{1}{2}\mathbf{I} - \mathbf{K}_K\right)\mathbf{D}_K^{-1}\left(\tfrac{1}{2}\mathbf{I} - \mathbf{K}_K'\right) \tag{4.15}$$

follows. To compute the unknown Dirichlet data $g = \gamma_0^K u \in H_*^{1/2}(\partial K)$ from given Neumann data $t = \gamma_1^K u \in H^{-1/2}(\partial K)$, we apply a Galerkin formulation once more, namely

$$\begin{aligned} &\text{Find } g \in H_*^{1/2}(\partial K) : \\ &(\mathbf{D}_K g, \xi)_{L_2(\partial K)} = \left((\tfrac{1}{2}\mathbf{I} - \mathbf{K}_K')t, \xi\right)_{L_2(\partial K)} \qquad \forall \xi \in H_*^{1/2}(\partial K)\,. \end{aligned} \tag{4.16}$$

This problem is reformulated into a saddle point formulation, which reads

$$\begin{aligned} &\text{Find } (g,\lambda) \in H^{1/2}(\partial K) \times \mathbb{R} : \\ &(\mathbf{D}_K g, \xi)_{L_2(\partial K)} + \lambda(\xi, 1)_{L_2(\partial K)} = \left((\tfrac{1}{2}\mathbf{I} - \mathbf{K}_K')t, \xi\right)_{L_2(\partial K)} \qquad \forall \xi \in H^{1/2}(\partial K)\,, \\ &\mu(g,1)_{L_2(\partial K)} = 0 \qquad \forall \mu \in \mathbb{R}\,. \end{aligned}$$

For $g \in H^{1/2}(\partial K) \setminus H_*^{1/2}(\partial K)$, we write $\mu = \lambda/(g,1)_{L_2(\partial K)} - \alpha$ with $\alpha \in \mathbb{R}$ and obtain from the second equation $\lambda = \alpha(g,1)_{L_2(\partial K)}$. The expression for the Lagrange multiplier λ also holds for $g \in H_*^{1/2}(\partial K)$, since testing the first equation with $\xi_0 = 1$ yields

$$\lambda(1,1)_{L_2(\partial K)} = 0$$

and thus $\lambda = 0$. Here, we employed

$$\ker \mathbf{D}_K = \ker\left(\tfrac{1}{2}\mathbf{I} + \mathbf{K}_K\right) = \operatorname{span}\{1\}\,,$$

$\mathbf{D}_K$ is self-adjoint and the solvability condition (4.12), such that

$$(\mathbf{D}_K g, \xi_0)_{L_2(\partial K)} = (g, \mathbf{D}_K \xi_0)_{L_2(\partial K)} = 0$$

and

$$\left((\tfrac{1}{2}\mathbf{I} - \mathbf{K}_K')t, \xi_0\right)_{L_2(\partial K)} = (t, \xi_0)_{L_2(\partial K)} - \left(t, (\tfrac{1}{2}\mathbf{I} + \mathbf{K}_K)\xi_0\right)_{L_2(\partial K)} = 0\,.$$

Inserting $\lambda = \alpha(g,1)_{L_2(\partial K)}$ into the first equation of the saddle point formulation, we obtain for fixed α the Galerkin formulation:

$$\begin{aligned} &\text{Find } g \in H^{1/2}(\partial K): \\ &\left(\widetilde{\mathbf{D}}_K g, \xi\right)_{L_2(\partial K)} = \left((\tfrac{1}{2}\mathbf{I} - \mathbf{K}_K')t, \xi\right)_{L_2(\partial K)} \quad \forall \xi \in H^{1/2}(\partial K)\,, \end{aligned} \tag{4.17}$$

where

$$\left(\widetilde{\mathbf{D}}_K \vartheta, \xi\right)_{L_2(\partial K)} = (\mathbf{D}_K \vartheta, \xi)_{L_2(\partial K)} + \alpha(\vartheta, 1)_{L_2(\partial K)}(\xi, 1)_{L_2(\partial K)}\,.$$

For $\alpha > 0$, the operator $\widetilde{\mathbf{D}}_K$ is bounded and elliptic on $H^{1/2}(\partial K)$ and consequently, the Galerkin formulation has a unique solution $g \in H^{1/2}(\partial K)$. This solution even belongs to $H_*^{1/2}(\partial K)$ since plugging $\xi_0 = 1$ into (4.17) yields with the same arguments as above

$$(g, 1)_{L_2(\partial K)}(1, 1)_{L_2(\partial K)} = 0\,.$$

Hence, the formulation (4.17) is equivalent to the initial variational formulation and the solution g is independent of α because of the unique solvability.

4.3 Boundary Element Method

The aim of this section is to introduce discrete Galerkin formulations for the direct approaches of the Dirichlet and Neumann problems derived in the previous section. Thus, we discretize the variational formulations (4.9) and (4.17). For this reason, we have to introduce approximation spaces for $H^{1/2}(\partial K)$ and $H^{-1/2}(\partial K)$ as well as in particular a discretization of ∂K. We follow standard approaches as described in [144, 151, 159], for instance.

First, the boundary ∂K of the domain K is decomposed into non-overlapping line segments in two-dimensions and triangles in three-dimensions, see Fig. 4.1, such that the resulting boundary mesh, which is denoted by $\mathscr{B}_h$, is regular. More precisely, we assume that $\mathscr{B}_h$ is shape-regular in the sense of Ciarlet such that neighbouring elements either share a common node or edge and the aspect ratio

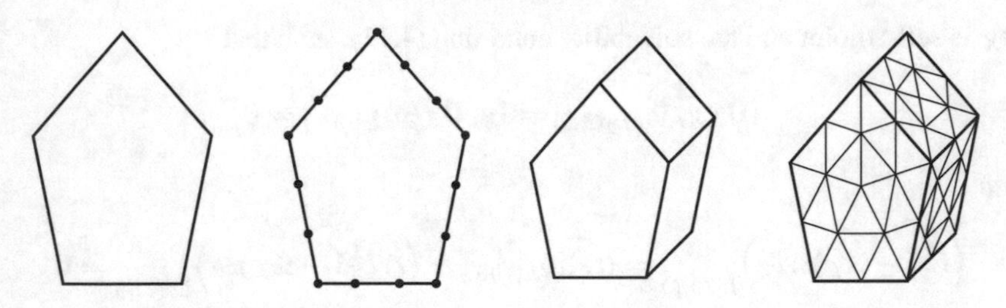

Fig. 4.1 A domain and its boundary mesh for $d = 2$ (left) and $d = 3$ (right)

of each triangle is uniformly bounded. In order to present approximation estimates later on, we additionally assume a uniform boundary mesh in the sense that all elements have comparable size. The elements of the mesh $\mathscr{B}_h$ are denoted by T. For the approximations, we utilize standard spaces of piecewise polynomials. Let $k \in \mathbb{N}$ be the desired approximation order in the boundary element method. We discretize $H^{-1/2}(\partial K)$ and thus the Neumann traces by piecewise polynomials of degree smaller or equal $k-1$ which might be discontinuous over element interfaces. This approximation space is given by

$$\mathscr{P}^{k-1}_{\mathrm{pw,d}}(\mathscr{B}_h) = \left\{\zeta \in L_2(\partial K) : \zeta\big|_T \in \mathscr{P}^{k-1}(T)\ \forall T \in \mathscr{B}_h\right\}. \tag{4.18}$$

The space $H^{1/2}(\partial K)$ and thus the Dirichlet traces are discretized by piecewise polynomials of degree smaller or equal k which are continuous over element interfaces. This approximation space is given by

$$\mathscr{P}^{k}_{\mathrm{pw}}(\mathscr{B}_h) = \mathscr{P}^{k}_{\mathrm{pw,d}}(\mathscr{B}_h) \cap C^0(\partial K)\,. \tag{4.19}$$

The choice of spaces yields conforming Galerkin approximations since

$$\mathscr{P}^{k}_{\mathrm{pw}}(\mathscr{B}_h) \subset H^{1/2}(\partial K) \qquad \text{and} \qquad \mathscr{P}^{k-1}_{\mathrm{pw,d}}(\mathscr{B}_h) \subset H^{-1/2}(\partial K)\,.$$

These spaces are equipped with the usual Lagrangian bases used in finite element methods. In two-dimensions we might also apply the polynomial basis defined in Sect. 2.3.1, cf. Fig. 2.9. The set of basis functions for $\mathscr{P}^{k}_{\mathrm{pw}}(\mathscr{B}_h)$ and $\mathscr{P}^{k-1}_{\mathrm{pw,d}}(\mathscr{B}_h)$ are fixed once and they are denoted in the following by Φ_D and Φ_N, respectively.

4.3.1 Dirichlet Problem

In order to treat the Dirichlet problem (4.1) for the Laplace equation, we utilize the direct approach and approximate the Galerkin formulation (4.9) for the unknown

Neumann trace $t = \gamma_1^K u \in H^{-1/2}(\partial K)$ from Sect. 4.2.1. This yields the discrete Galerkin formulation

$$\begin{aligned} &\text{Find } t_h \in \mathscr{P}^{k-1}_{\mathrm{pw,d}}(\mathscr{B}_h) : \\ &(\mathbf{V}_K t_h, \zeta)_{L_2(\partial K)} = \left(\left(\tfrac{1}{2}\mathbf{I} + \mathbf{K}_K\right) g, \zeta\right)_{L_2(\partial K)} \quad \forall \zeta \in \mathscr{P}^{k-1}_{\mathrm{pw,d}}(\mathscr{B}_h) , \end{aligned} \tag{4.20}$$

where g is the given Dirichlet data. Since the bilinear form induced by the single-layer potential operator is $H^{-1/2}(\partial K)$-elliptic as well as continuous on $H^{-1/2}(\partial K)$ and $\mathscr{P}^{k-1}_{\mathrm{pw,d}}(\mathscr{B}_h) \subset H^{-1/2}(\partial K)$, the variational formulation (4.20) admits a unique solution according to the Lax–Milgram Lemma. Furthermore, Céa's Lemma yields

$$\|t - t_h\|_{H^{-1/2}(\partial K)} \le C \inf_{\zeta \in \mathscr{P}^{k-1}_{\mathrm{pw,d}}(\mathscr{B}_h)} \|t - \zeta\|_{H^{-1/2}(\partial K)} .$$

From known approximation properties of polynomials, see [151, Theorem 4.3.20], we obtain

$$\|t - t_h\|_{H^{-1/2}(\partial K)} \le C\, h^{s+1/2} |t|_{H^s_{\mathrm{pw}}(\partial K)} , \tag{4.21}$$

when assuming $t \in H^s_{\mathrm{pw}}(\partial K)$ and $0 \le s \le k$. Here, h denotes the mesh size in the boundary element mesh $\mathscr{B}_h$. After the computation of t_h, we utilize it for approximating the solution $u(\mathbf{x})$ of the Dirichlet problem in an interior point $\mathbf{x} \in K$ by the representation formula (4.3). This yields

$$\widetilde{u}(\mathbf{x}) = \int_{\partial K} U^*(\mathbf{x}, \mathbf{y}) t_h(\mathbf{y}) \,\mathrm{d}s_{\mathbf{y}} - \int_{\partial K} \gamma^K_{1,\mathbf{y}} U^*(\mathbf{x}, \mathbf{y}) g(\mathbf{y}) \,\mathrm{d}s_{\mathbf{y}} , \tag{4.22}$$

and under the assumption of sufficient regularity we obtain for $k = 1$ the pointwise error estimate

$$|u(\mathbf{x}) - \widetilde{u}(\mathbf{x})| \le C(\mathbf{x})\, h^3\, |t|_{H^1_{\mathrm{pw}}(\partial K)} \tag{4.23}$$

for $\mathbf{x} \in K$ and in the $H^1(K)$-norm

$$\|u - \widetilde{u}\|_{H^1(K)} \le C\, h^{3/2}\, |t|_{H^1_{\mathrm{pw}}(\partial K)} .$$

Because of $\mathbf{x} \in K$, the integrands in (4.3) are non-singular and consequently approximation formulas for the derivatives of u can be derived by simply differentiating (4.22). These pointwise approximations of the derivatives converge with the same order as the approximation $\widetilde{u}(\mathbf{x})$ of $u(\mathbf{x})$. We point out, that the integrals in (4.22) can be evaluated analytically for piecewise polynomial data t_h and g. In our later application g is already piecewise polynomial. In the general case, however, the Dirichlet data is approximated by its L_2-projection $g_h \in \mathscr{P}^k_{\mathrm{pw}}(\mathscr{B}_h)$

and the error analysis additionally relies on Strang-type arguments. Utilizing the basis functions in Φ_D and Φ_N for $\mathscr{P}^k_{\mathrm{pw}}(\mathscr{B}_h)$ and $\mathscr{P}^{k-1}_{\mathrm{pw,d}}(\mathscr{B}_h)$, respectively, we make the ansatz

$$g_h(\mathbf{x}) = \sum_{\varphi\in\Phi_D} g_\varphi \varphi(\mathbf{x}) \quad \text{and} \quad t_h(\mathbf{x}) = \sum_{\tau\in\Phi_N} t_\tau \tau(\mathbf{x}) , \tag{4.24}$$

where $\mathbf{x} \in \partial K$. Furthermore, we identify the approximations g_h and t_h with their vectors $\underline{g}_h = (g_\varphi)_{\varphi\in\Phi_D}$ and $\underline{t}_h = (t_\tau)_{\tau\in\Phi_N}$ containing the expansion coefficients. Due to the L_2-projection, the coefficients in $\underline{g}_h$ are given as solution of

$$\sum_{\varphi\in\Phi_D} g_\varphi (\varphi, \xi)_{L_2(\partial K)} = (g, \xi)_{L_2(\partial K)} \quad \forall \xi \in \mathscr{P}^k_{\mathrm{pw}}(\mathscr{B}_h) . \tag{4.25}$$

The system of linear equations (4.25) involves the symmetric, positive definite mass matrix

$$\mathbf{M}^{DD}_{K,h} = \big((\varphi, \xi)_{L_2(\partial K)}\big)_{\xi\in\Phi_D,\varphi\in\Phi_D} .$$

Inserting the ansatz (4.24) into the discrete Galerkin formulation (4.20) yields a system of linear equations for $\underline{t}_h$, namely

$$\mathbf{V}_{K,h}\underline{t}_h = \left(\tfrac{1}{2}\mathbf{M}_{K,h} + \mathbf{K}_{K,h}\right) \underline{g}_h , \tag{4.26}$$

where the matrices are defined as

$$\mathbf{V}_{K,h} = \big((\mathbf{V}_K\tau, \vartheta)_{L_2(\partial K)}\big)_{\vartheta\in\Phi_N,\tau\in\Phi_N}$$

and

$$\mathbf{M}_{K,h} = \big((\varphi, \vartheta)_{L_2(\partial K)}\big)_{\vartheta\in\Phi_N,\varphi\in\Phi_D} , \quad \mathbf{K}_{K,h} = \big((\mathbf{K}_K\varphi, \vartheta)_{L_2(\partial K)}\big)_{\vartheta\in\Phi_N,\varphi\in\Phi_D} .$$

The system (4.26) is uniquely solvable since the matrix $\mathbf{V}_{K,h}$ is symmetric and positive definite due to the properties of the integral operator $\mathbf{V}_K$.

Remark 4.2 In the computational realization the matrices can be set up in different ways. Either a semi-analytic integration scheme is utilized, which evaluates the boundary integral operators applied to piecewise polynomial functions analytically and approximates the outer integrals by numerical quadrature, or a fully numerical integration scheme is applied. The semi-analytic scheme as well as the analytic formulas are given in [144] and an appropriate fully numerical quadrature is presented in [151] for the three-dimensional case. Corresponding formulas are also available for the two-dimensional case.

The approximation of the Dirichlet to Neumann map with the help of the system of linear equations (4.26) corresponds to the representation (4.7) of the Steklov–Poincaré operator. But, we have also derived a symmetric representation (4.8) which can be utilized to define a symmetric approximation of the Steklov–Poincaré operator. Since the Neumann trace satisfies

$$t = \mathbf{V}_K^{-1}\left(\tfrac{1}{2}\mathbf{I} + \mathbf{K}_K\right) g \,,$$

we use the previously derived approximation t_h and define

$$\widetilde{\mathbf{S}}_K g = \mathbf{D}_K g + \left(\tfrac{1}{2}\mathbf{I} + \mathbf{K}_K'\right) t_h \,. \tag{4.27}$$

This yields the symmetric discretization of the Steklov–Poincaré operator

$$\mathbf{S}_{K,h} = \mathbf{D}_{K,h} + \left(\tfrac{1}{2}\mathbf{M}_{K,h}^\top + \mathbf{K}_{K,h}^\top\right)\mathbf{V}_{K,h}^{-1}\left(\tfrac{1}{2}\mathbf{M}_{K,h} + \mathbf{K}_{K,h}\right) \tag{4.28}$$

with the matrix entries

$$\mathbf{S}_{K,h} = \left(\left(\widetilde{\mathbf{S}}_K \varphi, \phi\right)_{L_2(\partial K)}\right)_{\phi\in\Phi_D,\varphi\in\Phi_D} ,$$

where

$$\mathbf{D}_{K,h} = \left((\mathbf{D}_K \varphi, \phi)_{L_2(\partial K)}\right)_{\phi\in\Phi_D,\varphi\in\Phi_D} .$$

The matrix entries of $\mathbf{D}_{K,h}$ can be assembled with the help of the single-layer potential matrix $\mathbf{V}_{K,h}$. For piecewise smooth functions one can show that

$$(\mathbf{D}_K \varphi, \phi)_{L_2(\partial K)} = (\mathbf{V}_K \operatorname{curl}_{\partial K} \varphi, \operatorname{curl}_{\partial K} \phi)_{L_2(\partial K)} ,$$

where $\operatorname{curl}_{\partial K}$ denotes the surface curl of a scalar valued function on ∂K. For more details, we refer the interested reader to [144, 159].

Example 4.3 We demonstrate the performance of the boundary element method and give the numerical orders of convergence, cf. (1.7), for a model problem. Let K be a regular octagon centered at the origin with diameter 0.8, and consider the boundary value problem

$$-\Delta u = 0 \quad \text{in } K\,, \qquad u = g \quad \text{on } \partial K\,,$$

where the Dirichlet data g is chosen such that the unique solution of the problem is given as $u(\mathbf{x}) = \exp(2\pi(x_1 - 0.3))\cos(2\pi(x_2 - 0.3))$. The boundary element method is applied on a sequence of meshes for the approximation orders $k = 1, 2, 3$. The first mesh is defined to be the eight sides of the octagon and the following meshes are constructed by subdividing each line segment of the previous mesh into two new line segments of the same length. In the following tables we distinguish the meshes and the approximation orders by the number of degrees of freedom (DoF) used to approximate the Neumann trace. In Table 4.1, we present the convergence of the Neumann data in the L_2-norm. The observed numerical order of convergence (noc) is k that reflects the theoretical considerations, cf. (4.21). Furthermore, we evaluate the approximation and its gradient with the help of the representation formula (4.22) in the point $(0.2, 0)^\top \in K$ and present the relative errors as well as the numerical orders of convergence in Tables 4.2 and 4.3, respectively. We observe that the pointwise evaluation of the approximation as well as the evaluation of its gradient converge till numerical saturation is reached. For $k = 1$, the pointwise errors converge with cubic order for the function evaluation as well as for the gradient. This coincides with the estimate (4.23). For $k = 2, 3$, the tables indicate numerical convergence orders of 4 and 5, respectively.

Table 4.1 Degrees of freedom (DoF), error $\|t - t_h\|_{L_2(\Omega)}$ (err) and numerical order of convergence (noc) for $k = 1, 2, 3$ in Example 4.3

$k = 1$			$k = 2$			$k = 3$		
DoF	err	noc	DoF	err	noc	DoF	err	noc
8	$3.22 \times 10^{+0}$	–	16	8.06×10^{-1}	–	24	1.53×10^{-1}	–
16	9.90×10^{-1}	1.70	32	2.21×10^{-1}	1.87	48	1.76×10^{-2}	3.13
32	5.37×10^{-1}	0.88	64	5.98×10^{-2}	1.89	96	2.01×10^{-3}	3.12
64	2.63×10^{-1}	1.03	128	1.56×10^{-2}	1.93	192	2.42×10^{-4}	3.05
128	1.29×10^{-1}	1.03	256	3.96×10^{-3}	1.98	384	2.97×10^{-5}	3.03
256	6.40×10^{-2}	1.02	512	9.89×10^{-4}	2.00	768	3.68×10^{-6}	3.02
512	3.18×10^{-2}	1.01	1024	2.47×10^{-4}	2.00	1536	4.57×10^{-7}	3.01
1024	1.59×10^{-2}	1.00	2048	6.17×10^{-5}	2.00	3072	5.70×10^{-8}	3.00
2048	7.92×10^{-3}	1.00	4096	1.54×10^{-5}	2.00	6144	8.01×10^{-9}	2.83
Theory		1			2			3

Table 4.2 Degrees of freedom (DoF), relative error $|u(\mathbf{x}) - \widetilde{u}(\mathbf{x})|/|u(\mathbf{x})|$ for the point evaluation in $\mathbf{x} = (0.2, 0)^\top$ (err) and numerical order of convergence (noc) for $k = 1, 2, 3$ in Example 4.3

$k = 1$			$k = 2$			$k = 3$		
DoF	err	noc	DoF	err	noc	DoF	err	noc
8	3.46×10^{-2}	–	16	2.55×10^{-4}	–	24	1.57×10^{-4}	–
16	9.94×10^{-4}	5.12	32	3.80×10^{-5}	2.75	48	4.45×10^{-6}	5.14
32	1.12×10^{-4}	3.15	64	9.06×10^{-7}	5.39	96	1.19×10^{-7}	5.22
64	1.62×10^{-5}	2.79	128	1.21×10^{-7}	2.91	192	3.02×10^{-9}	5.31
128	2.03×10^{-6}	2.99	256	8.17×10^{-9}	3.89	384	7.46×10^{-11}	5.34
256	2.54×10^{-7}	2.99	512	4.70×10^{-10}	4.12	768	1.83×10^{-12}	5.35
512	3.22×10^{-8}	2.98	1024	2.53×10^{-11}	4.22	1536	3.52×10^{-14}	5.70
1024	4.09×10^{-9}	2.98	2048	1.28×10^{-12}	4.31	3072	4.51×10^{-14}	–
2048	5.20×10^{-10}	2.98	4096	1.69×10^{-13}	2.92	6144	1.21×10^{-13}	–

Table 4.3 Degrees of freedom (DoF), relative error $|\nabla u(\mathbf{x}) - \nabla \widetilde{u}(\mathbf{x})|/|\nabla u(\mathbf{x})|$ for the point evaluation of the gradient in $\mathbf{x} = (0.2, 0)^\top$ (err) and numerical order of convergence (noc) for $k = 1, 2, 3$ in Example 4.3

$k=1$			$k=2$			$k=3$		
DoF	err	noc	DoF	err	noc	DoF	err	noc
8	2.29×10^{-1}	–	16	6.68×10^{-3}	–	24	3.56×10^{-4}	–
16	1.53×10^{-2}	3.90	32	5.64×10^{-5}	6.89	48	2.24×10^{-6}	7.31
32	5.23×10^{-4}	4.87	64	2.09×10^{-6}	4.75	96	5.15×10^{-8}	5.45
64	5.58×10^{-5}	3.23	128	8.13×10^{-8}	4.69	192	1.46×10^{-9}	5.14
128	6.84×10^{-6}	3.03	256	5.14×10^{-9}	3.98	384	3.72×10^{-11}	5.29
256	8.42×10^{-7}	3.02	512	3.06×10^{-10}	4.07	768	9.43×10^{-13}	5.30
512	1.04×10^{-7}	3.01	1024	1.66×10^{-11}	4.20	1536	5.04×10^{-14}	4.22
1024	1.30×10^{-8}	3.01	2048	7.57×10^{-13}	4.46	3072	1.13×10^{-13}	–
2048	1.62×10^{-9}	3.00	4096	7.73×10^{-13}	–	6144	8.21×10^{-13}	–

4.3.2 Neumann Problem

The Neumann problem (4.2.3) is treated along the same lines as the Dirichlet problem in the previous section. We utilize the direct approach and approximate the Galerkin formulation (4.17) for the unknown Dirichlet trace $g = \gamma_0^K u \in H^{1/2}(\partial K)$ from Sect. 4.2.3. This yields the discrete Galerkin formulation

$$\begin{aligned}&\text{Find } g_h \in \mathscr{P}^k_{\text{pw}}(\mathscr{B}_h) :\\ &\big(\widetilde{\mathbf{D}}_K g_h, \xi\big)_{L_2(\partial K)} = \Big(\big(\tfrac{1}{2}\mathbf{I} - \mathbf{K}'_K\big)t, \xi\Big)_{L_2(\partial K)} \qquad \forall \xi \in \mathscr{P}^k_{\text{pw}}(\mathscr{B}_h)\,,\end{aligned} \tag{4.29}$$

where t is the given Neumann data with $\int_{\partial K} t \,\mathrm{d}s_{\mathbf{x}} = 0$. Since $\widetilde{\mathbf{D}}_K$ is bounded as well as elliptic on $H^{1/2}(\partial K)$ and $\mathscr{P}^k_{\text{pw}}(\mathscr{B}_h) \subset H^{1/2}(\partial K)$, the discrete Galerkin formulation has a unique solution according to the Lax–Milgram Lemma. Furthermore, Céa's Lemma yields

$$\|g - g_h\|_{H^{1/2}(\partial K)} \leq C \inf_{\xi \in \mathscr{P}^k_{\text{pw}}(\mathscr{B}_h)} \|g - \xi\|_{H^{1/2}(\partial K)}\,,$$

where known approximation properties of polynomials, see [151, Theorem 4.3.22], can be applied once more, such that

$$\|g - g_h\|_{H^{1/2}(\partial K)} \leq C\, h^{s-1/2} \|g\|_{H^s(\partial K)} \tag{4.30}$$

for $1/2 \leq s \leq k+1$, when assuming $g \in H^s(\partial K)$. Arguing as in Sect. 4.2.3, we even see that $g_h \in H^{1/2}_*(\partial K)$ since $\xi_0 \in \mathscr{P}^k_{\text{pw}}(\partial K)$. Inserting g_h into the representation formula (4.3), we obtain for $\mathbf{x} \in K$ the pointwise approximation

$$\widetilde{u}(\mathbf{x}) = \int_{\partial K} U^*(\mathbf{x}, \mathbf{y}) t(\mathbf{y}) \, \mathrm{d}s_{\mathbf{y}} - \int_{\partial K} \gamma_{1,\mathbf{y}}^K U^*(\mathbf{x}, \mathbf{y}) g_h(\mathbf{y}) \, \mathrm{d}s_{\mathbf{y}}$$

for the solution $u(\mathbf{x})$ of the Neumann problem. Under the assumption of sufficient regularity the approximation satisfies for $k = 1$ and $\mathbf{x} \in K$ the error estimate

$$|u(\mathbf{x}) - \widetilde{u}(\mathbf{x})| \le C(\mathbf{x}) \, h^3 \, |g|_{H^2(\partial K)} \, . \tag{4.31}$$

As in the Dirichlet problem we may approximate the given Neumann trace t by its L_2-projection $t_h \in \mathscr{P}^{k-1}_{\mathrm{pw,d}}(\partial K)$. Utilizing the ansatz (4.24) yields the coefficient vector $\underline{t}_h$ and therefore the approximation t_h as unique solution of

$$\sum_{\tau \in \Phi_N} t_\tau (\tau, \zeta)_{L_2(\partial K)} = (t, \zeta)_{L_2(\partial K)} \quad \forall \zeta \in \mathscr{P}^{k-1}_{\mathrm{pw,d}}(\mathscr{B}_h) \, . \tag{4.32}$$

The system of linear equations (4.32) involves the symmetric, positive definite mass matrix

$$\mathbf{M}^{NN}_{K,h} = \Big((\tau, \zeta)_{L_2(\partial K)} \Big)_{\zeta \in \Phi_N, \tau \in \Phi_N} \, .$$

The solvability condition $\int_{\partial K} t_h \mathrm{d}s_{\mathbf{x}} = 0$ is retained since $\operatorname{span}\{1\} \subset \mathscr{P}^{k-1}_{\mathrm{pw,d}}(\mathscr{B}_h)$. Inserting the ansatz (4.24) into the discrete Galerkin formulation (4.29) yields a system of linear equations for $\underline{g}_h$, namely

$$\widetilde{\mathbf{D}}_{K,h} \underline{g}_h = \left(\tfrac{1}{2} \mathbf{M}^\top_{K,h} - \mathbf{K}^\top_{K,h} \right) \underline{t}_h \, , \tag{4.33}$$

where

$$\widetilde{\mathbf{D}}_{K,h} = \mathbf{D}_{K,h} + \alpha \, \mathbf{d}_{K,h} \mathbf{d}^\top_{K,h} \, ,$$

with $\alpha > 0$ and

$$\mathbf{D}_{K,h} = \Big((\mathbf{D}_K \varphi, \phi)_{L_2(\partial K)} \Big)_{\phi \in \Phi_D, \varphi \in \Phi_D} \, , \qquad \mathbf{d}_{K,h} = \Big((\varphi, 1)_{L_2(\partial K)} \Big)_{\varphi \in \Phi_D} \, .$$

The system (4.33) is uniquely solvable since the matrix $\widetilde{\mathbf{D}}_{K,h}$ is symmetric and positive definite due to the properties of the integral operator $\mathbf{D}_K$.

Example 4.4 We demonstrate the performance of the boundary element method for the Neumann problem. Let K be a regular octagon centered at the origin with diameter 0.8 as in Example 4.3, and consider the boundary value problem

$$-\Delta u = 0 \quad \text{in } K \, , \qquad \gamma_1^K u = t \quad \text{on } \partial K \, ,$$

where the Neumann data t is chosen such that the unique solution of the problem in $H^1_*(K)$ is given as $u(\mathbf{x}) = \exp(2\pi(x_1 - 0.3))\cos(2\pi(x_2 - 0.3)) - C$ with $C \in \mathbb{R}$ such that $(\gamma_1^K u, 1)_{L_2(\partial K)} = 0$. The boundary element method is applied on a sequence of meshes for the approximation orders $k = 1, 2$. The first mesh is defined to be the eight sides of the octagon and the following meshes are constructed by subdividing each line segment of the previous mesh into two new line segments of the same length. In Table 4.4 we give the results for $k = 1$ and in Table 4.5 those for $k = 2$. The meshes are distinguished by the number of degrees of freedom (DoF) used to approximate the Dirichlet trace. From (4.30) we expect that the error $\|g - g_h\|_{L_2(\partial K)}$ of the Dirichlet trace in the L_2-norm converges with order $k+1$. This is verified by the numerical order of convergence (noc) in the tables. For the point evaluation, it has been shown in [159] that for $k = 1$ the optimal error estimate (4.31) is not achieved when the Neumann data has to be approximated

Table 4.4 Degrees of freedom (DoF) and errors $\|g - g_h\|_{L_2(\Omega)}$, $|u(\mathbf{x}) - \widetilde{u}(\mathbf{x})|/|u(\mathbf{x})|$ as well as $|\nabla u(\mathbf{x}) - \nabla\widetilde{u}(\mathbf{x})|/|\nabla u(\mathbf{x})|$ for $\mathbf{x} = (0.2, 0)^\top$ and numerical orders of convergence (noc) for $k = 1$ in Example 4.4

	$\|g - g_h\|_{L_2(\Omega)}$		$\|u(\mathbf{x}) - \widetilde{u}(\mathbf{x})\|/\|u(\mathbf{x})\|$		$\|\nabla u(\mathbf{x}) - \nabla\widetilde{u}(\mathbf{x})\|/\|\nabla u(\mathbf{x})\|$	
DoF	err	noc	err	noc	err	noc
8	3.23×10^{-1}	–	2.25×10^{-1}	–	1.60×10^{-1}	–
16	7.57×10^{-2}	2.09	4.49×10^{-2}	2.32	3.90×10^{-2}	2.04
32	1.67×10^{-2}	2.18	9.92×10^{-3}	2.18	8.97×10^{-3}	2.12
64	3.85×10^{-3}	2.12	2.41×10^{-3}	2.04	2.21×10^{-3}	2.02
128	9.19×10^{-4}	2.07	5.98×10^{-4}	2.01	5.51×10^{-4}	2.01
256	2.24×10^{-4}	2.04	1.49×10^{-4}	2.01	1.38×10^{-4}	2.00
512	5.52×10^{-5}	2.02	3.71×10^{-5}	2.00	3.44×10^{-5}	2.00
1024	1.37×10^{-5}	2.01	9.28×10^{-6}	2.00	8.59×10^{-6}	2.00
2048	3.41×10^{-6}	2.01	2.32×10^{-6}	2.00	2.15×10^{-6}	2.00

Table 4.5 Degrees of freedom (DoF) and errors $\|g - g_h\|_{L_2(\Omega)}$, $|u(\mathbf{x}) - \widetilde{u}(\mathbf{x})|/|u(\mathbf{x})|$ as well as $|\nabla u(\mathbf{x}) - \nabla\widetilde{u}(\mathbf{x})|/|\nabla u(\mathbf{x})|$ for $\mathbf{x} = (0.2, 0)^\top$ and numerical orders of convergence (noc) for $k = 2$ in Example 4.4

	$\|g - g_h\|_{L_2(\Omega)}$		$\|u(\mathbf{x}) - \widetilde{u}(\mathbf{x})\|/\|u(\mathbf{x})\|$		$\|\nabla u(\mathbf{x}) - \nabla\widetilde{u}(\mathbf{x})\|/\|\nabla u(\mathbf{x})\|$	
DoF	err	noc	err	noc	err	noc
16	3.46×10^{-2}	–	2.57×10^{-3}	–	2.45×10^{-2}	–
32	4.52×10^{-3}	2.94	4.88×10^{-4}	2.40	5.24×10^{-4}	5.55
64	6.06×10^{-4}	2.90	9.06×10^{-5}	2.43	2.06×10^{-5}	4.67
128	8.02×10^{-5}	2.92	2.18×10^{-5}	2.06	9.46×10^{-7}	4.45
256	1.04×10^{-5}	2.95	5.61×10^{-6}	1.96	5.56×10^{-8}	4.09
512	1.34×10^{-6}	2.95	1.44×10^{-6}	1.96	4.15×10^{-9}	3.74
1024	1.79×10^{-7}	2.91	3.65×10^{-7}	1.98	3.16×10^{-10}	3.72
2048	2.66×10^{-8}	2.75	8.99×10^{-8}	2.02	2.30×10^{-11}	3.78
4096	4.58×10^{-9}	2.54	2.04×10^{-8}	2.14	8.61×10^{-13}	4.74

by (4.32). Instead, only

$$|u(\mathbf{x}) - \widetilde{u}(\mathbf{x})| \leq C(\mathbf{x})\, h^2\, |g|_{H^2(\partial K)}$$

is obtained. This theoretical result is confirmed in Table 4.4. Furthermore, the quadratic convergence also holds for $k = 2$ in the numerical experiment, see Table 4.5. For the pointwise convergence of the gradient, the tables indicate numerical convergence orders of 2 and 4 for $k = 1$ and $k = 2$, respectively.

4.4 Nyström Approach

The Nyström method is an alternative approach for the approximation of integral equations. It was initially designed for domains with globally parametrized and smooth boundaries and was later adapted to domains with corners. Here, we restrict ourselves to the two-dimensional case and we utilize the indirect approach for the Dirichlet problem discussed in Sect. 4.2.2. The main idea is to replace the integral by a suitable quadrature formula and to approximate the resulting equation by means of collocation.

First of all, we seek the solution of the Laplace equation in the form (4.10) such that the unknown density ξ has to be approximated. In the case of the two-dimensional Laplace equation we have

$$\gamma_{1,\mathbf{y}}^K U^*(\mathbf{x}, \mathbf{y}) = \frac{(\mathbf{x} - \mathbf{y}) \cdot \mathbf{n}_K(\mathbf{y})}{2\pi\, |\mathbf{x} - \mathbf{y}|^2} \tag{4.34}$$

almost everywhere, where $\mathbf{n}_K(\mathbf{y})$ denotes the outer normal vector of K in the boundary point $\mathbf{y} \in \partial K$. The density ξ satisfies the boundary integral equation (4.11).

4.4.1 Domains with Smooth Boundary

If the boundary of the domain is smooth, i.e. C^2, and there is a global parametrization $\mathbf{x}(\theta)$ such that

$$\partial K = \left\{ \mathbf{x}(\theta) \in \mathbb{R}^2 : \theta_0 \leq \theta \leq \theta_1 \right\}$$

with $|\mathbf{x}'(\theta)| \neq 0$ for all $\theta \in [\theta_0, \theta_1]$, then (4.34) holds for all $\mathbf{x}, \mathbf{y} \in \partial K$ with a removable singularity at $\mathbf{x} = \mathbf{y}$. Furthermore, let the parametric curve be given as $\mathbf{x}(\theta) = (x_1(\theta), x_2(\theta))^\top$ in counter-clockwise orientation. Hence, the outer normal vector can be expressed as $\mathbf{n}_K(\mathbf{x}(\theta)) = (x_2'(\theta), -x_1'(\theta))^\top / |\mathbf{x}'(\theta)|$. Respecting

Remark 4.1, the integral equation (4.11) for the density ξ now reads

$$\tfrac{1}{2}\xi(\mathbf{x}(\theta))+\int_{\theta_0}^{\theta_1}\kappa(\mathbf{x}(\theta),\mathbf{x}(\tau))\xi(\mathbf{x}(\tau))\,\mathrm{d}\tau=-g(\mathbf{x}(\theta))\quad\text{for }\theta\in[\theta_0,\theta_1]\,,\tag{4.35}$$

with the integral kernel

$$\kappa(\mathbf{x}(\theta),\mathbf{x}(\tau))=\begin{cases}\dfrac{1}{2\pi}\dfrac{x_1'(\tau)(x_2(\theta)-x_2(\tau))-x_2'(\tau)(x_1(\theta)-x_1(\tau))}{(x_1(\theta)-x_1(\tau))^2+(x_2(\theta)-x_2(\tau))^2}\,, & \text{for }\theta\neq\tau\,,\\[2ex] \dfrac{1}{4\pi}\dfrac{x_1'(\tau)x_2''(\tau)-x_2'(\tau)x_1''(\tau)}{(x_1'(\tau))^2+(x_2'(\tau))^2}\,, & \text{for }\theta=\tau\,.\end{cases}$$

Next, we apply the composite trapezoidal rule to the integral with $N+1$ uniformly placed quadrature points $\mathbf{x}_j$, $j=0,\ldots,N$ and weights. Since ξ is periodic on the closed boundary, this quadrature rule is especially suited for the integration, see [62]. Furthermore, we have $\xi(\mathbf{x}_N)=\xi(\mathbf{x}_0)$. The resulting equation cannot hold for all $\theta\in[\theta_0,\theta_1]$ and therefore we enforce its validity by collocation in the quadrature notes. Consequently, we find due to the periodicity the following system of linear equations for the unknown values $\xi_i=\xi(\mathbf{x}_i)$ of the density:

$$\tfrac{1}{2}\xi_i+\sum_{j=1}^{N}\kappa(\mathbf{x}_i,\mathbf{x}_j)\xi_j\omega_j=-g(\mathbf{x}_i)\quad\text{for }i=1,\ldots,N\,,\tag{4.36}$$

where the quadrature points and weights are given by

$$\mathbf{x}_j=\mathbf{x}\left(\theta_0+j\frac{\theta_1-\theta_0}{N}\right)\qquad\text{and}\qquad\omega_j=\frac{\theta_1-\theta_0}{N}\,.$$

It is known that this trapezoidal Nyström method converges with order $\mathscr{O}(1/N^q)$ in the maximum norm, where $q\geq 0$ is related to the smoothness of the boundary ∂K as well as to the smoothness of ξ, see, e.g. [118]. Having the values ξ_i at hand we can approximate $u(\mathbf{x})$ for $\mathbf{x}\in K$ with the help of (4.10) and the trapezoidal rule by

$$\widetilde{u}(\mathbf{x})=-\sum_{j=1}^{N}\kappa(\mathbf{x},\mathbf{x}_j)\xi_j\omega_j\quad\text{for }\mathbf{x}\in K\,.\tag{4.37}$$

Since the integrand in (4.10) is smooth for $\mathbf{x}\in K$, differentiation and integration can be interchanged. Thus, we obtain an approximation of the gradient of the solution $\nabla u(\mathbf{x})$ as

$$\nabla\widetilde{u}(\mathbf{x})=-\sum_{j=1}^{N}\nabla\kappa(\mathbf{x},\mathbf{x}_j)\xi_j\omega_j\quad\text{for }\mathbf{x}\in K\,.\tag{4.38}$$

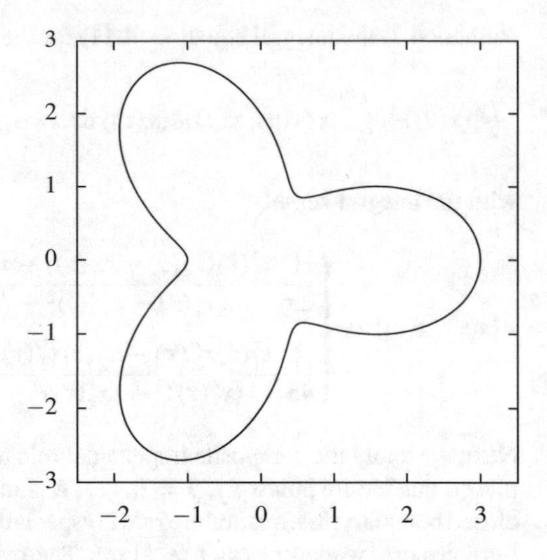

Fig. 4.2 Domain in Example 4.5 which is given by a globally smooth curve describing its boundary

The gradient of the integral kernel can be computed analytically by a small exercise and does not involve any difficulties.

Example 4.5 We consider the boundary value problem

$$-\Delta u = 0 \quad \text{in } K , \qquad u = g \quad \text{on } \partial K ,$$

where g is chosen such that $u(\mathbf{x}) = \ln |\mathbf{x} - \mathbf{x}^*|$ with $\mathbf{x}^* = (3, 3)^\top \notin K$ is the exact solution. The domain K is given by its boundary that is defined as a globally smooth curve with parametrization

$$\mathbf{x}(\theta) = (2 + \cos(3\theta)) \begin{pmatrix} \cos(\theta) \\ \sin(\theta) \end{pmatrix} \quad \text{for} \quad 0 \leq \theta < 2\pi ,$$

see Fig. 4.2. In Table 4.6, the convergence of the approximation (4.37) as well as of its gradient (4.38) is presented in the point $(1.5, 0)^\top$ for an increasing number of quadrature points (QP) which is equal N in this setting. Furthermore, the numerical order of convergence (noc) is given with respect to $1/N$. Obviously, the Nyström approach converges very fast till machine precision for domains with smooth boundaries.

4.4.2 Domains with Corners

Often, boundary value problems are considered on domains whose boundaries are not given as a globally smooth parametric curves and which may contain corners.

Table 4.6 Number of quadrature points (QP = N), pointwise error of approximation and its gradient in the point $\mathbf{x} = (1.5, 0)^\top$ as well as numerical order of convergence (noc) for Example 4.5

QP	$\lvert u(\mathbf{x}) - \widetilde{u}(\mathbf{x})\rvert / \lvert u(\mathbf{x})\rvert$	noc	$\lvert\nabla u(\mathbf{x}) - \nabla\widetilde{u}(\mathbf{x})\rvert / \lvert\nabla u(\mathbf{x})\rvert$	noc
4	5.12×10^{-1}	–	$1.59 \times 10^{+0}$	–
8	1.71×10^{-1}	1.58	$2.47 \times 10^{+0}$	−0.64
16	6.80×10^{-3}	4.65	1.86×10^{-1}	3.73
32	1.92×10^{-5}	8.47	1.64×10^{-3}	6.83
64	2.69×10^{-8}	9.48	2.38×10^{-7}	12.75
128	5.80×10^{-14}	18.82	4.12×10^{-13}	19.14
256	1.83×10^{-16}	8.31	5.27×10^{-16}	9.61

Consequently, we consider domains with piecewise smooth boundaries next. Thus, the boundary ∂K is decomposed into boundary segments such that each can be parametrized by a smooth curve. Without loss of generality, we concatenate these parametrizations to a piecewise smooth and globally given parametrization which is oriented counter-clockwise. Therefore, let $M \in \mathbb{N}$ be the number of boundary segments, we write

$$\partial K = \left\{\mathbf{x}(\theta) \in \mathbb{R}^2 : \theta_\ell \le \theta \le \theta_{\ell+1}, \ell = 0, \ldots, M-1\right\} .$$

Here, $|\mathbf{x}'(\theta)| \neq 0$ for all $\theta \in (\theta_\ell, \theta_{\ell+1})$, $\ell = 0, \ldots, M-1$ and $\mathbf{z}_\ell = \mathbf{x}(\theta_\ell)$ are the corner points or vertices of the domain. Since the boundary is closed we obviously have $\mathbf{x}(\theta_0) = \mathbf{x}(\theta_M)$. A special case are the polygonal domains which are used throughout this book. In this situation the boundary segments are given as straight lines and $\mathbf{x}'(\theta)$ is constant on each interval $(\theta_\ell, \theta_{\ell+1})$. Furthermore, the vertices $\mathbf{z}_\ell$ coincide with the nodes of the polygonal elements.

In order to derive the Nyström approximation we consider once more the boundary integral equation (4.11). But, since the boundary of the domain is not smooth in the points $\mathbf{z}_\ell$, we have to take care on $\varsigma(\mathbf{z}_\ell)$ which depends on the interior angle of the domain. This dependency is resolved by using (4.6) and reformulating (4.11) to

$$\frac{\xi(\mathbf{x}) + \xi(\mathbf{z})}{2} - \int_{\partial K} \gamma_{1,\mathbf{y}}^K U^*(\mathbf{x}, \mathbf{y})(\xi(\mathbf{y}) - \xi(\mathbf{z}))\,\mathrm{d}s_{\mathbf{y}} = -g(\mathbf{x}) \quad \text{for } \mathbf{x} \in \partial K ,$$

where $\mathbf{z}$ is the closest vertex $\mathbf{z}_\ell$ to $\mathbf{x}$. Next, the boundary integral is split into its contributions over the single boundary segments. The parametrization and the outer normal vector are treated within each smooth segment as in the previous section. This yields

$$\frac{\xi(\mathbf{x}(\theta)) + \xi(\mathbf{z})}{2} + \sum_{\ell=0}^{M-1} \int_{\theta_\ell}^{\theta_{\ell+1}} \kappa(\mathbf{x}(\theta), \mathbf{x}(\tau))(\xi(\mathbf{x}(\tau)) - \xi(\mathbf{z}))\,\mathrm{d}\tau = -g(\mathbf{x}(\theta))$$

for $\theta \in [\theta_0, \theta_M]$. If $\theta = \theta_\ell$, $\ell = 0, \ldots, M$, i.e. $\mathbf{x}(\theta) = \mathbf{z}_\ell$, there is a singularity in the integral kernel and the formula of the previous section for $\kappa(\mathbf{z}_\ell, \mathbf{z}_\ell)$ is actually not well defined. Instead of applying the composite trapezoidal rule directly, Kress proposed to perform a sigmoidal change-of-variables first, which copes with the singularity, see [117]. This variable transformation $\eta^{(\ell)} : [0, 1] \to [\theta_\ell, \theta_{\ell+1}]$ is strictly monotonic increasing and it is defined by

$$\eta^{(\ell)}(t) = \theta_\ell + \frac{(c(t))^p(\theta_{\ell+1} - \theta_\ell)}{(c(t))^p + (1 - c(t))^p},$$

where $c : [0, 1] \to [0, 1]$ with

$$c(t) = \left(\frac{1}{2} - \frac{1}{p}\right)(2t - 1)^3 + \frac{1}{p}(2t - 1) + \frac{1}{2}$$

and $p \geq 2$ is an integer. It is straight-forward to see that $\left(\eta^{(\ell)}\right)'$ has a root of order $p-1$ at each endpoint of the interval $[0, 1]$. Thus, we obtain with sufficiently large p and the composite trapezoidal rule on each boundary segment

$$\int_{\theta_\ell}^{\theta_{\ell+1}} \kappa(\mathbf{x}(\theta), \mathbf{x}(\tau))(\xi(\mathbf{x}(\tau)) - \xi(\mathbf{z}))\,\mathrm{d}\tau \approx \sum_{j=1}^{N-1} \kappa\left(\mathbf{x}(\theta), \mathbf{x}_j^{(\ell)}\right)\left(\xi\left(\mathbf{x}_j^{(\ell)}\right) - \xi(\mathbf{z})\right)\omega_j^{(\ell)}$$

where the quadrature points and weights are given by

$$\mathbf{x}_j^{(\ell)} = \mathbf{x}\left(\eta^{(\ell)}(j/N)\right) \quad \text{and} \quad \omega_j^{(\ell)} = \frac{\left(\eta^{(\ell)}\right)'(j/N)}{N},$$

$j = 0, \ldots, N$. The summands for $j = 0, N$ vanish because of the roots of $\left(\eta^{(\ell)}\right)'$. A careful convergence analysis of this quadrature is given in [117], showing that it is convergent for the kinds of integrands we encounter here of increasingly higher order in N as p is increased.

Applying the sigmoidal transform and the trapezoidal rule as above to the modified boundary integral equation and using collocation in the quadrature points yields the following system of linear equations with unknowns $\xi_i^{(k)} = \xi(\mathbf{x}_i^{(k)})$:

$$\frac{\xi_i^{(k)} + \tilde{\xi}_i^{(k)}}{2} + \sum_{\ell=0}^{M-1}\sum_{j=1}^{N-1} \kappa\left(\mathbf{x}_i^{(k)}, \mathbf{x}_j^{(\ell)}\right)\left(\xi_j^{(\ell)} - \tilde{\xi}_i^{(k)}\right)\omega_j^{(\ell)} = -g\left(\mathbf{x}_i^{(k)}\right) \tag{4.39}$$

for $i = 0, \ldots, N-1$, $k = 0, \ldots, M-1$, where $\tilde{\xi}_i^{(k)}$ is either $\xi_0^{(k)}$ or $\xi_N^{(k)}$ depending which point $\mathbf{x}_0^{(k)}$ or $\mathbf{x}_N^{(k)}$ is closer to $\mathbf{x}_i^{(k)}$. After we have solved the system, the function value $u(\mathbf{x})$ can be approximated for $\mathbf{x} \in K$ with the help of (4.10) and the quadrature by

$$\widetilde{u}(\mathbf{x}) = -\sum_{\ell=0}^{M-1}\sum_{j=1}^{N} \kappa\big(\mathbf{x}, \mathbf{x}_j^{(\ell)}\big)\xi_j^{(\ell)}\omega_j^{(\ell)} \quad \text{for } \mathbf{x} \in K\,. \tag{4.40}$$

Analogously, we obtain for the gradient $\nabla u(\mathbf{x})$ the approximation

$$\nabla\widetilde{u}(\mathbf{x}) = -\sum_{\ell=0}^{M-1}\sum_{j=1}^{N} \nabla\kappa\big(\mathbf{x}, \mathbf{x}_j^{(\ell)}\big)\xi_j^{(\ell)}\omega_j^{(\ell)} \quad \text{for } \mathbf{x} \in K\,. \tag{4.41}$$

Example 4.6 We consider the boundary value problem

$$\Delta u = 0 \quad \text{in } K\,, \qquad u = g \quad \text{on } \partial K\,,$$

where g is chosen such that $u(\mathbf{x}) = \ln|\mathbf{x} - \mathbf{x}^*|$ with $\mathbf{x}^* = (6, 8)^\top \notin K$ is the exact solution. The domain K is given by its boundary that is defined globally as a curve which is piecewise smooth. More precisely, we use the parametrization of an epicycloid that is

$$\mathbf{x}(\theta) = \begin{pmatrix} 8\cos(\theta) - \cos(8\theta) \\ 8\sin(\theta) - \sin(8\theta) \end{pmatrix} \quad \text{for} \quad 0 \le \theta < 2\pi\,,$$

with corners for $\theta = \theta_\ell = 2\pi\ell/7, \ell = 0, \ldots, 7$, see Fig. 4.3. The Nyström approach is applied with the parameter $p = 4$ in the sigmoidal change-of-variable. Each of the $M = 7$ smooth parts of the boundary is decomposed into N segments and thus $N + 1$ quadrature points. In Table 4.7, the convergence of the approximation (4.40)

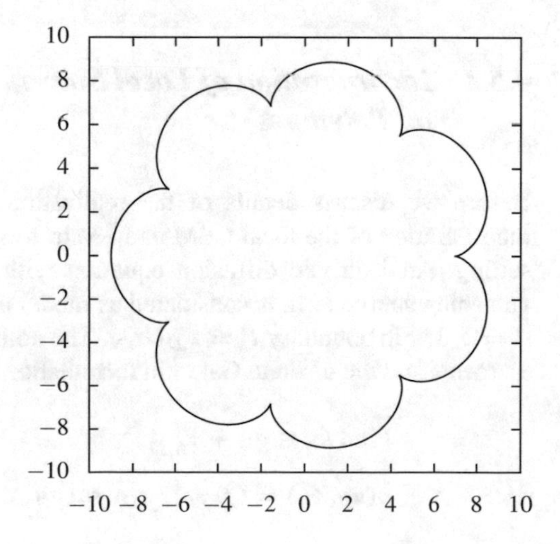

Fig. 4.3 Domain in Example 4.6 with piecewise smooth boundary which is given as parametrization of an epicycloid

Table 4.7 Number of quadrature points (QP= MN), pointwise error of approximation and its gradient in the point $\mathbf{x} = (5.6, 0)^\top$ as well as numerical order of convergence (noc) for Example 4.6 with parameter $p = 4$

QP	$\lvert u(\mathbf{x}) - \widetilde{u}(\mathbf{x})\rvert / \lvert u(\mathbf{x})\rvert$	noc	$\lvert\nabla u(\mathbf{x}) - \nabla\widetilde{u}(\mathbf{x})\rvert / \lvert\nabla u(\mathbf{x})\rvert$	noc
28	3.79×10^{-2}	–	3.41×10^{-1}	–
56	7.18×10^{-5}	9.04	3.19×10^{-2}	3.42
112	9.03×10^{-6}	2.99	6.54×10^{-5}	8.93
224	8.00×10^{-10}	13.46	2.26×10^{-7}	8.18
448	1.73×10^{-12}	8.85	6.89×10^{-10}	8.36
896	1.28×10^{-15}	10.40	3.43×10^{-13}	10.97
1792	8.54×10^{-16}	0.58	1.36×10^{-15}	7.98

as well as of its gradient (4.41) is presented in the point $(5.6, 0)^\top$ for an increasing number of quadrature points (QP) which is equal MN in total. Furthermore, the numerical order of convergence (noc) is given with respect to $1/\mathrm{QP}$. Obviously, the Nyström approach also converges very fast till machine precision for domains with piecewise smooth boundaries when the singularities at the corners are treated with an appropriate quadrature scheme.

4.5 Application in BEM-Based FEM

Throughout this book all numerical experiments and test have been performed with the help of a local BEM solver as described in the following. However, we also give a brute-force application of a local Nyström solver and discuss its potential advantageous and disadvantageous in the next sections for a test problem.

4.5.1 Incorporation of Local Solvers and Quadrature on Polytopes

Before we discuss details of the realization of the BEM-based FEM and the incorporation of the local BEM and Nyström solvers, we recapitulate the problem setting. An isotropic diffusion equation with mixed boundary data and a non-vanishing source term is considered as model problem (2.1) on a domain $\Omega \subset \mathbb{R}^d$, $d = 2, 3$ with boundary $\Gamma = \Gamma_D \cup \Gamma_N$. The domain Ω is decomposed into polytopal elements and the discrete Galerkin formulation (2.28) reads:

$$\begin{aligned}
&\text{Find } u_h \in g_D + V_{h,D}^k : \\
&b(u_h, v_h) = (f, v_h)_{L_2(\Omega)} + (g_N, v_h)_{L_2(\Gamma_N)} \quad \forall v_h \in V_{h,D}^k ,
\end{aligned}$$

where $b(u_h, v_h) = (a\nabla u_h, \nabla v_h)_{L_2(\Omega)}$, and the approximation space admits the decomposition $V_{h,D}^k = V_{h,H,D}^k \oplus V_{h,B}^k$ into two types of functions. The first ones are harmonic on each element and have piecewise polynomial data on the skeleton of the discretization, whereas the second ones vanish on the skeleton and have a polynomial Laplacian on each element. In particular, the discrete Galerkin formulation decouples for these kinds of functions as given in (2.31) and (2.32) in the case of a piecewise constant diffusion coefficient.

It remains to discuss the realization of the different terms in the discrete Galerkin formulation with the help of the local boundary element method and the local Nyström solver from this chapter. First of all, we recognize that every function $v_h \in V_{h,D}^k$ is given uniquely over each element $K \in \mathcal{K}_h$ by the local boundary value problem

$$-\Delta v_h = p_K \quad \text{in } K \quad \text{and} \quad v_h = p_{\partial K} \quad \text{on } \partial K$$

with prescribed data $p_K \in \mathcal{P}^{k-2}(K)$ and $p_{\partial K} \in \mathcal{P}_{\text{pw}}^k(\partial K)$, cf. (2.14) and (2.15). Since p_K is a polynomial, we can write $v_h = v_{h,H} + q$ with $q \in \mathcal{P}^k(K)$ such that

$$-\Delta v_{h,H} = 0 \quad \text{in } K \quad \text{and} \quad v_{h,H} = p_{\partial K} - q \quad \text{on } \partial K \tag{4.42}$$

with $p_{\partial K} - q \in \mathcal{P}_{\text{pw}}^k(\partial K)$. Therefore, it was sufficient to consider the pure Laplace problem in the previous sections for the local solvers. A constructive approach for finding q is presented in [113]. For a homogeneous polynomial $p \in \mathcal{P}^m(\mathbb{R}^d)$ of degree m, i.e. $p(c\mathbf{x}) = c^m p(\mathbf{x})$, the polynomial

$$q(\mathbf{x}) = \sum_{\ell=0}^{\lfloor m/2 \rfloor} \frac{(-1)^\ell \Gamma(d/2 + m - \ell)}{\Gamma(d/2 + m + 1)(\ell + 1)!} \left(\frac{|\mathbf{x}|^2}{4} \right)^{\ell+1} \Delta^\ell p(\mathbf{x}) \in \mathcal{P}^{m+2}(K)$$

satisfies $\Delta q = p$, where $\lfloor m/2 \rfloor$ denotes the integer part of $m/2$ and $\Gamma(\cdot)$ the gamma function, see [113, Theorem 2]. For non-homogeneous polynomials p the construction can be applied on the representation of p in the monomial basis, whose basis functions are homogeneous.

Let us focus on the two terms $(f, v_h)_{L_2(\Omega)}$ and $(g_N, v_h)_{L_2(\Gamma_N)}$. The latter one does not cause any difficulties. The Neumann boundary Γ_N is given as collection of line segments ($d = 2$) or triangles ($d = 3$) and the restrictions of the functions $v_h \in V_{h,D}^k$ onto Γ_N are piecewise polynomials. Consequently, we apply Gaussian quadrature on each segment or standard numerical integration on each triangle in order to approximate the integral value of the product of the given Neumann data g_N and the piecewise polynomial data of v_h over Γ_N. We also have to apply a quadrature scheme to approximate $(f, v_h)_{L_2(\Omega)}$ since no additional information on f is given in general. For this reason, we decompose the integral first into its contribution over the elements $K \in \mathcal{K}_h$ and afterwards we decompose it even further to its contributions over the triangles ($d = 2$) and tetrahedra ($d = 3$) of the auxiliary triangulation $\mathscr{T}_h(K)$, i.e.

$$(f, v_h)_{L_2(\Omega)} = \sum_{K\in\mathscr{K}_h} (f, v_h)_{L_2(K)} = \sum_{K\in\mathscr{K}_h} \sum_{T\in\mathscr{T}_h(K)} (f, v_h)_{L_2(T)} . \tag{4.43}$$

Now, we apply Gaussian quadrature over each triangle/tetrahedron T. Instead of utilizing the auxiliary discretization $\mathscr{T}_h(K)$ introduces in Sect. 2.2, we may also triangulate the elements with some software tool like `triangle` or `TetGen`, see [155, 157]. This strategy has been especially performed in the numerical tests in order to compute accurate errors in the L_2- and H^1-norm. The evaluation of v_h inside the elements is realized by means of the reformulation $v_h = v_{h,H} + q$ with (4.42) and the approximated representation formulas (4.22) and (4.40), respectively. This natural idea to apply a quadrature rule over a subtriangulation in order to approximate an integral over a polygonal domain has been applied in [164] for instance. An alternative approach is presented in [131], where quadrature points and weights for fixed polygonal domains are precomputed.

In order to treat the bilinear form we have different possibilities. We assume here that the diffusion coefficient is constant on each element and we split the integral into its contributions over the single elements

$$b(u_h, v_h) = (a\nabla u_h, \nabla v_h)_{L_2(\Omega)} = \sum_{K\in\mathscr{K}_h} a_K(\nabla u_h, \nabla v_h)_{L_2(K)} .$$

A brute-force approach would be to approximate the term $(\nabla u_h, \nabla v_h)_{L_2(K)}$ by a quadrature as described above using the representation formulas for the evaluation of ∇u_h and ∇v_h in the quadrature points. Alternatively, we may use Green's first identity (4.2) on each element such that

$$(\nabla u_h, \nabla v_h)_{L_2(K)} = (\gamma_1^K u_h, \gamma_0^K v_h)_{L_2(\partial K)} - (\Delta u_h, v_h)_{L_2(K)} .$$

Obviously, if either u_h or v_h is harmonic, the volume integral vanishes and we end up with a boundary integral solely, where the product of a Dirichlet and a Neumann trace has to be integrated. Since the approximation space $V_{h,D}^k = V_{h,H,D}^k \oplus V_{h,B}^k$ is given as a direct sum, we distinguish three cases:

1. $u_h, v_h \in V_{h,H,D}^k$: We end up with solely boundary integrals

$$(\nabla u_h, \nabla v_h)_{L_2(K)} = (\gamma_1^K u_h, \gamma_0^K v_h)_{L_2(\partial K)} .$$

2. $u_h, v_h \in V_{h,B}^k$: Let $v_h = v_{h,H} + q$ with (4.42), where $p_{\partial K} = 0$, then

$$(\nabla u_h, \nabla v_h)_{L_2(K)} = -(\Delta u_h, v_h)_{L_2(K)} = (\gamma_1^K u_h, \gamma_0^K v_{h,H})_{L_2(\partial K)} - (\Delta u_h, q)_{L_2(K)} .$$

3. $u_h \in V_{h,H,D}^k, v_h \in V_{h,B}^k$ or vice versa: As we have seen in (2.30), it is

$$(\nabla u_h, \nabla v_h)_{L_2(K)} = 0 \,.$$

The only volume integral which is left is $(\Delta u_h, v_h)_{L_2(K)}$. Since $\Delta u_h \in \mathscr{P}^{k-2}(K)$, the integrand is a polynomial of degree smaller or equal $2k-2$. The integral can thus be computed exactly by the quadrature over the auxiliary discretization or alternatively by applying the divergence theorem followed by a quadrature over the boundary of the element.

Local BEM Solver

The local BEM solver, which is used throughout this book, makes use of the reformulation of the bilinear form in order to reduce the volume integrals to integrals over the skeleton of the domain. More precisely, we end up with integrals over the element boundaries, where we have to integrate the product of a Dirichlet and Neumann trace of functions in V_h^k. This setting nicely fits into the boundary element strategy. The Dirichlet trace of the functions is known whereas their Neumann trace have to be approximated. Here, we proceed as described in Sect. 4.3.

Let $K \in \mathscr{K}_h$ and $\mathscr{B}_h$ be an appropriate boundary element mesh of ∂K consisting of line segments ($d=2$) or triangles ($d=3$). Furthermore, we denote by Φ_D and Φ_N the basis of $\mathscr{P}_{\mathrm{pw}}^k(\mathscr{B}_h)$ and $\mathscr{P}_{\mathrm{pw,d}}^{k-1}(\mathscr{B}_h)$, which are used as approximation spaces for $H^{1/2}(\partial K)$ and $H^{-1/2}(\partial K)$, respectively. Since $\gamma_0^K u_h$ is already polynomial of degree k over each edge/face of K, the trace is represented exactly in the basis Φ_D such that in the notation of Sect. 4.3 it is $g_h^{(u)} = \gamma_0^K u_h \in \mathscr{P}_{\mathrm{pw}}^k(\mathscr{B}_h)$. The Neumann trace $\gamma_1^K u_h$ is approximated by $t_h^{(u)} \in \mathscr{P}_{\mathrm{pw,d}}^{k-1}(\mathscr{B}_h)$ according to the discrete Galerkin formulation (4.20) with the ansatz (4.24). With the help of the Steklov–Poincaré operator we write

$$(\gamma_1^K u_h, \gamma_0^K v_h)_{L_2(\partial K)} = (\mathbf{S}_K \gamma_0^K u_h, \gamma_0^K v_h)_{L_2(\partial K)} \,.$$

Next, we may either use the non-symmetric, see (4.7), or the symmetric, see (4.8), representation of the Steklov–Poincaré operator. According to (4.26), the non-symmetric representation leads to

$$(\gamma_1^K u_h, \gamma_0^K v_h)_{L_2(\partial K)} \approx (t_h^{(u)}, g_h^{(v)})_{L_2(\partial K)} = (\underline{g}_h^{(v)})^\top \mathbf{M}_{K,h}^\top \, \underline{t}_h^{(u)} = (\underline{g}_h^{(v)})^\top \mathbf{S}_{K,h}^{\mathrm{unsym}} \, \underline{g}_h^{(u)}$$

with

$$\mathbf{S}_{K,h}^{\mathrm{unsym}} = \mathbf{M}_{K,h}^\top \mathbf{V}_{K,h}^{-1} \left(\tfrac{1}{2} \mathbf{M}_{K,h} + \mathbf{K}_{K,h} \right) \,.$$

On the other hand, the symmetric representation with (4.27) leads to

$$(\gamma_1^K u_h, \gamma_0^K v_h)_{L_2(\partial K)} \approx (\widetilde{\mathbf{S}}_K g_h^{(u)}, g_h^{(v)})_{L_2(\partial K)} = (\underline{g}_h^{(v)})^\top \mathbf{S}_{K,h} \, \underline{g}_h^{(u)} \,.$$

Both matrices $\mathbf{S}_{K,h}^{\text{unsym}}$ and $\mathbf{S}_{K,h}$ originate from the same symmetric bilinear form. Whereas $\mathbf{S}_{K,h}^{\text{unsym}}$ is a non-symmetric matrix, $\mathbf{S}_{K,h}$ retains the symmetry because of its Definition (4.28).

Without saying, we have already linked the approximation order in the BEM with the one of the global FEM formulation naturally. For V_h^k in the BEM-based FEM, we have chosen $\mathscr{P}_{\text{pw}}^k(\mathscr{B}_h)$ and $\mathscr{P}_{\text{pw,d}}^{k-1}(\mathscr{B}_h)$ with the same degree k in the local BEM solver. This choice is appropriate, since the trace of functions in V_h^k on the boundary of an element lies in the used boundary element space. Furthermore, we point out that $\mathscr{P}^k(K) \subset V_h^k\big|_K$ and for a polynomial $p \in \mathscr{P}^k(K)$ it is $\gamma_0^K p \in \mathscr{P}_{\text{pw}}^k(\partial K)$ and, in particular, $\gamma_1^K p \in \mathscr{P}_{\text{pw,d}}^{k-1}(\partial K)$. Thus, the local BEM solver is exact, up to quadrature errors, for all polynomials contained in V_h^k. The choice of the boundary mesh $\mathscr{B}_h$, however, is still open. It turns out that $\mathscr{B}_h = \mathscr{T}_h(\partial K)$ is an adequate choice, i.e., the naturally given boundary mesh consisting of edges ($d = 2$) and triangular faces ($d = 3$) of the polytopal elements. This mesh is also the coarsest possible one to discretize ∂K.

The boundary element matrices only depend on the geometry and on the discretization of ∂K, but they are independent of the basis functions of the BEM-based FEM. Thus, the matrices are precomputed once per element and they are used throughout the simulation for the setup of the global FEM matrix as well as for the evaluation of all functions of V_h^k insight elements and for the approximation of their Neumann traces on the skeleton of the domain. If the mesh $\mathscr{K}_h$ consists of a few element types only, it is possible to compute the BEM matrices solely for the representative elements since they are invariant under translation and rotation. Consequently, a kind of lookup table can be used to reduce the computational cost by using the same matrices for several elements, see Sect. 6.2.6. Beside this improvement, we point out that the boundary element matrices in our application are rather small because of the coarse meshes $\mathscr{B}_h = \mathscr{T}_h(\partial K)$ and the fact that the number of nodes and the number of edges/faces is uniformly bounded, cf. Sect. 2.2.

Finally, the assembling of the global FEM matrix is performed as usual by adding up the local element-wise contributions. Here, the matrix

$$\mathbf{S}_{K,h}^{\text{unsym}} \qquad \text{or} \qquad \mathbf{S}_{K,h}$$

serves as a local stiffness matrix in the BEM-based FEM simulation.

Remark 4.7 The 2D implementation of the BEM-based FEM, used in all numerical examples in this book, utilizes $\mathbf{S}_{K,h}$ as local stiffness matrix. The entries of the boundary element matrices are computed by means of a fully numeric integration routine involving adaptive quadratures techniques. The 3D implementation, in contrary, is set up on a semi-analytic integration technique for the computation of the boundary element matrices and the assembling of the global FEM matrix is performed using $\mathbf{S}_{K,h}^{\text{unsym}}$ as local stiffness matrix. In both cases, the representation formulas are evaluated with analytic expressions. As already mentioned, the boundary element matrices are rather small. Therefore, no additional matrix compression techniques like the Adaptive Cross Approximation (ACA) have been applied, cf. [20, 144, 148]. Furthermore, the inversion of the single-layer potential matrix $\mathbf{V}_{K,h}$ is done with the help of an efficient LAPACK [6] routine.

Local Nyström Solver

We restrict ourselves to two-dimensions and proceed as in Sect. 4.4. The boundary ∂K of a polygonal element is prescribed as a union of straight lines and can thus be parametrized by a piecewise smooth curve easily. The number of boundary segments M corresponds to the number of edges, and the vertices $\mathbf{z}_\ell$ are given by the nodes of the element. For the approximation of the bilinear form of the global FEM formulation, we proceed with the local volume integrals $(\nabla u_h, \nabla v_h)_{L_2(K)}$, since the Neumann trace $\gamma_1^K u_h$ is not accessible directly by the Nyström approximation. Consequently, the elements are subdivided into triangles and a quadrature rule is applied on each of them, where the evaluations of ∇u_h and ∇v_h are realized by means of (4.41).

The Nyström approximation has to be performed for each basis function. This involves the solution of a system of linear equations (4.39) each. In contrast, the local BEM solver only inverts one matrix per element. Furthermore, an effective generalization of the Nyström approximation to 3D is not obvious and the use of volume quadrature is unpleasant. But, the implementation of a Nyström code is much easier than the appropriate numerical approximation of the BEM matrices. Furthermore, due to the sigmoidal change-of-variables the Nyström approximation copes with singularities that appear at reentrant corners. The BEM, which is applied on the coarsest possible mesh, might need additional attention on this, see Sect. 4.5.2.2.

In order to bypass the unpleasant volume integration in the evaluation of the global FEM bilinear form using the local Nyström solver, it is possible to apply an advanced strategy. In [135], the authors have proposed a Nyström discretization using harmonic conjugates which directly gives access to the Neumann trace of the approximation. Consequently, this approach can be applied to approximate the boundary integrals $(\gamma_1^K u_h, \gamma_0^K v_h)_{L_2(\partial K)}$ in the reformulation of the FEM bilinear form such that volume integrals are avoided. Beside this, the Nyström approach relies on a piecewise smooth boundary curve only and therefore, it opens the developments towards polygonal elements with curved edges, see [5].

4.5.2 *Numerical Examples and Comparison*

In this section we substantiate our considerations on the local solvers. Numerical examples for the Nyström approximation as well as for the BEM have been presented in the previous sections. Furthermore, the whole book contains examples for the BEM-based FEM using the local BEM solver. Therefore, we restrict ourselves here to demonstrate the applicability of the local Nyström solver and discuss a comparison of the two solvers in the case of the presence of singularities in the approximation space V_h^k.

4.5.2.1 Interpolation with Local Nyström Solver

To demonstrate the interpolation properties, we first interpolate the smooth function

$$v(\mathbf{x}) = \sin(2\pi x_1)\sin(2\pi x_2) \qquad \text{for } \mathbf{x} \in \Omega = (0,1)^2$$

on two different families of meshes which have already been used in a former example. The first family has been generated by the software package PolyMesher [167], and consists of convex polygons that are primarily pentagons and hexagons. The second family consist of rectangles and L-shaped elements, and has been chosen to illustrate that the presence of non-convex elements does not negatively impact the interpolation properties of the associated local spaces. The meshes with the convex and L-shaped elements are depicted in Figs. 2.14 and 2.16, respectively. The relative interpolation errors for

$$\mathfrak{I}_h^k : H^2(\Omega) \to V_h^k$$

in the H^1- as well as in the L_2-norm are presented with respect to the mesh size h in logarithmic scale in Fig. 4.4 for different approximation orders k. The results for the meshes generated by PolyMesher are visualized with a solid line whereas the results for the meshes with L-shaped elements are given with a dashed line. We observe optimal rates of convergence for both families of meshes as expected from the theory developed in Chap. 2.

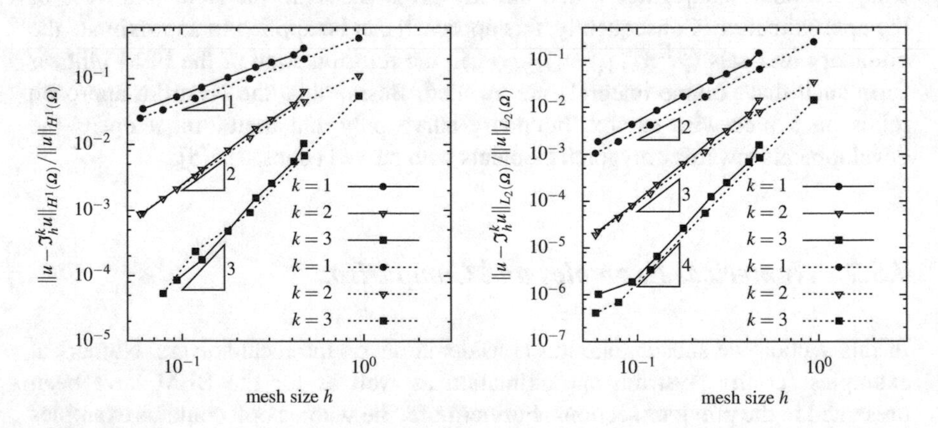

Fig. 4.4 Relative interpolation error in H^1-norm (left) and L_2-norm (right) with respect to the mesh size h on meshes produced by PolyMesher (lines, cf. Fig. 2.14) and meshes with L-shaped elements (dashed, cf. Fig. 2.16), for $k = 1, 2, 3$ and local Nyström solver

4.5.2.2 Comparison of Local Solvers for L-Domain

In a second example, we make use of polar coordinates $\mathbf{x} = (r\cos\phi, r\sin\phi)^\top$. The function

$$v(\mathbf{x}) = r^{2/3}\sin(2(\phi - \pi/2)/3) \qquad \text{for } \mathbf{x} \in \Omega = (-1, 1)^2 \setminus [0, 1]^2$$

is interpolated in the space V_h^1. This function exhibits the typical singularity at the reentrant corner, which is located in the origin of the coordinate system. We compare the L_2-interpolation error for three families of meshes, see Fig. 4.5, using the local Nyström solver. Afterwards, we compare the Nyström solver with a naive application of the local BEM solver and we discuss improvements.

We specify the meshes by a discretization parameter $n \sim h^{-1}$ instead of the mesh size h. The first family is denoted by $\mathcal{K}_n^1$ and the nth mesh consists of one L-shaped element, $(-1/3, 1/3)^2 \setminus [0, 1/3]^2$, and $24n^2$ squares of size $(3n)^{-1} \times (3n)^{-1}$. Thus, $\mathcal{K}_n^1$ has $(2n + 1)(12n + 1) + 1$ vertices. The second family $\mathcal{K}_n^2$ solely consists of congruent squares of size $(3n)^{-1} \times (3n)^{-1}$ such that, the nth mesh has $27n^2$ elements. The third family $\mathcal{K}_n^3$ is obtained from $\mathcal{K}_n^2$ by agglomerating the three squares that have the origin as a vertex. The vertices in the meshes coincide with the nodes in the corresponding FEM discretization. For all square elements in either mesh, the local spaces are the bilinear functions. If K is the L-shaped element in $\mathcal{K}_n^1$, then it contains $6n + 2$ nodes, i.e. degrees of freedom, on its boundary and $V_h^1|_K$ contains functions having the correct singular behaviour at the reentrant

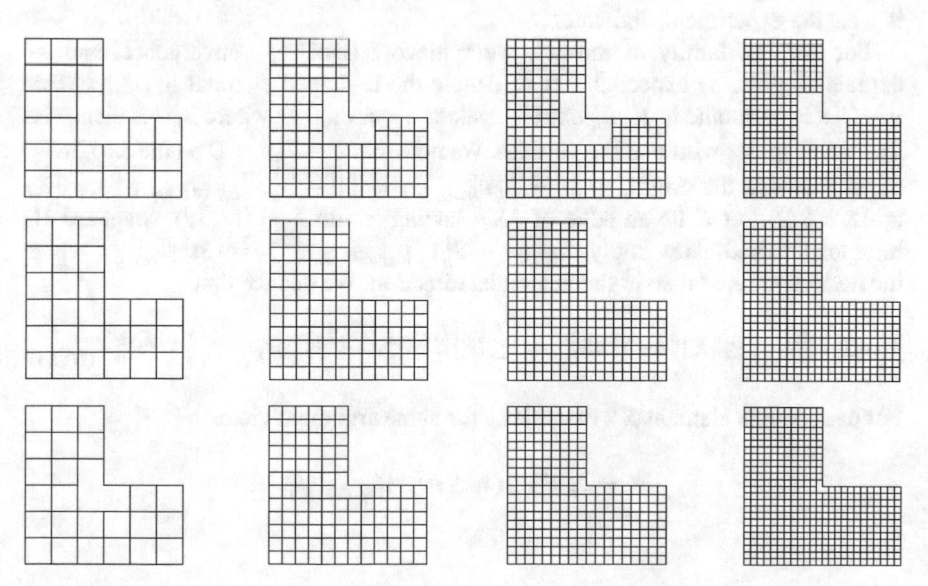

Fig. 4.5 First four meshes of first family $\mathcal{K}_n^1$ (top), second family $\mathcal{K}_n^2$ (middle), and third family $\mathcal{K}_n^3$ (bottom)

Table 4.8 Relative L_2-error (err) for interpolation in V_h^1 for the three families of meshes depicted in Fig. 4.5, and numerical order of convergence (noc) with respect to the number of degrees of freedom (DoF)

First family			Second family			Third family		
DoF	err	noc	DoF	err	noc	DoF	err	noc
40	3.24×10^{-3}	–	40	1.26×10^{-2}	–	40	3.24×10^{-3}	–
126	8.02×10^{-4}	1.22	133	4.00×10^{-3}	0.96	133	1.12×10^{-3}	0.89
260	3.55×10^{-4}	1.12	280	2.04×10^{-3}	0.90	280	5.87×10^{-4}	0.86
442	1.99×10^{-4}	1.09	408	1.46×10^{-3}	0.89	408	4.25×10^{-4}	0.86
672	1.27×10^{-4}	1.07	833	7.85×10^{-4}	0.87	833	2.32×10^{-4}	0.85
950	8.79×10^{-5}	1.06	1045	6.45×10^{-4}	0.86	1045	1.92×10^{-4}	0.84
1276	6.44×10^{-5}	1.05	1281	5.42×10^{-4}	0.86	1281	1.61×10^{-4}	0.84

corner. If K is the L-shaped element in $\mathscr{K}_n^3$, then it contains only 8 nodes, i.e. degrees of freedom, on the boundary and $V_h^1\big|_K$ also contains functions having the correct singular behaviour at the reentrant corner.

In Table 4.8, the relative interpolation error in the L_2-norm is given for all three sequences of meshes, maintaining comparable numbers of degrees of freedom between the spaces. Furthermore, the numerical order of convergence (noc) is given. This is an estimate of the exponent q in the error model $C\,\mathrm{DoF}^{-q}$. Standard bilinear interpolation theory for functions $v \in H^{1+s}(\Omega)$ on the second family of meshes yields $\|v - \mathfrak{I}_h^1 v\|_{L_2(\Omega)} = \mathscr{O}(\mathrm{DoF}^{-(1+s)/2})$, i.e. $q = (1+s)/2$. Since $v \in H^{1+s}(\Omega)$ for any $s < 2/3$, we expect to see essentially $q = 5/6$ for the second family, which is what the experiments indicate.

For the first family of meshes, we achieve $\mathscr{O}(\mathrm{DoF}^{-1})$ convergence, and we explain why this is expected. Let K denote the L-shaped element in $\mathscr{K}_{n,1}$. Since $v - \mathfrak{I}_h^1 v$ is harmonic in K, its extreme values occur on ∂K, where $\mathfrak{I}_h^1 v$ is piecewise linear and agrees with v at the vertices. We note that $v - \mathfrak{I}_h^1 v = 0$ on the two (long) edges touching the origin, so $\|v - \mathfrak{I}_h^1 v\|_{L_\infty(K)} = \|v - \mathfrak{I}_h^1 v\|_{L_\infty(\partial K_I)}$, where ∂K_I is $\partial K \setminus \partial\Omega$. Let E be an edge of ∂K_I, having length $h = 1/(3n)$. Standard 1D interpolation estimates imply that $\|v - \mathfrak{I}_h^1 v\|_{L_\infty(E)} \leq h^2 \|\partial^2 v/\partial \mathbf{t}^2\|_{L_\infty(E)}$, where the derivatives are taken in the tangential direction. We deduce that

$$\|v-\mathfrak{I}_h^1 v\|_{L_2(K)}^2 \leq |K| \|v-\mathfrak{I}_h^1 v\|_{L_\infty(K)}^2 \leq h^4 |K| \|\partial^2 v/\partial \mathbf{t}^2\|_{L_\infty(\partial K_I)}^2 \leq \tfrac{1}{3} h^4 |v|_{W_\infty^2(\Omega\setminus K)}^2 .$$

For each square element K', essentially the same argument yields

$$\|v - \mathfrak{I}_h^1 v\|_{L_2(K')}^2 \leq h^6 |v|_{W_\infty^2(\Omega\setminus K)}^2 .$$

From this, we determine that

$$\|v-\mathfrak{I}_h^1 v\|_{L_2(\Omega)}^2 \le \left(\tfrac{1}{3}h^4 + h^6(|\mathscr{K}_n^1|-1)\right)|v|_{W_\infty^2(\Omega\setminus K)}^2 = \left(\tfrac{1}{3}h^4 + \tfrac{8}{3}h^4\right)|v|_{W_\infty^2(\Omega\setminus K)}^2 .$$

In other words, $\|v - \mathfrak{I}_h^1 v\|_{L_2(\Omega)} = \mathscr{O}(h^2) = \mathscr{O}(\mathrm{DoF}^{-1})$.

Now suppose that K is the L-shaped element in $\mathscr{K}_n^3$ with edge length h. Since the maximal interpolation error happens on the boundary ∂K, we estimate it directly by computing the linear interpolant on each edge, and comparing it with v on that edge. We determine that

$$\|v-\mathfrak{I}_h^1 v\|_{L_2(K)}^2 \le 3h^2\|v-\mathfrak{I}_h^1 v\|_{L_\infty(K)}^2 \le 3h^2(0.023201h^{2/3})^2 \le 0.0016149h^{10/3} .$$

For comparison, we estimate the interpolation error for the three $h \times h$ squares in $\mathscr{K}_n^2$ that touch the origin, namely for $K_1 = [-h, 0] \times [0, h]$, $K_2 = [0, h] \times [-h, 0]$ and for $K_3 = [-h, 0] \times [-h, 0]$, noting that $K = K_1 \cup K_2 \cup K_3$. Since v is most naturally expressed in polar coordinates, we convert the bilinear interpolant $\mathfrak{I}_h^1 v$ to polar coordinates on each square, and compute upper and lower bounds on $\|v - \mathfrak{I}_h^1 v\|_{L_2(K_j)}$,

$$\|v - \mathfrak{I}_h^1 v\|_{L_2(D_j)} \le \|v - \mathfrak{I}_h^1 v\|_{L_2(K_j)} \le \|v - \mathfrak{I}_h^1 v\|_{L_2(\widehat{D}_j)} ,$$

where D_j, $\widehat{D}_j$ are sectors of disks centered at the origin, having radii h and $\sqrt{2}h$ respectively, and satisfying $D_j \subset K_j \subset \widehat{D}_j$. These bounds are

$$0.172h^{5/3} \le \|v - \mathfrak{I}_h^1 v\|_{L_2(K_1)} = \|v - \mathfrak{I}_h^1 v\|_{L_2(K_2)} \le 0.364h^{5/3} ,$$
$$0.571h^{5/3} \le \|v - \mathfrak{I}_h^1 v\|_{L_2(K_3)} \le 0.941h^{5/3} .$$

This explains why the interpolation error for the third family, while being of the same order as that of the second family, is slightly smaller.

Finally, for the first family of meshes, only the local interpolant on the (fixed) L-shaped element K, has to be approximated numerically. For the results above in Table 4.8, we used the Nyström approach. For comparison, we repeat the interpolation error experiment for the first family, using instead three versions of the local BEM solver to treat $\mathfrak{I}_h^1 v$ on K. For the first version (large edges), the boundary element mesh is precisely that suggested by K itself, consisting of two edges of length $1/3$ touching the origin, and $6n$ edges of length $h = 1/(3n)$, cf. Fig. 4.6 (left). For the second version, the two long edges are each partitioned into n subedges of length h, cf. Fig. 4.6 (middle). As seen in Table 4.9, the convergence for the first version stagnates almost immediately, whereas the convergence for the second version is similar to what has been seen for the second and third families above. So, we see that the BEM discretization error dominates the interpolation error in these cases, but the Nyström discretization error does not. Recalling that the BEM integral formulation is attempting to compute $\gamma_1^K v$ on ∂K, and $\gamma_1^K v(\mathbf{x}) = -\frac{2}{3}r^{-1/3}$ on both

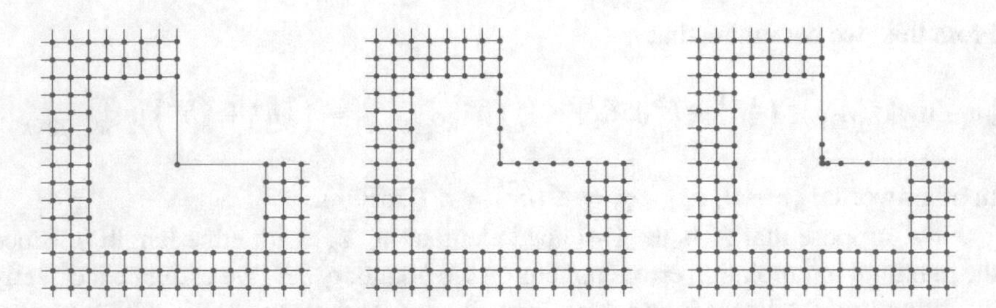

Fig. 4.6 Zoom of L-shaped element for different BEM discretizations with two large edges (left), small edges (middle) and a graded mesh (right)

Table 4.9 Relative L_2-error (err) for interpolation in V_h^1 for the first family of meshes depicted in Fig. 4.5, and numerical order of convergence (noc) with respect to the number of degrees of freedom (DoF) with three versions of the local BEM solver

Version 1 (large edges)			Version 2 (small edges)			Version 3 (graded mesh)		
DoF	err	noc	DoF	err	noc	DoF	err	noc
40	6.65×10^{-3}	–	40	6.65×10^{-3}	–	40	6.65×10^{-3}	–
126	4.88×10^{-3}	0.27	128	2.16×10^{-3}	0.97	128	1.29×10^{-3}	1.41
260	4.56×10^{-3}	0.09	264	1.10×10^{-3}	0.93	264	5.26×10^{-4}	1.23
442	4.45×10^{-3}	0.05	448	7.09×10^{-4}	0.83	448	2.64×10^{-4}	1.31
672	4.39×10^{-3}	0.03	680	5.06×10^{-4}	0.81	680	1.60×10^{-4}	1.20
950	4.36×10^{-3}	0.02	960	3.83×10^{-4}	0.81	960	1.08×10^{-4}	1.15
1276	4.34×10^{-3}	0.02	1288	3.02×10^{-4}	0.82	1288	7.75×10^{-5}	1.12

of the edges touching the origin, it is not surprising that the BEM struggles in its discretization. This challenge is mitigated for the local BEM solver by employing the a priori knowledge of the singular behaviour or by a self-adaptive procedure. We prescribe an appropriate graded mesh along the two large edges, cf. Fig. 4.6 (right), which copes with the singularity in the Neumann trace. The underlying regularity theory and the construction of graded meshes for the boundary element method has been studied in [88, 161, 171]. We repeat the convergence test with the adapted BEM discretization on the first family and retrieve the optimal rates of convergence for the interpolation error in Table 4.9.

Chapter 5
Adaptive BEM-Based Finite Element Method

As long as the solutions of boundary value problems are sufficiently regular, the refinement of the mesh size h and the increase of the approximation order k in the discretization space V_h^k yields an improvement in the accuracy. In particular, this yields optimal convergence rates. But, in most applications the regularity of the solution is restricted due to corners of the domain or jumping physical quantities. Therefore, it is essential to adapt the solution process to the underlying problem in order to retrieve optimal approximation properties. In this chapter, we deal with a posteriori error estimates which can be used to drive an adaptive mesh refinement procedure and we recover optimal rates of convergence for the adaptive methods in the numerical experiments in the presence of singularities. For the error estimation, we cover the classical residual based error estimator as well as goal-oriented techniques on general polytopal meshes. Whereas, we derive reliability and efficiency estimates for the first mentioned estimator, we discuss the benefits and potentials of the second one for general meshes.

5.1 Preliminaries

In the following derivations we restrict ourselves to the model problem and the BEM-based FEM formulation given in Chap. 2. Therefore, let $\Omega \subset \mathbb{R}^d$, $d = 2, 3$ be a polygonal or polyhedral domain. Its boundary $\Gamma = \Gamma_D \cup \Gamma_N$ is split into a Dirichlet and a Neumann part, where we assume $|\Gamma_D| > 0$. Given a source term $f \in L_2(\Omega)$, a Dirichlet datum $g_D \in H^{1/2}(\Gamma_D)$ as well as a Neumann datum $g_N \in L_2(\Gamma_N)$, the

© Springer Nature Switzerland AG 2019

S. Weißer, *BEM-based Finite Element Approaches on Polytopal Meshes*,
Lecture Notes in Computational Science and Engineering 130,
https://doi.org/10.1007/978-3-030-20961-2_5

problem reads

$$
\begin{aligned}
-\operatorname{div}(a\nabla u) &= f \quad \text{in } \Omega\ ,\\
u &= g_D \quad \text{on } \Gamma_D\ ,\\
a\nabla u \cdot \mathbf{n} &= g_N \quad \text{on } \Gamma_N\ .
\end{aligned}
$$

Furthermore, we restrict ourselves for the presentation in this chapter to piecewise constant diffusion coefficients which are aligned with the mesh, i.e.

$$a(\mathbf{x}) = a_K \quad \text{for } \mathbf{x} \in K \text{ and } K \in \mathscr{K}_h$$

for the initial mesh and consequently for all meshes in the refinement process. The Galerkin as well as the corresponding discrete Galerkin formulation are given in Sect. 2.5. We assume for simplicity, that the extension of the Dirichlet data g_D can be chosen in V_h^k. The Galerkin formulations thus read

$$
\begin{aligned}
&\text{Find } u \in g_D + H_D^1(\Omega):\\
&b(u,v) = (f,v)_{L_2(\Omega)} + (g_N,v)_{L_2(\Gamma_N)} \quad \forall v \in H_D^1(\Omega)\ ,
\end{aligned}
\tag{5.1}
$$

and

$$
\begin{aligned}
&\text{Find } u_h \in g_D + V_{h,D}^k:\\
&b(u_h,v_h) = (f,v_h)_{L_2(\Omega)} + (g_N,v_h)_{L_2(\Gamma_N)} \quad \forall v_h \in V_{h,D}^k\ .
\end{aligned}
\tag{5.2}
$$

In Chap. 2, the approximation spaces are defined and we have derived a priori error estimates for the Galerkin approximation $u_h \in V_h^k$ of the form

$$\|u - u_h\|_{H^1(\Omega)} \leq c\, h^k\, |u|_{H^{k+1}(\Omega)} \quad \text{for } u \in H^{k+1}(\Omega)\ . \tag{5.3}$$

As already mentioned, the convergence rate k in these estimates is linked to, and restricted by the regularity of the solution $u \in H^{k+1}(\Omega)$. Furthermore, the estimate cannot be evaluated for computational purposes since it contains the unknown solution u in the right hand side. The aim of an adaptive FEM is to retrieve the convergence $\mathcal{O}(h^k)$, $k \in \mathbb{N}$ of the error although the exact solution is not regular at all, i.e. $u \notin H^2(\Omega)$. In order to achieve this, we need an error estimator that is computable and can serve as an indicator for local refinement. We consider estimates of the form

$$\|u - u_h\| \leq c\,\eta \quad \text{for} \quad \eta^2 = \sum_{K\in\mathscr{K}_h} \eta_K^2\ ,$$

where $\eta = \eta(u_h)$ is desirable for practical considerations. Here, $\|\cdot\|$ denotes some norm and η is a computable error estimator, which depends on the current approximation u_h but not on the unknown solution u explicitly. Therefore, the inequality is called a posteriori error estimate. The values η_K, which are assigned to the elements $K \in \mathscr{K}_h$, serve as error indicator over the corresponding elements. With their help, we can monitor the approximation quality over the single elements and we can use this information for local mesh refinement.

The preceding considerations lead to an adaptive finite element strategy, which is often abbreviated to AFEM in the literature. This scheme can be sketched as

$$\textbf{SOLVE} \to \textbf{ESTIMATE} \to \textbf{MARK} \to \textbf{REFINE} \to \textbf{SOLVE} \to \cdots .$$

First, the discrete boundary value problem is solved on a given mesh and the error estimator η and the error indicators η_K are computed for all elements. If the desired accuracy is reached according to η, we are done. If not, some elements are marked for refinement. These elements are chosen on the basis of the error indicators η_K. Next, the marked elements are refined, and thus we obtain a new mesh which is adapted to the problem. Afterwards, we can solve the boundary value problem on the refined mesh and continue this procedure until the desired accuracy is reached.

For triangular meshes and piecewise linear trial functions, the first convergence proof for the adaptive finite element method applied to the Poisson problem can be found in [67]. Here, the mesh has to satisfy some fineness assumption. In [129], this condition is removed and the notion of data oscillation is introduced. A general convergence result for conforming adaptive finite elements, which is valid for several error estimates and for a class of problems, has been published 7 years later in [130]. The first convergence rates are proven in [36], where an additional coarsening step is introduced and the refinement is done in such a way that a new node lies inside each marked element of the previous mesh. In [55], the authors show a decay rate of the energy error plus data oscillation in terms of the number of degrees of freedom without the additional assumptions on coarsening and refining. A state of the art discussion and an axiomatic presentation of the proof of optimal convergence rates of adaptive finite element methods can be found in [53].

Whereas the cited theory is done for triangular meshes, we state an adaptive finite element method on regular and stable polygonal meshes. In the **SOLVE** step, we approximate the solution of the boundary value problem on the current mesh $\mathscr{K}_h$. This is done as described in Chap. 2. Solving the discrete problem, we obtain an approximation $u_h \in V_h^k$ on the current mesh for a fixed order k.

The **ESTIMATE** part serves for the computation of the a posteriori error estimator η and local error indicators η_K. There is a great variety of estimators in the literature. The most classical one is the residual error estimate which goes back to [15]. This estimator measures the jumps of the conormal derivative of the approximation u_h over the element boundaries. Other estimators are obtained by solving local Dirichlet [16] or Neumann [19] problems on element patches. The engineering community came up with an error indicator that uses the difference between ∇u_h and its continuous approximation, see [183]. The equilibrated residual error estimator [39] is obtained by post-processing of the approximation and belongs to the more general class of functional analytic error estimates [142]. Finally, we mention the hierarchical [64] and the goal oriented [18] error estimates. For a comparison of all these strategies see for instance [54].

After the computation of the estimator and the local error indicators, we have to **MARK** several elements for refinement. There are different strategies in the literature for this task. The most classical one is the *maximum strategy* which has been proposed already in [16]. Here, all elements $K \in \mathcal{K}_h$ are marked which satisfy

$$\eta_K \geq \theta\ \eta_{\max}$$

for a given parameter $0 \leq \theta \leq 1$ and $\eta_{\max} = \max\{\eta_K : K \in \mathcal{K}_h\}$. So, the elements with the largest error indicators are chosen for refinement. For large values of θ, the strategy becomes more selective, whereas for small θ, we obtain almost a uniform refinement. A similar idea is used by the modified *equidistribution strategy*. For a given parameter $0 \leq \theta \leq 1$ and the global error estimator η, all elements $K \in \mathcal{K}_h$ are marked which satisfy

$$\eta_K \geq \theta \frac{\eta}{\sqrt{|\mathcal{K}_h|}} .$$

In this strategy one tries to reach a state where the error is distributed equally over all elements. The parameter θ controls again the selectivity. This kind of strategy has been used in Sect. 3.4.6 for the generation of anisotropically adapted meshes. Finally, we mention *Dörfler's strategy*, see [67]. Here, a set of elements $\mathcal{K}_M \subset \mathcal{K}_h$ is marked such that

$$\left(\sum_{K \in \mathcal{K}_M} \eta_K^2 \right)^{1/2} \geq (1-\theta)\,\eta ,$$

where $0 \leq \theta < 1$ is again a given parameter and η the global estimator. It is advantageous to choose the set $\mathcal{K}_M$ as small as possible. This can be achieved by sorting the elements $K \in \mathcal{K}_h$ according to the value of their error indicators η_K. Since every sorting algorithm is computationally expensive, Dörfler proposed in [67] the following procedure with given parameter $0 < \nu < 1$, which is chosen to be small.

```
sum = 0.0
μ = 1.0
while (sum < (1-θ)² η²)
do
    μ = μ - ν
    for all K ∈ 𝒦_h
      if (K is not marked)
        if (η_K > μ η_max)
          mark K
          sum = sum + η_K²
```

Dörfler's marking strategy was one of the key points in the proofs of convergence and convergence rates of AFEM in the literature mentioned above.

As the name of the last step **REFINE** already indicates, this is the time where the marked elements are refined. Usually, this step is more complicated for standard methods working on triangular or quadrangular meshes because of the strict admissibility conditions on the mesh. In such cases, it has to be guaranteed that no hanging nodes appear. Therefore, the mesh has to be completed in the sense that neighbouring elements are refined until all hanging nodes disappear. In the literature, one can find several strategies like red-green refinement or newest vertex bisection with completion algorithms, see [36, 170]. Another possibility to handle hanging nodes is to treat them as conditional degrees of freedom, i.e., to fix the value of the finite element functions in these points to be a suitable interpolation of their neighbouring regular nodes. Nevertheless, the first idea with completion spreads the local refinement into a neighbourhood and the second one produces artificial nodes. Both scenarios are somehow unpleasant for the numerical realization. Due to the use of the BEM-based FEM, we are in the fortune situation to cope with arbitrary polytopal meshes. Therefore, we do not have to worry about hanging nodes because they are incorporated as ordinary nodes in the strategy and thus contribute to the approximation accuracy. This behaviour is discussed more precisely in Sect. 5.2.3. The refinement only affects the marked elements and is done as described in Sect. 2.2.3. During this refinement process with the discussed bisection algorithm, the stability of the sequence of meshes is not preserved automatically. Thus, we might want to enforce this property explicitly in the mesh refinement.

In certain algorithms and applications an additional **COARSEN** step is necessary which reverses the local mesh refinement in some areas of the domain. This has been introduced in [36] for theoretical reasons in order to prove convergence rates for the adaptive algorithm. But also in time-dependent problems, this additional step is meaningful if, for instance, the singularity of the solution travels through the spacial domain. The coarsening often relies on the hierarchy of adaptive meshes obtained during the refinement. For polytopal meshes, however, one might agglomerate almost arbitrary elements in this step since the union of polytopes is a polytope. This demonstrates once more the flexibility of these general meshes.

5.2 Residual Based Error Estimator

In this section, we consider one of the most classical a posteriori error estimators, namely the residual based error estimator, and formulate it on polytopal meshes. For the classical results on simplicial meshes see, e.g., [4, 170]. This a posteriori error estimate bounds the difference of the exact solution and the Galerkin approximation in the energy norm $\|\cdot\|_b$ associated to the symmetric and positive definite bilinear form, i.e. $\|\cdot\|_b^2 = b(\cdot,\cdot)$. Among others, the estimate contains the jumps of the conormal derivatives over the element interfaces. Since we are dealing with the three- and two-dimensional case simultaneously, $F \in \mathscr{F}_h$ denotes a face ($d = 3$) or edge ($d = 2$), respectively. Such a jump over an internal face $F \in \mathscr{F}_{h,\Omega}$ is defined by

$$[\![u_h]\!]_F = \left(a_K \gamma_1^K u_h + a_{K'} \gamma_1^{K'} u_h \right)\Big|_F ,$$

where $K, K' \in \mathscr{K}_h$ are the neighbouring elements of F with $F \in \mathscr{F}(K) \cap \mathscr{F}(K')$. The element residual is given by

$$R_K = f + a_K \Delta u_h \quad \text{for } K \in \mathscr{K}_h ,$$

and the face/edge residual by

$$R_F = \begin{cases} 0 & \text{for } F \in \mathscr{F}_{h,D} , \\ g_N - a_K \gamma_1^K u_h & \text{for } F \in \mathscr{F}_{h,N} \text{ with } F \in \mathscr{F}(K) , \\ -\frac{1}{2} [\![u_h]\!]_F & \text{for } F \in \mathscr{F}_{h,\Omega} . \end{cases}$$

We can proceed as for the two-dimensional case in [174, 180] in order to formulate the residual based error estimator and to prove its reliability and efficiency on polytopal meshes. This estimator involves the previously defined element and face residuals and gives an upper bound for the Galerkin error in the energy norm which does not contain any unknown quantity.

Theorem 5.1 (Reliability) *Let $\mathscr{K}_h$ be a regular and stable mesh. Furthermore, let $u \in g_D + H^1_D(\Omega)$ and $u_h \in g_D + V^k_{h,D}$ be the solutions of* (5.1) *and* (5.2)*, respectively. The residual based error estimate is reliable, i.e.*

$$\|u - u_h\|_b \le c\, \eta_R \quad \textit{with} \quad \eta_R^2 = \sum_{K \in \mathscr{K}_h} \eta_K^2 ,$$

where the error indicator is defined by

$$\eta_K^2 = h_K^2 \|R_K\|^2_{L_2(K)} + \sum_{F \in \mathscr{F}(K)} h_F \|R_F\|^2_{L_2(F)} .$$

The constant $c > 0$ only depends on the regularity and stability parameters of the mesh, see Sect. 2.2, the approximation order k, the space dimension d and on the diffusion coefficient a.

In this presentation it is assumed that we can compute $\gamma_1^K u_h$ analytically. However, in the realization these terms are treated by means of boundary element methods as discussed in Chap. 4. This approximation of the Neumann traces has been incorporated in [180] and yields an additional term in the estimate.

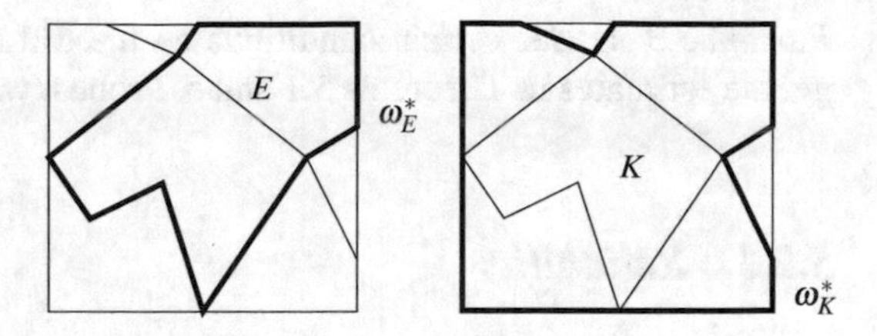

Fig. 5.1 Example of modified neighbourhoods of edges and elements in two space dimensions, cf. Fig. 3.1

Whereas the reliability gives an upper bound for the error, the efficiency states a local upper bound for the error indicator in terms of the approximation error and the problem data. Beside of the neighbourhoods (3.2) of nodes, edges, faces and elements defined in Sect. 3.1, we additionally need the following modified versions

$$\overline{\omega}_E^* = \bigcup_{K'\in\mathcal{K}_h:E\in\mathcal{E}(K')} \overline{K'}\,, \qquad \overline{\omega}_F^* = \bigcup_{K'\in\mathcal{K}_h:F\in\mathcal{F}(K')} \overline{K'}\,, \qquad \overline{\omega}_K^* = \bigcup_{F\in\mathcal{F}(K)} \overline{\omega}_F^*\,, \tag{5.4}$$

cf. Fig. 5.1. Furthermore, we introduce the notation $\|\cdot\|_{b,\omega}$ for $\omega\subset\Omega$, which means that the energy norm is only computed over the subset ω. More precisely, it is $\|v\|_{b,\omega}^2 = (a\nabla v,\nabla v)_{L_2(\omega)}$ for our model problem.

Theorem 5.2 (Efficiency) *Under the assumptions of Theorem 5.1, the residual based error indicator is efficient, i.e.*

$$\eta_K \le c\Bigg(\|u-u_h\|_{b,\omega_K^*}^2 + h_K^2\|f-\widetilde{f}\|_{L_2(\omega_K^*)}^2 \\ + \sum_{F\in\mathcal{F}(K)\cap\mathcal{F}_{h,N}} h_F\|g_N-\widetilde{g}_N\|_{L_2(F)}^2 + h_K\sum_{K'\subset\omega_K^*}\|\gamma_1^{K'}u_h - \widetilde{\gamma_1^{K'}u_h}\|_{L_2(\partial K')}^2\Bigg)^{1/2},$$

where $\widetilde{f}$, $\widetilde{g}_N$ and $\widetilde{\gamma_1^{K'}u_h}$ are piecewise polynomial approximations of f, g_N and $\gamma_1^{K'}u_h$, respectively. The constant $c>0$ only depends on the regularity and stability parameters of the mesh, see Sect. 2.2, the approximation order k, the space dimension d and on the diffusion coefficient a.

The terms involving the data approximation $\|f-\widetilde{f}\|_{L_2(\omega_K^*)}$ and $\|g_N-\widetilde{g}_N\|_{L_2(F)}$ are often called data oscillations. They are usually of higher order. Additionally, we have the term $\|\gamma_1^{K'}u_h - \widetilde{\gamma_1^{K'}u_h}\|_{L_2(\partial K')}$ measuring oscillations in the Neumann trace of the approximation u_h on the boundaries of the elements. The piecewise polynomial function $\widetilde{\gamma_1^{K'}u_h}$ might be chosen as the approximation of the Neumann trace obtained as solution of the derived boundary integral equation from Chap. 4. Thus, known error estimates from the boundary element method can be applied in order to bound this term further if needed.

Remark 5.3 Under certain conditions on the diffusion coefficient it is possible to get the estimates in Theorems 5.1 and 5.2 robust with respect to a, see, e.g., [139].

5.2.1 Reliability

We follow the classical lines in the proof of the reliability, see, e.g., [170]. However, we have to take care on the polytopal elements and the quasi-interpolation operators.

Proof (Theorem 5.1) The bilinear form $b(\cdot,\cdot)$ is a inner product on $V = H^1_D(\Omega)$ due to its boundedness and ellipticity, and thus, V is a Hilbert space together with $b(\cdot,\cdot)$ and $\|\cdot\|_b$. The Riesz representation theorem yields

$$\|u - u_h\|_b = \sup_{v\in V\setminus\{0\}} \frac{|\mathfrak{R}(v)|}{\|v\|_b} \quad \text{with} \quad \mathfrak{R}(v) = b(u - u_h, v)\,. \tag{5.5}$$

Thus, in order to prove the theorem, we reformulate and estimate the term $|\mathfrak{R}(v)|$ in the following. Let $v_h \in V^1_{h,D}$, the Galerkin orthogonality $b(u - u_h, v_h) = 0$ and integration by parts over each element lead to

$$\mathfrak{R}(v) = \sum_{K\in\mathcal{K}_h} \Big((R_K, v - v_h)_{L_2(K)} + \sum_{F\in\mathcal{F}(K)} (R_F, v - v_h)_{L_2(F)} \Big)\,. \tag{5.6}$$

The Cauchy–Schwarz inequality yields

$$|\mathfrak{R}(v)| \le \sum_{K\in\mathcal{K}_h} \Big(\|R_K\|_{L_2(K)} \|v - v_h\|_{L_2(K)} + \sum_{F\in\mathcal{F}(K)} \|R_F\|_{L_2(F)} \|v - v_h\|_{L_2(F)} \Big)\,.$$

We choose $v_h = \mathfrak{I}_C v$, where $\mathfrak{I}_C$ is the Clément interpolation operator from Sect. 3.3, which preserves the homogeneous boundary data on Γ_D. Estimating the L_2-norms of $v - \mathfrak{I}_C v$ over the elements and faces with the help of Theorem 3.7, we find

$$\begin{aligned}|\mathfrak{R}(v)| &\le c \sum_{K\in\mathcal{K}_h} \Big(h_K \|R_K\|_{L_2(K)} |v|_{H^1(\omega_K)} + \sum_{F\in\mathcal{F}(K)} h_F^{1/2} \|R_F\|_{L_2(F)} |v|_{H^1(\omega_F)} \Big) \\ &\le c \Big(\sum_{K\in\mathcal{K}_h} \eta_K^2 \Big)^{1/2} \Big(\sum_{K\in\mathcal{K}_h} |v|^2_{H^1(\omega_K)} \Big)^{1/2} \le c\, \eta_R\, |v|_{H^1(\Omega)}\,,\end{aligned}$$

where in the last two estimates we utilized several times Cauchy–Schwarz inequality and the facts, that each element has a bounded number of faces, see Lemmata 2.7 and 2.16, and that it is covered by a uniformly bounded number of patches only, see Lemma 3.1. Because of $\sqrt{a/a_{\min}} > 1$, it is $|v|_{H^1(\Omega)} \le \|v\|_b/\sqrt{a_{\min}}$ and thus (5.5) together with the previous inequality completes the proof. □

5.2.2 *Efficiency*

The classical proof of efficiency for the residual based error estimator makes use of special bubble functions over the simplicial meshes. These functions have support over single elements and are used to localize the residuals. We adapt the bubble function technique to polytopal meshes. Therefore, let ϕ_T and ϕ_F be the usual polynomial bubble functions over the auxiliary discretization $\mathscr{T}_h(\mathscr{K}_h)$ consisting of triangles ($d = 2$) or tetrahedra ($d = 3$), see [4, 170]. Here, ϕ_T is a cubic ($d = 2$) or quartic ($d = 3$) polynomial over the triangle/tetrahedron $T \in \mathscr{T}_h(\mathscr{K}_h)$, which vanishes on $\Omega \setminus T$ and in particular on ∂T. It is usually defined as the product of the barycentric coordinates of the triangle and tetrahedron, respectively, and scaled such that its maximum is one. The edge bubble ϕ_F is a piecewise quadratic ($d = 2$) or cubic ($d = 3$) polynomial over the adjacent triangles/tetrahedra in $\mathscr{T}_h(\mathscr{K}_h)$, sharing the common edge/face F, and it vanishes elsewhere. This bubble function can also be defined as scaled product of barycentric coordinates.

At first glance, we might define the bubble functions over polytopes as product of the first order basis functions defined in Sect. 2.3.1 or one might use the element bubble functions defined in Sect. 2.3.2. However, in these cases the functions are no polynomials that complicates their treatment in the analysis. In contrast, we define the new bubble functions over the polytopal mesh with the help of the bubble functions over the auxiliary discretization, namely

$$\varphi_K = \sum_{T \subset \mathscr{T}_h(K)} \phi_T \qquad \text{and} \qquad \varphi_F = \phi_F$$

for $K \in \mathscr{K}_h$ and $F \in \mathscr{F}_h$.

Lemma 5.4 *Let $K \in \mathscr{K}_h$ and $F \in \mathscr{F}(K)$ of a regular and stable mesh $\mathscr{K}_h$. The bubble functions satisfy*

$$\begin{aligned} \operatorname{supp} \varphi_K &= K\,, \quad 0 \le \varphi_K \le 1\,,\\ \operatorname{supp} \varphi_F &\subset \omega_F^*\,, \quad 0 \le \varphi_F \le 1\,, \end{aligned}$$

and fulfil for $p \in \mathscr{P}^k(K)$ the estimates

$$\begin{aligned} \|p\|_{L_2(K)}^2 &\le c\,(\varphi_K p, p)_{L_2(K)}\,, & |\varphi_K p|_{H^1(K)} &\le c h_K^{-1}\,\|p\|_{L_2(K)}\,,\\ \|p\|_{L_2(F)}^2 &\le c\,(\varphi_F p, p)_{L_2(F)}\,, & |\varphi_F p|_{H^1(K)} &\le c h_F^{-1/2}\,\|p\|_{L_2(F)}\,,\\ & & \|\varphi_F p\|_{L_2(K)} &\le c h_F^{1/2}\,\|p\|_{L_2(F)}\,. \end{aligned}$$

The constants $c > 0$ only depend on the regularity and stability parameters of the mesh, see Sect. 2.2, the approximation order k and the space dimension d.

Proof Similar estimates are valid for ϕ_T and ϕ_F on triangular and tetrahedral meshes, see [4, 170]. By the use of Cauchy–Schwarz inequality and the properties of the auxiliary discretization $\mathscr{T}_h(\mathscr{K}_h)$ the estimates translate to the new bubble functions. The details of the proof are omitted. □

With these ingredients the proof of Theorem 5.2 can be addressed. The arguments follow the line of [4].

Proof (Theorem 5.2) Let $\widetilde{R}_K \in \mathscr{P}^k(K)$ be a polynomial approximation of the element residual R_K for $K \in \mathscr{K}_h$. For $v = \varphi_K \widetilde{R}_K \in H_0^1(K)$ and $v_h = 0$ Eq. (5.6) yields

$$b(u - u_h, \varphi_K \widetilde{R}_K) = \mathfrak{R}(\varphi_K \widetilde{R}_K) = (R_K, \varphi_K \widetilde{R}_K)_{L_2(K)} .$$

Lemma 5.4 gives

$$\begin{aligned}\|\widetilde{R}_K\|_{L_2(K)}^2 &\le c\,(\varphi_K \widetilde{R}_K, \widetilde{R}_K)_{L_2(K)}\\ &= c\left((\varphi_K \widetilde{R}_K, \widetilde{R}_K - R_K)_{L_2(K)} + (\varphi_K \widetilde{R}_K, R_K)_{L_2(K)}\right)\\ &\le c\left(\|\widetilde{R}_K\|_{L_2(K)}\|\widetilde{R}_K - R_K\|_{L_2(K)} + b(u - u_h, \varphi_K \widetilde{R}_K)\right),\end{aligned}$$

and furthermore,

$$b(u-u_h, \varphi_K \widetilde{R}_K) \le c\,|u-u_h|_{H^1(K)}|\varphi_K \widetilde{R}_K|_{H^1(K)} \le c h_K^{-1}\,\|u-u_h\|_{b,K}\|\widetilde{R}_K\|_{L_2(K)} .$$

We thus get

$$\|\widetilde{R}_K\|_{L_2(K)} \le c\left(h_K^{-1}\|u - u_h\|_{b,K} + \|\widetilde{R}_K - R_K\|_{L_2(K)}\right),$$

and by the reverse triangle inequality

$$\|R_K\|_{L_2(K)} \le c\left(h_K^{-1}\|u - u_h\|_{b,K} + \|\widetilde{R}_K - R_K\|_{L_2(K)}\right).$$

Next, we consider the face residual. Let $\widetilde{R}_F \in \mathscr{P}^k(F)$ be an approximation of R_F, with $F \in \mathscr{F}_{h,\Omega}$. The case $F \in \mathscr{F}_{h,N}$ is treated analogously. For $v = \varphi_F \widetilde{R}_F \in H_0^1(\omega_F^*)$ and $v_h = 0$ Eq. (5.6) yields in this case

$$b(u-u_h, \varphi_F \widetilde{R}_F) = \mathfrak{R}(\varphi_F \widetilde{R}_F) = \sum_{K\subset\omega_F^*}\left((R_K, \varphi_F \widetilde{R}_F)_{L_2(K)} + (R_F, \varphi_F \widetilde{R}_F)_{L_2(F)}\right) .$$

Applying Lemma 5.4 and the previous formula leads to

$$\begin{aligned}\|\widetilde{R}_F\|^2_{L_2(F)} &\leq c\,(\varphi_F\widetilde{R}_F, \widetilde{R}_F)_{L_2(F)}\\ &= c\left((\varphi_F\widetilde{R}_F, \widetilde{R}_F - R_F)_{L_2(F)} + (\varphi_F\widetilde{R}_F, R_F)_{L_2(F)}\right)\\ &\leq c\left(\|\widetilde{R}_F\|_{L_2(F)}\|\widetilde{R}_F - R_F\|_{L_2(F)} + (\varphi_F\widetilde{R}_F, R_F)_{L_2(F)}\right),\end{aligned}$$

and

$$\begin{aligned}|(\varphi_F\widetilde{R}_F, R_F)_{L_2(F)}| &= \tfrac{1}{2}\left|b(u-u_h, \varphi_F\widetilde{R}_F) - \sum_{K\subset\omega_F^*}(R_K, \varphi_F\widetilde{R}_F)_{L_2(K)}\right|\\ &\leq c\left(|u-u_h|_{H^1(\omega_F^*)}|\varphi_F\widetilde{R}_F|_{H^1(\omega_F^*)} + \sum_{K\subset\omega_F^*}\|R_K\|_{L_2(K)}\|\varphi_F\widetilde{R}_F\|_{L_2(K)}\right)\\ &\leq c\left(h_F^{-1/2}\|u-u_h\|_{b,\omega_F^*} + \sum_{K\subset\omega_F^*}h_F^{1/2}\|R_K\|_{L_2(K)}\right)\|\widetilde{R}_F\|_{L_2(F)}.\end{aligned}$$

Therefore, it is

$$\|\widetilde{R}_F\|_{L_2(F)} \leq c\left(h_F^{-1/2}\|u-u_h\|_{b,\omega_F^*} + \sum_{K\subset\omega_F^*}h_F^{1/2}\|R_K\|_{L_2(K)} + \|\widetilde{R}_F - R_F\|_{L_2(F)}\right).$$

By the reverse triangle inequality, $h_K^{-1} \leq h_F^{-1}$ and the previous estimate for $\|R_K\|_{L_2(K)}$, we obtain

$$\|R_F\|_{L_2(F)} \leq c\left(h_F^{-1/2}\|u-u_h\|_{b,\omega_F^*} + \sum_{K\subset\omega_F^*}h_F^{1/2}\|\widetilde{R}_K - R_K\|_{L_2(K)} + \|\widetilde{R}_F - R_F\|_{L_2(F)}\right).$$

Let $\widetilde{f}$, $\widetilde{g}_N$ and $\widetilde{\gamma_1^K u_h}$ be piecewise polynomial approximations of f, g_N and $\gamma_1^K u_h$, respectively. We choose $\widetilde{R}_K = \widetilde{f} + a_K\Delta u_h$ for $K \in \mathscr{K}_h$ and

$$\widetilde{R}_F = \begin{cases} 0 & \text{for } F \in \mathscr{F}_{h,D},\\ \widetilde{g}_N - a_K\widetilde{\gamma_1^K u_h} & \text{for } F \in \mathscr{F}_{h,N} \text{ with } F \in \mathscr{F}(K),\\ -\frac{1}{2}\left(a_K\widetilde{\gamma_1^K u_h} + a_{K'}\widetilde{\gamma_1^{K'} u_h}\right) & \text{for } F \in \mathscr{F}_{h,\Omega} \text{ with } F \subset \mathscr{F}(K)\cap\mathscr{F}(K'). \end{cases}$$

Consequently, we have $\widetilde{R}_K \in \mathscr{P}^k(K)$ and $\widetilde{R}_F \in \mathscr{P}^k(F)$. Finally, the estimates for $\|R_K\|_{L_2(K)}$ and $\|R_F\|_{L_2(F)}$ yield after some applications of the Cauchy–Schwarz inequality and due to $h_F \le h_K$ and $|\mathscr{F}(K)| \le c$, see Lemmata 2.7 and 2.16,

$$\begin{aligned}\eta_K^2 &\le c\left(\|u-u_h\|_{b,\omega_K^*}^2 + h_K^2 \sum_{K'\subset\omega_K^*} \|\widetilde{R}_{K'} - R_{K'}\|_{L_2(K')}^2 + \sum_{F\in\mathscr{F}(K)} h_F \|\widetilde{R}_F - R_F\|_{L_2(F)}^2 \right)\\ &\le c\Bigg(\|u-u_h\|_{b,\omega_K^*}^2 + h_K^2 \|f-\widetilde{f}\|_{L_2(\omega_K^*)}^2\\ &\qquad + \sum_{F\in\mathscr{F}(K)\cap\mathscr{F}_{h,N}} h_F \|g_N - \widetilde{g}_N\|_{L_2(F)}^2 + h_K \sum_{K'\subset\omega_K^*} \|\gamma_1^{K'} u_h - \widetilde{\gamma_1^{K'} u_h}\|_{L_2(\partial K')}^2 \Bigg) .\end{aligned}$$

□

5.2.3 Numerical Experiments

The residual based error estimate can be used as stopping criteria to check if the desired accuracy is reached in a simulation on a sequence of meshes. However, it is well known that residual based estimators overestimate the true error a lot. But, because of the equivalence of the norms $\|\cdot\|_{1,\Omega}$ and $\|\cdot\|_b$ on $H_D^1(\Omega)$, we can still use η_R to verify numerically the convergence rates for uniform mesh refinement when $h \to 0$. On the other hand, we can utilize the error indicators in order to gauge the approximation quality over the single elements and drive an adaptive mesh refinement strategy with this information. The adaptive algorithm discussed in Sect. 5.1, has been implemented with Dörfler's marking strategy. During the refinement of the mesh we enforce the stability condition. This is done by refining elements that do not satisfy $h_K < c_T h_E$ for a threshold parameter c_T.

In the following we present numerical examples in 2-dimensions on uniformly and adaptively refined meshes. For the convergence analysis, we consider the error with respect to the mesh size $h = \max\{h_K : K \in \mathscr{K}_h\}$ for uniform refinement. For the adaptive BEM-based FEM, the convergence is studied with respect to the number of degrees of freedom (DoF). On uniform meshes the relation $\mathrm{DoF} = \mathscr{O}(h^{-2})$ holds, whereas on adaptive meshes the mesh size does not decrease uniformly.

Experiment 1: Uniform Refinement Strategy
Consider the Dirichlet boundary value problem

$$-\Delta u = f \quad \text{in } \Omega = (0,1)^2 , \qquad u = 0 \quad \text{on } \Gamma ,$$

where $f \in L_2(\Omega)$ is chosen in such a way that $u(\mathbf{x}) = \sin(\pi x_1)\sin(\pi x_2)$ for $\mathbf{x} \in \Omega$ is the exact solution. The solution is smooth, and thus, we expect optimal rates of convergence for uniform mesh refinement. The problem is treated with the BEM-based FEM for different approximation orders $k = 1, 2, 3$ on a sequence of meshes with L-shaped elements of decreasing diameter, see Fig. 5.2 left. In Fig. 5.3, we

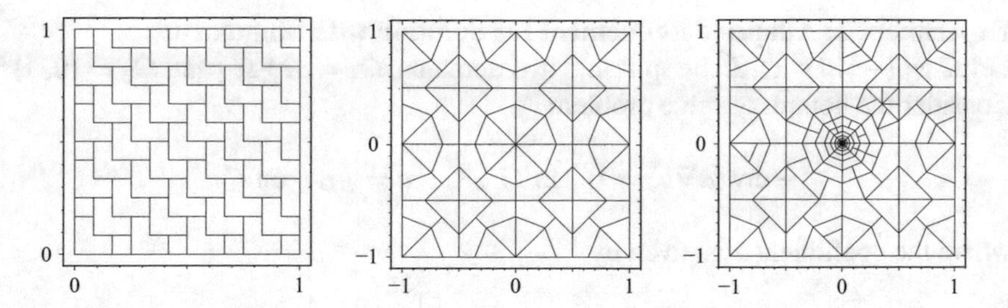

Fig. 5.2 Mesh with L-shaped elements for uniform refinement (left), initial mesh for adaptive refinement (middle), adaptive refined mesh after 30 steps for $k = 2$ with solution having a singularity in the origin of the coordinate system (right)

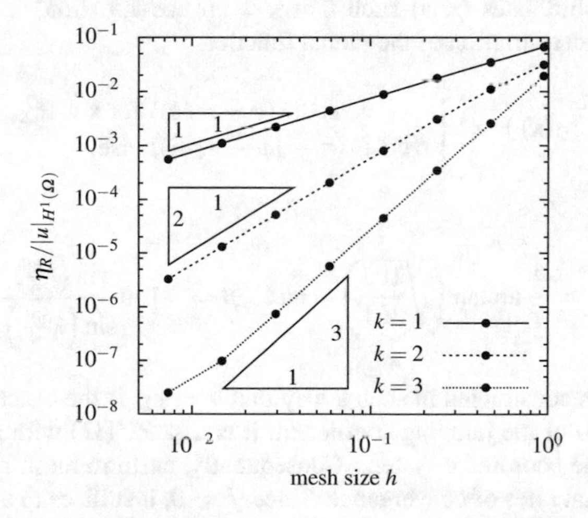

Fig. 5.3 Convergence graph for sequence of uniform meshes with L-shaped elements and V_h^k, $k = 1, 2, 3$, where $\eta_R/|u|_{H^1(\Omega)}$ is given with respect to h in logarithmic scale

give the convergence graphs in logarithmic scale for the value $\eta_R/|u|_{H^1(\Omega)}$, which behaves like the relative H^1-error, with respect to the mesh size h. The example confirms the theoretical rates of convergence stated in Sect. 2.5 on a sequence of meshes with non-convex elements. The highly accurate computations for V_h^3 involve approximately 690,000 degrees of freedom. Due to the decoupling of the variational formulation discussed in Sect. 2.5 into (2.31) and (2.32) the global system of linear equations has about 540,000 unknowns and the remaining degrees of freedom are determined by local projections.

Experiment 2: Adaptive Refinement for Solution with Singularity
Let $\Omega = (-1, 1)^2 \subset \mathbb{R}^2$ be split into two domains, $\Omega_1 = \Omega \setminus \overline{\Omega_2}$ and $\Omega_2 = (0, 1)^2$. Consider the boundary value problem

$$-\operatorname{div}(a\nabla u) = 0 \quad \text{in } \Omega\,, \qquad u = g_D \quad \text{on } \Gamma\,,$$

where the coefficient a is given by

$$a = \begin{cases} 1 \text{ in } \Omega_1\,, \\ 100 \text{ in } \Omega_2\,. \end{cases}$$

Using polar coordinates (r, ϕ) such that $\mathbf{x} = (r\cos\phi, r\sin\phi)^\top$, we choose the boundary data as restriction of the global function

$$g_D(\mathbf{x}) = r^\lambda \begin{cases} \cos(\lambda(\phi - \pi/4)) \text{ for } \mathbf{x} \in \mathbb{R}^2_+, \\ \beta\cos(\lambda(\pi - |\phi - \pi/4|)) \text{ else,} \end{cases}$$

with

$$\lambda = \frac{4}{\pi}\arctan\left(\sqrt{\frac{103}{301}}\right) \quad \text{and} \quad \beta = -100\frac{\sin\left(\lambda\frac{\pi}{4}\right)}{\sin\left(\lambda\frac{3\pi}{4}\right)}\,.$$

This problem is constructed in such a way that $u = g_D$ is the exact solution in Ω. Due to the ratio of the jumping coefficient, it is $u \notin H^2(\Omega)$ with a singularity in the origin of the coordinate system. Consequently, uniform mesh refinement does not yield optimal rates of convergence. Since $f = 0$, it suffices to approximate the solution in $V^k_{h,H}$ with the variational formulation (2.31). Starting from an initial polygonal mesh, see Fig. 5.2 middle, the adaptive BEM-based FEM produces a sequence of locally refined meshes. The approach detects the singularity in the origin of the coordinate system and polygonal elements appear naturally during the local refinement, see Fig. 5.2 right. In Fig. 5.4, the energy error $\|u - u_h\|_b$ as well as the error estimator η_R are plotted with respect to the number of degrees of freedom in logarithmic scale. As expected by the theory the residual based error estimate represents the behaviour of the energy error very well. Furthermore, the adaptive approach yields optimal rates of convergence in the presence of a singularity, namely a slope of $-k/2$ in the logarithmic plot.

Experiment 3: Adaptive Refinement, Closer Look
Using polar coordinates again, let $\Omega = \{\mathbf{x} \in \mathbb{R}^2 : |r| < 1 \text{ and } 0 < \varphi < 3\pi/2\}$ and the boundary data g_D be chosen in such a way that

$$u(r\cos\phi, r\sin\phi) = r^{2/3}\sin\left(\frac{2\phi}{3}\right)$$

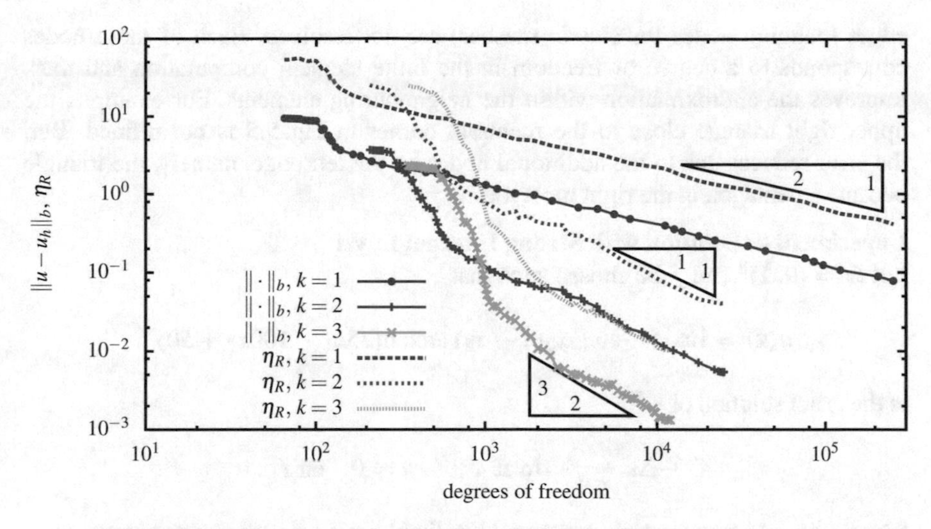

Fig. 5.4 Convergence graph for adaptive mesh refinement with $V^k_{h,H}$, $k = 1, 2, 3$, the energy error and the residual based error estimator are given with respect to the number of degrees of freedom in logarithmic sale

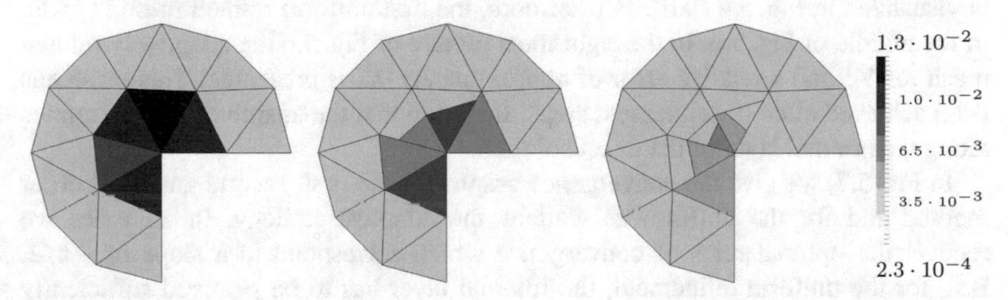

Fig. 5.5 Closer look at the error distribution $|u-u_h|^2_{H^1(K)}$ for the first three meshes in the adaptive refinement approach

is the solution of the boundary value problem

$$-\Delta u = 0 \quad \text{in } \Omega, \qquad u = g_D \quad \text{on } \Gamma.$$

The function u is constructed in such a way that its derivatives have a singularity at the origin of the coordinate system. The boundary value problem is discretized using the first order approximation space V^1_h and we analyse the first steps in the adaptive refinement strategy in more detail. This will stress the use and the flexibility of polygonal meshes in adaptive computations. For this purpose the error distribution is visualized in Fig. 5.5 for the first three meshes. Each element K is colored according to the value $|u - u_h|^2_{H^1(K)}$. The adaptive algorithm apparently marks and refines the elements with the largest error contribution. The introduced nodes on straight edges (hanging nodes for classic meshes) are not resolved. Each of these nodes corresponds to a degree of freedom in the finite element computation and thus, improves the approximation within the neighbouring elements. For example, the upper right triangle close to the reentrant corner in Fig. 5.5 is not refined. But, the error reduces due to the additional nodes on the left edge, namely, the triangle became a pentagon in the right most mesh.

Experiment 4: Solution with Strong Internal Layer
Let $\Omega = (0, 1)^2$ and f be chosen such that

$$u(\mathbf{x}) = 16x_1(1 - x_1)x_2(1 - x_2)\arctan(25x_1 - 100x_2 + 50) ,$$

is the exact solution of

$$-\Delta u = f \quad \text{in } \Omega , \qquad u = 0 \quad \text{on } \Gamma .$$

Since u is arbitrary smooth, we expect optimal rates of convergence in the case of uniform mesh refinement in an asymptotic regime. Although the solution u is smooth, it has a strong internal layer along the line $x_2 = 1/2+x_1/4$. The initial mesh is visualized in Fig. 5.6 (left). Furthermore, the first uniform refined mesh is given in the middle of Fig. 5.6. In the right most picture of Fig. 5.6 the adaptively refined mesh for V_h^1 and a relative error of approximately 0.2 is presented. This mesh has been achieved after 19 refinement steps. It is seen that the adaptive strategy refines along the internal layer of the exact solution.

In Fig. 5.7, we give the convergence graphs for the first, second and third order method and for the uniform as well as the adaptive strategy. In all cases we recover the optimal rates of convergence which correspond to a slope of $-k/2$. But, for the uniform refinement, the internal layer has to be resolved sufficiently before the optimal rates are achieved. Since the adaptive strategy resolves the layer

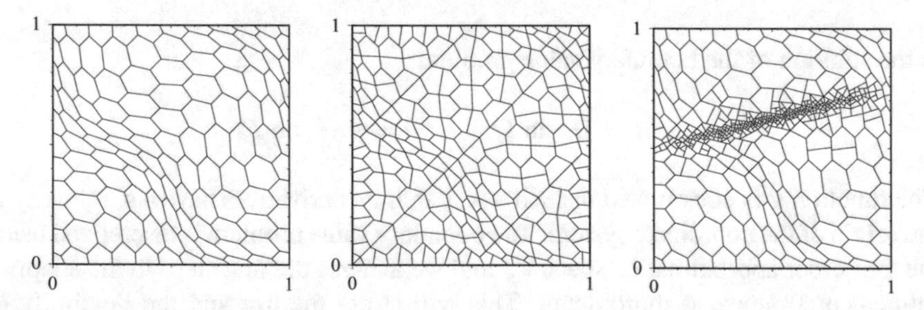

Fig. 5.6 Initial mesh (left), uniformly refined mesh (middle), adaptively refined mesh for $k = 1$ and $\|u - u_h\|_b/\|u\|_b \approx 0.2$ (right) for the solution with an internal layer

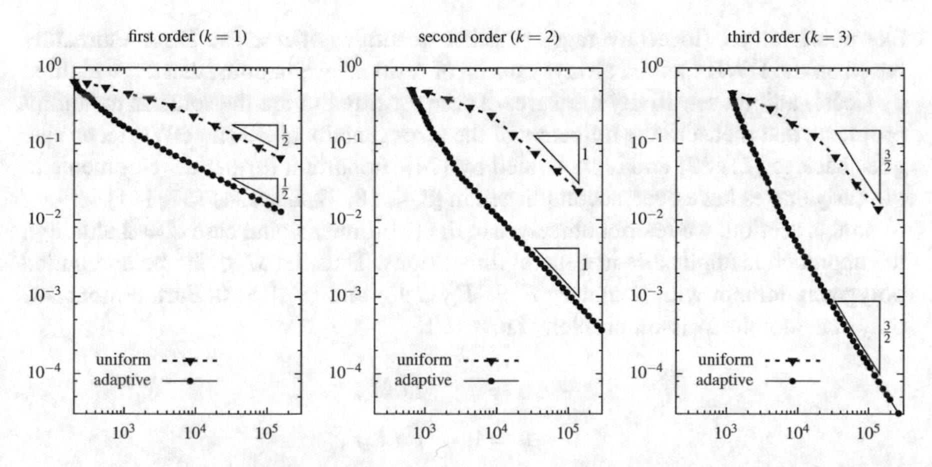

Fig. 5.7 Convergence of the relative energy error $\|u - u_h\|_b/\|u\|_b$ with respect to the number of degrees of freedom for the approximation orders $k = 1, 2, 3$ on uniformly and adaptively refined meshes

automatically, the adaptive BEM-based FEM is much more accurate for the same number of unknowns.

In Chap. 3, we have introduced the notion of anisotropic polytopal meshes. Such meshes are especially suited for the approximation of functions with strong layers. In Sect. 3.4.6, an algorithm has been given in order to adapt the mesh to the layers of a function sought to be approximated. In this algorithm, it has been assumed that the function and its derivatives are known. The anisotropic polytopal meshes clearly outperformed the uniformly and adaptively refined meshes with isotropic elements in that case. In this section, we have investigated an adaptive refinement method that does not need the knowledge of the exact solution and the refinement is done fully automatic. So the next step would be to combine the adaptive algorithm driven by a posteriori error estimates with anisotropic mesh refinement for problems containing strong layers in their solutions.

5.3 Goal-Oriented Error Estimation

In the previous section, the adaptive algorithm has been driven by an error indicator penalizing the error measured in the energy norm. In engineering applications, however, this quantity might not be of importance for the considered simulation. In this case, goal-oriented error estimation techniques are advantages that enable adaptive refinements with different emphases. The dual-weighted residual (DWR) method allows for estimating the error $u - u_h$ between the exact solution of the boundary value problem and its Galerkin approximation in terms of a general (error) functionals J. These functionals can be norms but also more general expressions, like point-values, (local) averages or other quantities of interest. Error estimators based on the DWR method always consist of residual evaluations, that are weighted by (local) adjoint sensitivity measures. These sensitivities are the solution to adjoint problems that measure the influence of the error functional J. The DWR technique goes back to [21, 22] and is motivated by [74]. Important further developments in the early stages have been accomplished in [3, 4, 18, 37, 83, 134, 137, 141].

In this section, we restrict ourselves to the two-dimensional case $d = 2$ although the approach is applicable in general dimensions. Thus, let $\Omega \subset \mathbb{R}^2$ be a bounded polygonal domain with boundary $\Gamma = \Gamma_D \cup \Gamma_N$ and $|\Gamma_D| > 0$. Furthermore, we only consider the Poisson problem, i.e. $a \equiv 1$,

$$\begin{aligned} -\Delta u &= f && \text{in } \Omega\,, \\ u &= 0 && \text{on } \Gamma_D\,, \\ \nabla u \cdot \mathbf{n} &= g_N && \text{on } \Gamma_N\,, \end{aligned}$$

with source function f, homogeneous Dirichlet condition and Neumann data g_N for simplicity. This setting is sufficient to highlight the key concepts which can be applied to more general problems.

5.3.1 DWR Method for Linear Goal Functionals

The DWR method is aimed to measure the error in an adaptive algorithm via certain quantities of interest, i.e., goal functionals $J(\cdot)$. Although the theory is applicable for non-linear goal functionals, see [22], we restrict ourselves to the linear case. Such quantities of interest can be mean values of the solution and its derivatives or more involved technical values such as drag or lift in fluid dynamics. These examples include, for instance,

$$J(u) = \int_\Omega u \,\mathrm{d}\mathbf{x}\,, \quad J(u) = \int_\Gamma \nabla u \cdot \mathbf{n} \,\mathrm{d}s\,, \quad J(u) = u(\mathbf{x}^*),\ \mathbf{x}^* \in \Omega\,, \tag{5.7}$$

that are a mean value, a line integral related to the stress values in elasticity and a point value. If the exact solution u is unknown and only its approximation by a discrete function u_h is given, the question arises, whether we can bound the error

$$J(u) - J(u_h)\,.$$

The DWR approach tackles this task by exploiting a dual problem

$$\text{Find } z \in H^1_D(\Omega): \quad b(v, z) = J(v) \quad \forall v \in H^1_D(\Omega)\,, \tag{5.8}$$

where the bilinear form coincides for the Poisson equation with the one in the primal problem. The boundary conditions are of homogeneous Dirichlet and Neumann type. The derivation of the dual (or better 'adjoint') problem follows the Lagrangian formalism that is well-known in optimization. The original motivation is provided in detail in [22]. The solvability and regularity theory for (5.8) follows standard arguments. Thus, we may recognize that the last functional in (5.7), the point evaluation, does not fall into this theory, since it is not defined for functions in $H^1(\Omega)$. Consequently, one may regularize the point evaluation by a convolution with a mollifier, i.e. with an appropriate smooth function having small local support.

Choosing $v = u - u_h$ in (5.8) and applying the Galerkin orthogonality, namely $b(u - u_h, v_h) = 0$ for all $v_h \in V^k_{h,D}$, yields

$$b(u - u_h, z - v_h) = J(u - u_h) .$$

This is a key point in the DWR method. Since v_h is an arbitrary discrete test function, we can, for instance, use an interpolation or projection $v_h = \mathrm{i}_h z$ to obtain an error representation.

Proposition 5.5 *For the Galerkin approximation of the above bilinear form, we have the a posteriori error identity:*

$$J(u - u_h) = b(u - u_h, z - \mathrm{i}_h z) . \tag{5.9}$$

We cannot simply evaluate the error identity because z is only analytically known in very special cases. Consequently, in order to obtain a computable error representation, z is approximated through a discrete function z_h^*, that is, as the primal problem itself, obtained from solving a discretized version of (5.8).

Proposition 5.6 *Let z_h^* be the discrete dual function. For the Galerkin approximation of the above bilinear form, we have the a posteriori error representation*

$$J(u - u_h) \approx b(u - u_h, z_h^* - \mathrm{i}_h z_h^*) .$$

The straight forward choice of $z_h^* = z_h \in V^k_{h,D}$ as solution of

$$b(v_h, z_h) = J(v_h) \quad \forall v_h \in V^k_{h,D}$$

is not applicable. Since $z_h - \mathrm{i}_h z_h \in V^k_{h,D}$ and due to the Galerkin orthogonality this choice yields

$$J(u - u_h) \approx b(u - u_h, z_h - \mathrm{i}_h z_h) = 0 .$$

For the evaluation of the error in the form (5.9), we have to calculate approximations $z_h^* - \mathrm{i}_h z_h^*$ of the interpolation errors $z - \mathrm{i}_h z$. This approximation is the critical part in the DWR framework that limits strict reliability [132]. A remedy is only given by spending sufficient effort on the estimation of these weights on fine meshes [22, 52] or an additional control of the approximation error in $z - \mathfrak{i}_h z$, see [132]. As just mentioned, it is well-known that the discrete approximation of $z - \mathfrak{i}_h z$ must be finer than the trial space for the primal variable as the residual is orthogonal on $V_{h,D}^k$. On triangular and quadrilateral meshes there are basically two main strategies in the literature.

- Global more accurate approximation: The dual problem is either treated on the same mesh with a higher order approximation space or on a finer mesh with the same order approximation space. Both variants are quite expensive [22].
- Local more accurate reconstruction: The primal and dual problem are treated on the same mesh with the same approximation space, but, the dual approximation is post-processed locally using a patch-mesh structure [22]. This is a cheaper alternative, but needs an agglomeration of elements.

Both strategies are applicable on polygonal meshes. However, we even propose a new approach which is based on a local post-processing using a single element. This enables the treatment of the primal and dual problem with the same mesh and approximation space followed by an element-wise higher order reconstruction in order to obtain z_h^*. Detailed explanations of this variant are provided in Sect. 5.3.3.2.

In order to obtain an error estimator, the right hand side of (5.9) is either estimated or approximated by some $\eta(u_h, z)$. The quality of this error estimator with respect to the true error is measured in terms of the effectivity index I_{eff} with

$$I_{\text{eff}}(u_h, z) = \left| \frac{\eta(u_h, z)}{J(u - u_h)} \right| \to 1 \quad \text{for } h \to 0 \,. \tag{5.10}$$

In many applications, the asymptotic sharpness 1 cannot be achieved, but it should be emphasized that even overestimations of a factor 2 or 4 still yield a significant reduction of the computational cost in order to obtain a desired accuracy for the goal functional $J(u)$. The residual based error estimator studied in Sect. 5.2 is known to have a bad effectivity. In the numerical experiments in Sect. 5.2.3 the error estimator η_R overestimated the true error by a factor between 5 and 10 in the problem with singular solution and by a factor between 7 and 22 in the problem with smooth solution containing an internal layer. The DWR method produces sharper estimates.

In the following sections, we explain the realization of the dual-weighted residual method for goal-oriented error estimation on polygonal discretizations. We first introduce special meshes and then recall various strategies to discretize the primal and dual problems. In particular, we introduce an element-based post-processing of the dual solution. The above mentioned error estimator $\eta(u_h, z)$ is the basis for the derivation of a posteriori error estimates. In order to use this formulation from Proposition 5.5 for mesh refinement, we need to localize the error contributions on each element. Therefore, two error representations are finally recapitulated: using the classical method with strong forms of the differential operator, and secondly, using a partition of unity for the variational form.

5.3.2 Preparation for Post-processing: Special Meshes

The local post-processing of the dual solution might be done on coarsened meshes obtained by agglomerating polygonal elements in a classical way. However, on these general meshes we might alternatively use novel kinds of hierarchies. In the later described post-processing of the dual solution, we exploit this possibility. Therefore, we do not allow general polygonal meshes $\mathcal{K}_h$ as described in Sect. 2.2. Here, we restrict ourselves to regular and stable meshes $\mathcal{K}_h$ with polygonal elements having an even number of nodes, such that every second node lies on an straight part of the boundary of the element. Furthermore, we assume that by removing these nodes from the mesh we obtain a coarsened polygonal mesh $\mathcal{K}_{2h}$ which is still regular. In Fig. 5.8, we visualize such meshes $\mathcal{K}_h$ in the middle column and their corresponding coarsened meshes $\mathcal{K}_{2h}$ in the left column. Using these meshes we define the approximation spaces V_h^k and V_{2h}^k, respectively.

The condition on the node count for $\mathcal{K}_h$ is not a real restriction. We can always introduce some additional nodes in the mesh to ensure the requirements. This is also done when we refine some given meshes. The middle column of Fig. 5.8 shows a sequence of uniform refined meshes which are used in later numerical experiments in Sect. 5.3.5. In the refinement procedure each element in the mesh $\mathcal{K}_h$ is bisected as described in Sect. 2.2.3. This yields a mesh which does not satisfy the requirement on the node count for each element in general, see Fig. 5.8 right. However, we can ensure the required structure of the mesh by introducing some additional nodes. This can be observed by comparing the refined, but inappropriate mesh, in the right column of Fig. 5.8 with the next mesh in the sequence depicted in the middle column.

5.3.3 Approximation of the Primal and Dual Solution

The primal and dual problems are approximated on polygonal meshes as described in Chap. 2. For this reason, let $\Omega \subset \mathbb{R}^2$ be a bounded polygonal domain meshed into polygonal elements satisfying the regularity and stability assumption and, in most experiments, the requirement on the node count as described above. The primal variable u is approximated by $u_h \in V_h^k$, which is given by the decoupled weak formulation (2.31) as well as (2.32) and reads in this setting for $u_h = u_{h,H} + u_{h,B}$ with $u_{h,H} \in V_{h,H}^k$ and $u_{h,B} \in V_{h,B}^k$:

$$\begin{aligned} &\text{Find } u_{h,H} \in V_{h,H,D}^k : \\ &b(u_{h,H}, v_h) = (f, v_h)_{L_2(\Omega)} + (g_N, v_h)_{L_2(\Gamma_N)} \quad \forall v_h \in V_{h,H,D}^k , \end{aligned} \tag{5.11}$$

Fig. 5.8 Each row corresponds to the mesh in one FEM simulation, the middle column corresponds to the actual mesh $\mathscr{K}_h$, the left column shows the mesh $\mathscr{K}_{2h}$ after coarsening and the right column shows the mesh after refinement before the nodes are added to ensure the condition on the node count

and

$$\text{Find } u_{h,B} \in V_{h,B}^k : \quad b(u_{h,B}, v_h) = (f, v_h)_{L_2(\Omega)} \quad \forall v_h \in V_{h,B}^k . \tag{5.12}$$

For the approximation of the dual solution z we have a corresponding decoupling and we focus on two strategies. Either we use globally a higher order for the approximation, which is, however, practically expensive, or we apply a local post-processing to $z_h \in V_h^k$. The local post-processing is especially attractive for the approximation space of the BEM-based FEM, since there is no need for local agglomerations of elements as we see in the next sections.

5.3.3.1 Dual Solution with Globally Higher Order Discretization

A brute-force strategy to obtain an approximation of the dual solution, which is suited for error estimation, is to solve the discrete variational formulation with higher accuracy. To track the approximation order, we write $u_h = u_h^{(k)} \in V_h^k$ for the approximation of the primal solution. The dual solution can be approximated by $z_h^{(k+1)} \in V_h^{k+1}$ on the same mesh. The choice $z_h^* = z_h^{(k+1)}$ is applicable for the error representation, cf. (5.9). Here, we do not need the restriction on the node count for the mesh $\mathcal{K}_h$. As we already mentioned, this strategy is computationally expensive in practical applications. However, it serves as a good starting point to verify the performance of the dual-weighted residual method on polygonal meshes.

5.3.3.2 Dual Solution Exploiting Local Post-Processing

A more convenient and efficient strategy is to approximate the dual solution by $z_h = z_h^{(k)} \in V_h^k$ on the same mesh with the same approximation order as the primal solution. Afterwards z_h^* is chosen as a post-processed version of z_h on a coarsened mesh with higher approximation order. This strategy is well discussed in the literature for simplicial meshes, see [143] and the references therein. In fact, this has already been introduced in the early studies [22]. The key point is, how the meshes, and especially the coarse meshes, are chosen. Since polygonal meshes are very flexible and inexpensive for coarsening and refining, they are well suited for this task. It is possible to just agglomerate two or more neighbouring elements to construct a coarsened mesh and to proceed in a classical way for the local post-processing.

In the following we describe a slightly different strategy that does not need the agglomeration of elements and is applicable on single elements. We use the meshes $\mathcal{K}_h$ and $\mathcal{K}_{2h}$ discussed in Sect. 5.3.2 satisfying the requirement on the node count. The approach relies on two key ingredients: the hierarchy of the discretization of the element boundaries ∂K in these two meshes and the decoupling of the dual problem analogously to (5.11) and (5.12) for the primal problem.

Let $z_h \in V_h^k$ be the approximation of the dual problem over the mesh $\mathscr{K}_h$. We construct $z_h^* \in V_{2h}^{k+1}$ as locally post-processed function over the mesh $\mathscr{K}_{2h}$. We write the mapping $z_h = z_h^{(k)} \mapsto z_h^* \in V_{2h}^{k+1}$ also in operator notation with $\mathfrak{P}_{2h}^{k+1} : V_h^k \to V_{2h}^{k+1}$ such that $z_h^* = \mathfrak{P}_{2h}^{k+1} z_h^{(k)}$. It is sufficient to define the post-processing on a single element $K \in \mathscr{K}_h$, since it directly generalizes to the entire mesh. By construction, each element $K = K_h \in \mathscr{K}_h$ has a corresponding element $K_{2h} \in \mathscr{K}_{2h}$, which is obtained by skipping every second node on the boundary ∂K_h. Thus, the shapes of these elements coincide and they only differ in the number of nodes on the boundary. Consequently, ∂K_h can be interpreted as a refinement of ∂K_{2h}, or in other words, ∂K_h and ∂K_{2h} are one-dimensional patched meshes of the element boundary. Therefore, it is $\mathscr{P}_{\text{pw}}^k(\partial K_{2h}) \subset \mathscr{P}_{\text{pw}}^k(\partial K_h)$. In terms of the approximation space we set $V_{2h}^k(K_h) = V_h^k(K_{2h}) \subset V_h^k(K_h)$. Since it is clear from the approximation space which element is meant, we skip the index h and $2h$ again.

Suppose we would approximate the dual problem globally in V_{2h}^{k+1}. Then, the weak formulation decouples into a global system of linear equations in order to compute the expansion coefficients of the harmonic basis functions and into a projection of the error functional into the space of element bubble functions. We similarly proceed with the post-processing. Exploiting the hierarchy of the boundary, we construct $z_h^* = z_{h,H}^* + z_{h,B}^* \in V_{2h}^{k+1}(K) = V_{2h,H}^{k+1}(K) \oplus V_{2h,B}^{k+1}(K)$ from the approximation $z_h = z_{h,H} + z_{h,B} \in V_h^k(K)$ in the following way: We set

$$z_{h,H}^* \in V_{2h,H}^{k+1}(K) \text{ as interpolation of } z_{h,H} \in V_{h,H}^k(K) \tag{5.13}$$

and

$$z_{h,B}^* \in V_{2h,B}^{k+1}(K) \text{ as solution of:} \quad (\nabla z_{h,B}^*, \nabla \varphi)_{L_2(K)} = J(\varphi) \quad \forall \varphi \in V_{2h,B}^{k+1}(K)\,. \tag{5.14}$$

The interpolation process in (5.13) is equivalent to an interpolation of a function in $\mathscr{P}_{\text{pw}}^k(\partial K_h)$ by a function in $\mathscr{P}_{\text{pw}}^{k+1}(\partial K_{2h})$. Thus, a standard point-wise interpolation procedure is applied. The definition of $z_{h,B}^*$ is exactly the projection of the error functional into the space of element bubble functions. Both operations are local over a single element and are thus suited for a computationally inexpensive post-processing.

Remark 5.7 The first idea might be to use the interpolation operator $\mathfrak{I}_h^k$ studied in Sect. 2.4 and to set $z_h^* = \mathfrak{I}_{2h}^{k+1} z_h^{(k)}$. But, this strategy does not work. The interpolation affecting the harmonic basis functions yields the same results as described above. However, the transition from the lower order element bubble functions to the higher order ones is not well suited. Since there is no agglomeration of elements and the process is kept on a single element, there is no additional information in the interpolation using higher order element bubble functions. This is reflected by the fact that $V_{h,B}^k(K) = V_{2h,B}^k(K)$. The choice (5.14) overcomes this deficit and includes the required information for the element bubble functions by exploiting the dual problem.

5.3.4 The Localized Error Estimators

In this section, we discuss the localization of the error representation derived in Sect. 5.3.1 on polygonal meshes. The representation involves the adjoint sensitivity measure $z - \mathfrak{i}_h z$ with $\mathfrak{i}_h : V \to V_h^k$. Since the dual solution is not known in general, it is approximated in the numerical tests as discussed in Sect. 5.3.3. In the realization, we replace z in the error estimates by z_h^*. The operator $\mathfrak{i}_h$ is realized in the following with the help of the interpolation operator $\mathfrak{I}_h^k$, which is given and studied in Sect. 2.4.

5.3.4.1 The Classical Way of Localization

The error identity in Proposition 5.5 is realized in the classical way by using the concrete problem, followed by integration by parts on every mesh element $K \in \mathscr{K}_h$, yielding

$$J(u - u_h) = \sum_{K \in \mathscr{K}_h} \Big(\big(f + \Delta u_h, z - \mathfrak{i}_h z\big)_{L_2(K)} - \big(\gamma_1^K u_h, (z - \mathfrak{i}_h z)\big)_{L_2(\partial K)} \Big) + \big(g_N, (z - \mathfrak{i}_h z)\big)_{L_2(\Gamma_N)} .$$

Following the usual procedure for residual based error estimators as in Sect. 5.2.1, we combine each two boundary integrals over element edges to a normal jump and proceed with the Cauchy–Schwarz inequality to derive an upper bound of the error.

Proposition 5.8 *For the BEM-based FEM approximation of the Poisson equation, we have the a posteriori error estimate based on the classical localization:*

$$|J(u - u_h)| \le \eta^{\mathrm{CL}} = \sum_{K \in \mathscr{K}_h} \eta_K^{\mathrm{CL}} \tag{5.15}$$

with

$$\eta_K^{\mathrm{CL}} = \|R_K\|_{L_2(K)} \|z - \mathfrak{i}_h z\|_{L_2(K)} + \sum_{E \in \mathscr{E}(K)} \|R_E\|_{L_2(E)} \|z - \mathfrak{i}_h z\|_{L_2(E)} , \tag{5.16}$$

where R_K and R_E are the element and edge residuals defined in Sect. 5.2, namely

$$R_K = f + \Delta u_h \quad \textit{for } K \in \mathscr{K}_h ,$$

and

$$R_E = \begin{cases} 0 & \textit{for } E \in \mathscr{E}_{h,D} , \\ g_N - \gamma_1^K u_h & \textit{for } E \in \mathscr{E}_{h,N} \textit{ with } E \in \mathscr{E}(K) , \\ -\frac{1}{2} [\![u_h]\!]_E & \textit{for } E \in \mathscr{E}_{h,\Omega} . \end{cases}$$

According to the definition of the trial space we have $\Delta u_h \in \mathscr{P}^{k-2}(K)$ in each $K \in \mathscr{K}_h$. Since most of the basis functions are harmonic, Δu_h is directly obtained by the expansion coefficients of u_h corresponding to the element bubble functions. The term $\gamma_1^K u_h$ is treated by means of boundary element methods in the realization and therefore it is approximated in $\mathscr{P}^{k-1}_{\mathrm{pw,d}}(\partial K)$.

The local error indicator (5.16) is usually estimated in order to separate it into two parts such that $\eta_K^{\mathrm{CL}} \leq \mathfrak{r}_K(u_h)\mathfrak{w}_K(z)$, see, e.g., [4, 18, 21, 22, 143]. The first part $\mathfrak{r}_K(u_h)$ contains the residual with the discrete solution u_h and the problem data and the second part $\mathfrak{w}_K(z)$ contains the adjoint sensitivity measure $z - \mathfrak{i}_h z$. The separation is obtained by further applications of the Cauchy–Schwarz inequality and reads in our notation

$$|J(u-u_h)| \leq \sum_{K\in\mathscr{K}_h} \underbrace{\left(\|R_K\|_{L_2(K)} + h_K^{-1/2}\|R_E\|_{L_2(\partial K)}\right)}_{=\mathfrak{r}_K(u_h)} \cdot \underbrace{\left(\|z-\mathfrak{i}_h z\|_{L_2(K)} + h_K^{1/2}\|z-\mathfrak{i}_h z\|_{L_2(\partial K)}\right)}_{=\mathfrak{w}_K(z)} . \tag{5.17}$$

In order to incorporate the polygonal structure of the elements and in particular the different numbers and lengths of their edges, we propose to split the L_2-norms over the boundaries of the elements. This refined manipulation yields

$$|J(u-u_h)| \leq \sum_{K\in\mathscr{K}_h} \underbrace{\left(\|R_K\|^2_{L_2(K)} + \sum_{E\in\mathscr{E}(K)} h_E^{-1}\|R_E\|^2_{L_2(E)}\right)^{1/2}}_{=\mathfrak{r}_K(u_h)} \cdot \underbrace{\left(\|z-\mathfrak{i}_h z\|^2_{L_2(K)} + \sum_{E\in\mathscr{E}(K)} h_E\|z-\mathfrak{i}_h z\|^2_{L_2(E)}\right)^{1/2}}_{=\mathfrak{w}_K(z)} . \tag{5.18}$$

The powers of h_K and h_E in (5.17) and (5.18) are chosen in such a way that the volume and boundary terms contribute in the right proportion. This weighting of the norms implicitly makes use of $h_E \sim h_K$, which is guaranteed by the stability of the polygonal meshes. For triangular and quadrilateral meshes the terms h_E and h_K only differ by a small multiplicative factor. (For quadrilaterals it is $h_K = \sqrt{2}h_E$.) In polygonal meshes, however, the ratio $h_K/h_E < c_{\mathscr{K}}$ can be large and it might even blow up in the numerical tests, if the stability is not enforced. Due to these reasons, it seems to be natural to weight directly the volume term $\|R_K\|_{L_2(K)}$ with $\|z-\mathfrak{i}_h z\|_{L_2(K)}$ and the edge term $\|R_E\|_{L_2(E)}$ with $\|(z-\mathfrak{i}_h z)\|_{L_2(E)}$ which gives rise to Proposition 5.8.

5.3.4.2 A Variational Error Estimator with PU Localization

We use a new localization approach [143] based on the variational formulation. Localization is simply based on introducing a partition of unity (PU) into the global error representation Proposition 5.5. In the case of triangular or quadrilateral meshes, the (bi-)linear basis functions are usually utilized, which are associated with nodes. The same is possible for polygonal meshes and the corresponding nodal basis functions, cf. Sect. 2.3.1, which satisfy the partition of unity property. However, this yields a node-wise error indicator, whereas the adaptive refinement is an element-wise procedure. Therefore, we define an element-wise partition of unity in order to obtain directly an element-wise indicator. For this reason, let $\mathfrak{n}(\mathbf{z}) = |\{K \in \mathcal{K}_h : \mathbf{z} \in \mathcal{N}(K)\}|$ be the number of neighbouring elements to the node $\mathbf{z} \in \mathcal{N}_h$. We can write

$$1 = \sum_{\mathbf{z}\in\mathcal{N}_h} \psi_{\mathbf{z}} = \sum_{K\in\mathcal{K}_h} \sum_{\mathbf{z}\in\mathcal{N}(K)} \frac{1}{\mathfrak{n}(\mathbf{z})}\psi_{\mathbf{z}} = \sum_{K\in\mathcal{K}_h} \chi_K \quad \text{on } \Omega\,,$$

and thus obtain a new partition of unity employing the element-wise functions

$$\chi_K = \sum_{\mathbf{z}\in\mathcal{N}(K)} \frac{1}{\mathfrak{n}(\mathbf{z})}\psi_{\mathbf{z}}\,. \tag{5.19}$$

The support of χ_K is local and covers the neighbouring elements of K, namely

$$\operatorname{supp}\chi_K = \overline{\{\mathbf{x} \in K' : K' \in \mathcal{K}_h, \overline{K} \cap \overline{K'} \neq \varnothing\}} = \overline{\omega}_K\,.$$

Inserting the partition of unity into the global error representation Proposition 5.5 yields

$$J(u-u_h) = \sum_{K\in\mathcal{K}_h} b\big(u-u_h, (z-\mathfrak{i}_h z)\chi_K\big)\,.$$

Consequently, when we refer from now on to the PU-based localization technique, we mean the following error representation.

Proposition 5.9 *For the BEM-based FEM approximation of the Poisson equation, we have the element-wise PU-DWR a posteriori error representation and estimate*

$$J(u-u_h) = \eta^{\mathrm{PU}} = \sum_{K\in\mathcal{K}_h} \eta_K^{\mathrm{PU}} \quad \text{and} \quad |J(u-u_h)| \le \eta_{\mathrm{abs}}^{\mathrm{PU}} = \sum_{K\in\mathcal{K}_h} |\eta_K^{\mathrm{PU}}|\,, \tag{5.20}$$

respectively, with

$$\eta_K^{\mathrm{PU}} = \big(f, (z-\mathfrak{i}_h z)\chi_K\big)_{L_2(\Omega)} + \big(g_N, (z-\mathfrak{i}_h z)\chi_K\big)_{L_2(\Gamma_N)} - \big(\nabla u_h, \nabla((z-\mathfrak{i}_h z)\chi_K)\big)_{L_2(\Omega)}\,.$$

We finish this section by a comment on the practical realization. Even if high-order approximations are used for the primal and dual problems, the PU can be realized using a lowest order method involving the nodal basis functions only.

5.3.5 Numerical Tests

In this section, we substantiate our formulations of the dual-weighted residual estimator and the treatment of the dual solution with several different numerical tests and various goal functionals. In the first example, we consider the standard Poisson problem with a regular goal functional. The second example considers a norm-based goal functional. In the third example we study adaptivity in detail. In all examples, we compare the classical and PU localization techniques. Moreover, we compare as previously mentioned different ways to approximate the dual solution.

In analyzing our results, we notice that the tables and graphs are given with respect to the number of degrees of freedom (DoF) in the following. This highlights the fact that the considered sequences of meshes may have the same shapes of elements, but have different numbers of degrees of freedom. This behaviour is due to the mesh requirement for the local post-processing involving additional nodes on the boundaries of the elements. The degrees of freedom are also the usual criterion for adaptive refined meshes.

The adaptive algorithm discussed in Sect. 5.1 has been realized in a slightly adjusted way. In the **SOLVE** step, we additionally have to compute the approximate (higher order) dual solution z_h^*. For the error estimator we now distinguish between $\eta = \sum_K \eta_K$ and $\eta_{\mathrm{abs}} = \sum_K |\eta_K|$ in **ESTIMATE**. Note that $\eta_{\mathrm{abs}}^{\mathrm{CL}} = \eta^{\mathrm{CL}}$ but $\eta_{\mathrm{abs}}^{\mathrm{PU}} \neq \eta^{\mathrm{PU}}$. This also influences the formulation in the marking later on since the error indicators are not squared here. In the **MARK** step, we utilize this time the equidistribution strategy such that all elements K are marked that have values $|\eta_K|$ above the average $\theta\eta_{\mathrm{abs}}/|\mathscr{K}_h|$. Furthermore, we point out that not all theoretical assumptions on the regularity and stability of the mesh from Sect. 2.2 are enforced in **REFINE** for the following tests. During the refinement, the edge lengths may degenerate with respect to the element diameter. If not otherwise stated, all appearing volume integrals are treated by numerical quadrature over polygonal elements as described in Sect. 4.5.1.

Problem 1: Verification in Terms of a Domain Goal Functional
Let $\Omega = (0, 1)^2$. We consider the boundary value problem

$$-\Delta u = 1 \quad \text{in } \Omega\ , \qquad u = 0 \quad \text{on } \Gamma\ ,$$

on two uniform sequences of meshes depicted in Fig. 5.8 (left and middle columns). With a little abuse of notation we denote the sequence of meshes by $\mathscr{K}_{2h}$ and $\mathscr{K}_h$ for the left and middle column in Fig. 5.8, respectively. The goal functional is chosen as

$$J(v) = \int_{\Omega} v\,\mathbf{dx}\,,$$

such that the dual and primal problems coincide. The regularity of the solutions is only limited by the corners of the domain and consequently, it is $u, z \in H^{3-\varepsilon}(\Omega)$ for arbitrary small $\varepsilon > 0$. We use the reference value $J(u) \approx 0.03514425375 \pm 10^{-10}$ taken from [143] for the convergence analysis.

In the first experiment, we compare the different representations of the classical localization technique given in Sect. 5.3.4.1. Here we detect a significant difference depending on the partition into residual terms and dual weights of the classical estimator. The primal solution is approximated in V_h^1 and the dual solution is treated by globally higher order, i.e. $z_h^* = z_h^{(2)}$. For this choice, we do not need the requirement on the node count for the meshes. Therefore, we perform the computations on the mesh sequence $\mathscr{K}_{2h}$ of the unite square Ω. The effectivity index I_{eff} is presented in Table 5.1. For comparisons, we also provide results computed on a sequence of structured meshes with rectangular elements. Obviously, the sharpened estimate (5.18) performs better than the usual form (5.17) of the estimator. We observe, however, that the effectivity index is indeed closest to one for the estimate (5.15) which does not separate the residual part from the sensitivity measure. Therefore, we only apply (5.15) in the following experiments for the classical localization. Furthermore, the comparison with structured meshes indicate that the polygonal shapes of the elements do not influence the effectivity on these uniform refined meshes.

Next, we compare the effectivity index for the PU-based and the classical localization with (5.15). The problems are approximated with $k = 1, 2$. In Table 5.2, we show I_{eff} for the choice $z_h^* = z_h^{(k+1)}$ on a sequence of meshes $\mathscr{K}_{2h}$. The

Table 5.1 Problem 1 approximated with $u_h \in V_h^1$, and dual solution treated by globally higher order, i.e. $z_h^* = z_h^{(2)}$; comparison of effectivity for different representations of the classic localization on a mesh sequence $\mathscr{K}_{2h}$ and on structured meshes

Polygonal-meshes					Quad-meshes		
DoF	$J(u - u_h^{(1)})$	$I_{\text{eff}}^{\text{CL}}$ (5.15)	$I_{\text{eff}}^{\text{CL}}$ (5.17)	$I_{\text{eff}}^{\text{CL}}$ (5.18)	DoF	$J(u - u_h^{(1)})$	$I_{\text{eff}}^{\text{CL}}$ (5.15)
4	5.52×10^{-3}	3.01	6.30	3.31	9	2.56×10^{-3}	2.91
8	3.48×10^{-3}	2.21	6.56	4.03	49	6.51×10^{-4}	2.93
13	4.30×10^{-3}	1.74	4.83	2.59	121	2.90×10^{-4}	2.93
25	2.33×10^{-3}	2.02	5.42	2.98	225	1.64×10^{-4}	2.92
57	1.32×10^{-3}	1.99	5.89	3.37	361	1.05×10^{-4}	2.92
129	5.36×10^{-4}	2.29	6.47	3.64	529	7.27×10^{-5}	2.92
289	2.63×10^{-4}	2.34	6.73	3.84	729	5.35×10^{-5}	2.92
620	1.17×10^{-4}	2.65	7.45	4.28	961	4.09×10^{-5}	2.92
1297	5.67×10^{-5}	2.66	7.48	4.24	1225	3.23×10^{-5}	2.92

Table 5.2 Problem 1 approximated with $u_h \in V_h^k$, $k = 1, 2$ and dual solution treated by globally higher order, i.e. $z_h^* = z_h^{(k+1)}$; comparison of effectivity for PU localization and classical localization with (5.15) on mesh sequence $\mathscr{K}_{2h}$

DoF	$J(u - u_h^{(1)})$	$I_{\text{eff}}^{\text{CL}}$ (5.15)	$I_{\text{eff}}^{\text{PU}}$	DoF	$J(u - u_h^{(2)})$	$I_{\text{eff}}^{\text{CL}}$ (5.15)	$I_{\text{eff}}^{\text{PU}}$
8	3.48×10^{-3}	2.21	1.16	35	1.76×10^{-4}	2.09	1.30
13	4.30×10^{-3}	1.74	0.99	65	7.36×10^{-5}	1.59	1.24
25	2.33×10^{-3}	2.02	1.00	129	1.41×10^{-5}	1.65	1.30
57	1.32×10^{-3}	1.99	1.01	273	4.00×10^{-6}	1.60	1.27
129	5.36×10^{-4}	2.29	1.03	577	7.80×10^{-7}	1.68	1.34
289	2.63×10^{-4}	2.34	1.04	1217	1.93×10^{-7}	1.66	1.34
620	1.17×10^{-4}	2.65	1.07	2519	3.63×10^{-8}	1.80	1.41
1297	5.67×10^{-5}	2.66	1.07	5153	4.95×10^{-9}	3.62	2.90

Table 5.3 Problem 1 approximated with $u_h \in V_h^k$, $k = 1, 2$ and dual solution treated by local post-processing, i.e. $z_h^* = \mathfrak{P}_{2h}^{k+1} z_h^{(k)}$; comparison of effectivity for classical with (5.15) and PU localization on mesh sequence $\mathscr{K}_h$

DoF	$J(u - u_h^{(1)})$	$I_{\text{eff}}^{\text{CL}}$ (5.15)	$I_{\text{eff}}^{\text{PU}}$	DoF	$J(u - u_h^{(2)})$	$I_{\text{eff}}^{\text{CL}}$ (5.15)	$I_{\text{eff}}^{\text{PU}}$
25	3.41×10^{-3}	1.40	0.92	69	1.21×10^{-5}	1.20	0.57
45	1.76×10^{-3}	1.96	0.96	129	1.07×10^{-5}	1.27	0.86
89	9.17×10^{-4}	2.05	0.95	257	8.69×10^{-7}	1.19	0.68
193	4.63×10^{-4}	2.36	0.96	545	6.76×10^{-7}	1.50	1.08
465	2.31×10^{-4}	1.99	0.93	1249	4.36×10^{-8}	1.34	0.79
953	1.14×10^{-4}	2.10	0.95	2545	2.65×10^{-8}	1.56	1.14
2069	5.66×10^{-5}	2.12	0.95	5417	1.50×10^{-9}	2.01	1.39
4269	2.83×10^{-5}	2.09	0.96	11,097	$<10^{-9}$	–	–

effectivity index for the PU localization is close to one whereas the classical localization lacks on effectivity for the first order approximation $k = 1$. For $k = 2$ the effectivity I_{eff}^{CL} is improved.

Furthermore, in Table 5.3, we applied the local post-processing of $z_h^{(k)}$ in order to construct $z_h^* = \mathfrak{P}_{2h}^{k+1} z_h^{(k)}$ and therefore the computations are done on the sequence of meshes $\mathscr{K}_h$, which satisfy the condition on the node count. Although the elements have the same shapes in the sequences of meshes, the number of degrees of freedom is larger in $\mathscr{K}_h$ than in $\mathscr{K}_{2h}$. Both localization strategies show good effectivity in Table 5.3. Due to the local post-processing instead of the globally higher order approximation for the dual solution, the computational cost is significantly reduced compared to the experiments for Table 5.2. We finally remark that for obtaining errors of similar order in the case of $k = 2$, the meshes in Table 5.2 are one times more refined in comparison to the method presented in Table 5.3. However, as just explained, the mesh itself is coarser but the number of degrees of freedom is higher on the other hand when using the local post-processing of $z_h^{(k)}$.

Problem 2: A Norm-Based Goal Functional
In our second example, let again $\Omega = (0,1)^2$. We consider the boundary value problem

$$-\Delta u = f \quad \text{in } \Omega\,, \qquad u = 0 \quad \text{on } \Gamma\,,$$

where f is chosen such that $u(\mathbf{x}) = \sin(\pi x_1)\sin(\pi x_2)$ is the analytical solution. As in the previous problem, we compare the different localization techniques and the two choices of z_h^*. The computations are done solely on the sequence of meshes satisfying the node count condition, which is depicted in Fig. 5.8 in the middle column. The error functional is chosen as

$$J(v) = \frac{(u - u_h, v)_{L_2(\Omega)}}{\|u - u_h\|_{L_2(\Omega)}}\,,$$

such that $J(u-u_h) = \|u-u_h\|_{L_2(\Omega)}$. Our results of the effectivity indices are shown in the Tables 5.4 and 5.5. All indices are close to one and behave similar to those of the previous Problem 1. I_{eff}^{PU} is hardly effected by the different approximations of the dual solution and also the classical localization shows comparable effectivity. Consequently, the computationally less expensive post-processing is to favor over the higher order approximation of the dual solution in practical applications.

Problem 3: Adaptivity
Finally, let $\Omega = (-1,1) \times (-1,1) \setminus [0,1] \times [-1,0]$ be an L-shaped domain and its boundary is split into $\Gamma_D = \{(x_1,x_2) \in \mathbb{R}^2 : x_1 \in [0,1],\ x_2 = 0 \text{ or } x_1 = 0,\ x_2 \in [-1,0]\}$ and $\Gamma_N = \partial\Omega \setminus \Gamma_D$. We consider the mixed boundary value problem

$$-\Delta u = 0 \quad \text{in } \Omega\,, \qquad u = 0 \quad \text{on } \Gamma_D\,, \qquad \nabla u \cdot \mathbf{n} = g_N \quad \text{on } \Gamma_N\,,$$

Table 5.4 Problem 2 approximated with $u_h \in V_h^k$, $k = 1, 2$ and dual solution treated by globally higher order, i.e. $z_h^* = z_h^{(k+1)}$; comparison of effectivity for PU localization and classical localization with (5.15) on mesh sequence $\mathscr{K}_h$

DoF	$J(u - u_h^{(1)})$	$I_{\text{eff}}^{\text{CL}}$ (5.15)	$I_{\text{eff}}^{\text{PU}}$	DoF	$J(u - u_h^{(2)})$	$I_{\text{eff}}^{\text{CL}}$ (5.15)	$I_{\text{eff}}^{\text{PU}}$
25	3.80×10^{-2}	1.76	0.92	69	5.64×10^{-3}	1.38	0.95
45	2.10×10^{-2}	1.99	0.98	129	2.90×10^{-3}	1.26	0.95
89	1.05×10^{-2}	1.91	0.81	257	8.48×10^{-4}	1.27	0.97
193	5.34×10^{-3}	2.05	0.83	545	3.57×10^{-4}	1.29	0.96
465	2.59×10^{-3}	1.98	0.82	1249	1.17×10^{-4}	1.46	0.93
953	1.35×10^{-3}	2.06	0.83	2545	4.04×10^{-5}	1.35	0.98
2069	6.75×10^{-4}	2.11	0.82	5417	1.59×10^{-5}	1.37	0.99
4269	3.38×10^{-4}	2.04	0.84	11,097	5.26×10^{-6}	1.36	1.05

Table 5.5 Problem 2 approximated with $u_h \in V_h^k$, $k = 1, 2$ and dual solution treated by local post-processing, i.e. $z_h^* = \mathfrak{P}_{2h}^{k+1} z_h^{(k)}$; comparison of effectivity for classical with (5.15) and PU localization on mesh sequence $\mathscr{K}_h$

DoF	$J(u - u_h^{(1)})$	$I_{\text{eff}}^{\text{CL}}$ (5.15)	$I_{\text{eff}}^{\text{PU}}$	DoF	$J(u - u_h^{(2)})$	$I_{\text{eff}}^{\text{CL}}$ (5.15)	$I_{\text{eff}}^{\text{PU}}$
25	3.80×10^{-2}	1.65	0.83	69	5.64×10^{-3}	1.26	0.95
45	2.10×10^{-2}	1.79	0.86	129	2.90×10^{-3}	1.28	0.96
89	1.05×10^{-2}	2.29	0.84	257	8.48×10^{-4}	1.28	0.97
193	5.34×10^{-3}	2.16	0.80	545	3.57×10^{-4}	1.30	0.95
465	2.59×10^{-3}	2.24	0.82	1249	1.17×10^{-4}	1.29	0.89
953	1.35×10^{-3}	2.20	0.82	2545	4.04×10^{-5}	1.32	0.97
2069	6.75×10^{-4}	2.25	0.82	5417	1.59×10^{-5}	1.33	0.97
4269	3.38×10^{-4}	2.19	0.82	11,097	5.26×10^{-6}	1.34	1.01

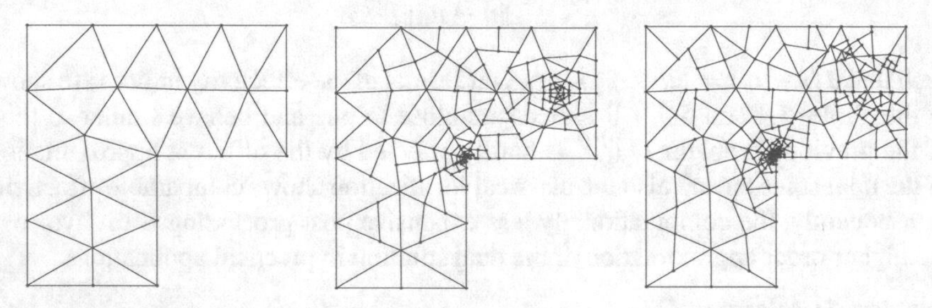

Fig. 5.9 Initial mesh of the L-shaped domain in Problem 3 with triangular elements (left) and adaptive meshes for $k = 2$ after 10 refinements for classical (middle) and PU (right) localization, where the dual problem is treated by globally higher order, i.e. $z_h^* = z_h^{(k+1)}$

where g_N is chosen with the help of polar coordinates (r, ϕ), such that

$$u(r\cos\phi, r\sin\phi) = r^{2/3}\sin\left(\tfrac{2}{3}\phi\right)$$

is the exact solution. This is a classical problem for mesh adaptivity, since the gradient of the solution inherits a singularity at the reentrant corner in the origin of the coordinate system. It holds $u \in H^{5/3}(\Omega)$. The considered goal functional is a point evaluation

$$J(v) = v(\mathbf{x}^*) ,$$

where $\mathbf{x}^*$ is chosen as the upper right node inside the domain, which is adjacent to six elements of the initial mesh, see Fig. 5.9 (left). We apply the adaptive strategy and compare the resulting meshes for the different localization techniques and approximations of the dual solution.

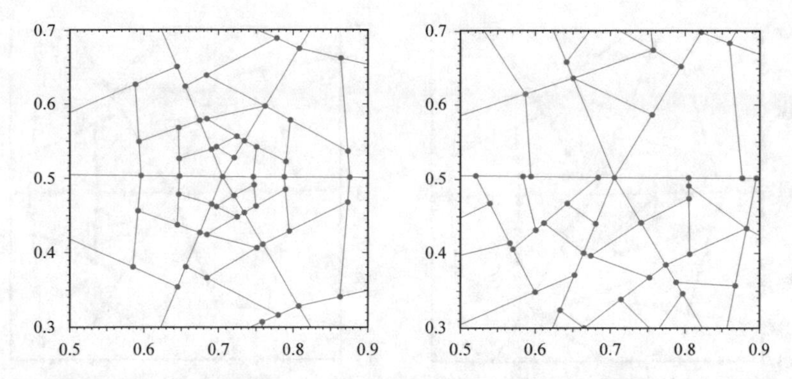

Fig. 5.10 Zoom into L-shaped domain in Problem 3 with adaptive meshes for $k = 2$ after 10 refinements for classical (left) and PU (right) localization, where the dual problem is treated by globally higher order, i.e. $z_h^* = z_h^{(k+1)}$

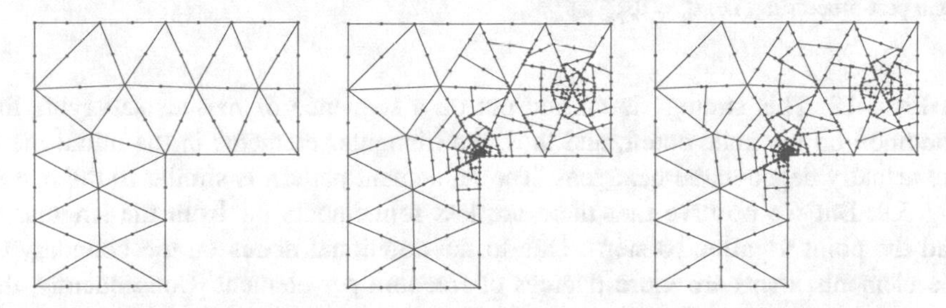

Fig. 5.11 Initial mesh of the L-shaped domain in Problem 3 with triangular elements (left), which are actually degenerated hexagons, and adaptive meshes for $k = 2$ after 10 refinements for classical (middle) and PU (right) localization, where the dual problem is treated by local post-processing, i.e. $z_h^* = \mathfrak{P}_{2h}^{k+1} z_h^{(k)}$

In Fig. 5.9, we display the initial mesh and the adaptively refined meshes for $k = 2$ after 10 refinement steps for the classical and the PU localization. A zoom-in highlighting the resulting shapes of adaptively refined elements is provided in Fig. 5.10. The dual problem is treated by a globally higher order discretization, i.e. $z_h^* = z_h^{(k+1)}$. This experiment has been carried out on sequences of meshes, which do not satisfy the condition on the node count. The elements in the initial mesh are triangles. The adaptive process, however, produces naturally polygonal elements during the local refinements. These refinements are located in the expected regions.

The resulting meshes for the experiments with local post-processing for the dual solution, i.e. $z_h^* = \mathfrak{P}_{2h}^{k+1} z_h^{(k)}$, are visualized in Fig. 5.11. As before, a zoom-in highlighting the resulting shapes of adaptively refined elements is provided

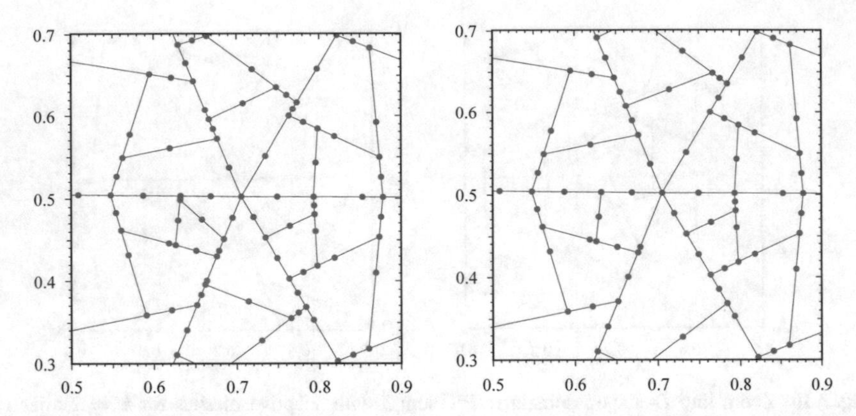

Fig. 5.12 Zoom into L-shaped domain in Problem 3 with adaptive meshes for $k = 2$ after 10 refinements for classical (left) and PU (right) localization, where the dual problem is treated by local post-processing, i.e. $z_h^* = \mathfrak{P}_{2h}^{k+1} z_h^{(k)}$

in Fig. 5.12. This strategy is carried out on a sequence of meshes satisfying the condition on the node count, and thus, the triangular elements in the initial mesh are actually degenerated hexagons. The refinement pattern is similar to the one in Fig. 5.9. But we observe that there are less refinements far from the singularity and the point $\mathbf{x}^*$ after 10 steps. Due to the additional nodes on the boundary of the elements, there are more degrees of freedom per element. Consequently, the approximation over the degenerated hexagonal elements (with triangular shape) is more accurate compared to the corresponding triangular elements in Fig. 5.9.

In order to study convergence, we plot the absolute values of the errors and the estimators with respect to the number of degrees of freedom on a logarithmic scale. The abbreviation $e = u - u_h$ is used in the key of the plots. If we run the computations on a sequence of uniform refined meshes, the convergence slows down due to the singularity located at the reentrant corner. The tests are performed on a uniform sequence $\mathcal{K}_{2h}$, which does not satisfy the condition on the node count, and on a uniform sequence $\mathcal{K}_h$, which satisfies this condition. The initial meshes are visualized in Figs. 5.9 and 5.11, respectively. The corresponding convergence graphs are given in Fig. 5.13 for $k = 1, 2$. In these graphs, the error estimator η^{PU} is given additionally, which clearly reflects the behaviour of the true error $J(e)$.

Next, we apply the adaptive refinement strategy. The following computations are run on meshes satisfying the condition on the node count only. We have performed 25 adaptive refinement steps for the different localization techniques and the two choices of z_h^*. Since $f = 0$ in this test, we directly obtain from (5.12) that $u_{h,B} = 0$ and thus $u_h = u_{h,H} \in V_{h,H}^k$. Consequently, we can reduce the

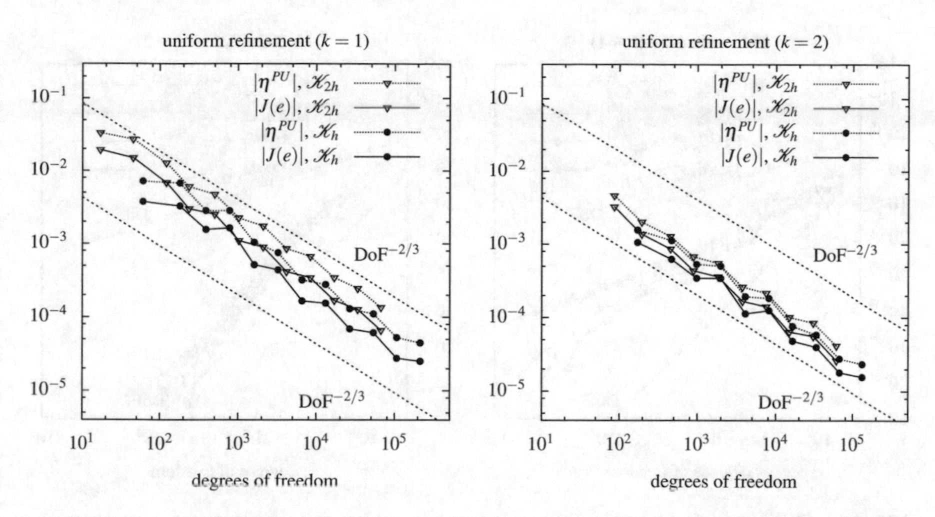

Fig. 5.13 Convergence of uniform refinement strategy with respect to the number of degrees of freedom for Problem 3 with PU localization and $z_h^* = z_h^{(k+1)}$

volume integral in η^{PU} to the boundaries of the elements. Let $K' \in \mathcal{K}_h$ with $K' \subset \omega_K = \operatorname{supp} \chi_K$, it is

$$\left(\nabla u_h, \nabla((z - i_h z)\chi_K)\right)_{L_2(K')} = \left(\gamma_1^{K'} u_h, \gamma_0^{K'}((z - i_h z)\chi_K)\right)_{L_2(\partial K')}$$

according to Green's first identity. This reformulation improved the accuracy of the numerical results. The convergence graphs are given in Fig. 5.14 for the PU localization and in Fig. 5.15 for the classical localization stated in Prop. 5.8. In contrast to the uniform refinement strategy, we recover higher convergence rates, which are not limited by the regularity of the primal solution. Both localization techniques show comparable performance in Figs. 5.14 and 5.15, respectively. The PU localization, however, has a better effectivity while less computational effort is spent for the dual problem. Furthermore, we point out that the convergence is actually faster than expected. Indeed for finite elements, L_∞ regularity results for irregular meshes have been established in [153] and further references to regular meshes are cited therein. In particular, assuming enough regularity, we would expect for $k = 2$ a behaviour like $\mathcal{O}(\mathrm{DoF}^{-3/2})$. For $k = 1$ we would expect $\mathcal{O}(\mathrm{DoF}^{-1})$ including a logarithm term [153]. However in our computations, we observe for $k = 2$ a behaviour like $\mathcal{O}(\mathrm{DoF}^{-3})$. For $k = 1$ the error $J(e)$ seems to converge with $\mathcal{O}(\mathrm{DoF}^{-2})$ rather than with $\mathcal{O}(\mathrm{DoF}^{-1})$ indicated by the estimators η^{PU} and η^{CL}. These effects might be caused by the special meshes, which include additional

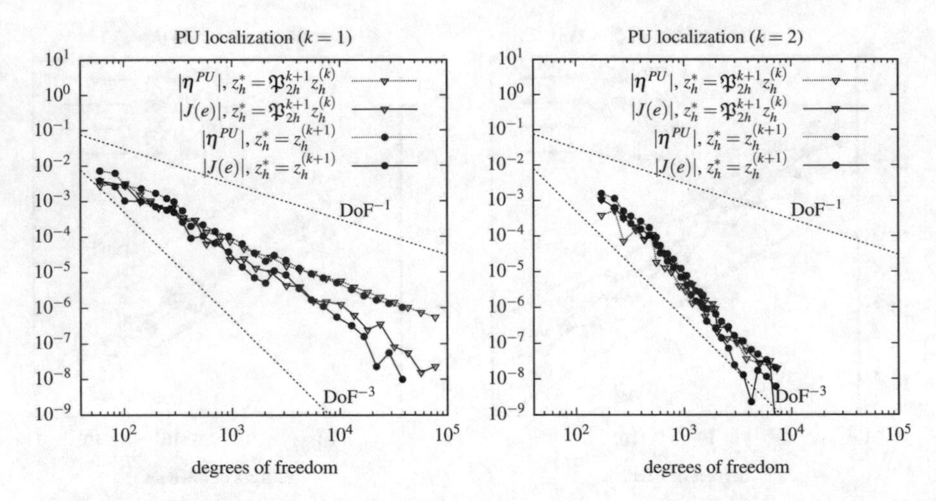

Fig. 5.14 Convergence of adaptive refinement strategy with respect to the number of degrees of freedom for Problem 3 with PU localization

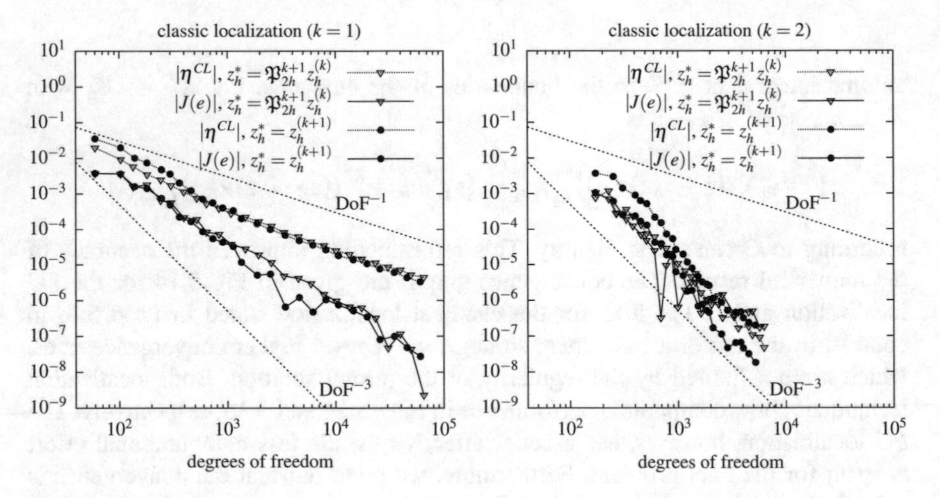

Fig. 5.15 Convergence of adaptive refinement strategy with respect to the number of degrees of freedom for Problem 3 with classic localization

nodes in order to satisfy the condition on the node count during the refinement. Furthermore, the implementation allows edge degeneration, that is excluded in the current theory of most polygonal discretization techniques, but which might be beneficial. These observations rise open questions for future research.

Chapter 6
Developments of Mixed and Problem-Adapted BEM-Based FEM

In the final chapter some extensions and improvements of the BEM-based FEM are discussed which have not been addressed so far. In particular, the focus lies on two topics: The use of the method within mixed finite element formulations and the generalization of the construction of basis functions to polyhedral elements with polygonal faces in 3D with an application to convection-dominated problems.

The challenge in the treatment of mixed formulations is the proper construction of a $\mathbf{H}(\mathrm{div})$-conforming, vector valued approximation space over polytopal discretizations. In contrast to the previous definitions of basis functions, the construction involves local Neumann problems, which are treated in the numerical realization by appropriate boundary element methods.

The forthcoming generalization to 3D gives a H^1-conforming discretization once more which makes use of a hierarchical construction of basis functions. This adapted construction shows in particular advantageous properties when applied to convection-diffusion-reaction problems in the convection-dominated regime. The experiments indicate an improved resolution of exponential layers at outflow boundaries for the proposed approach when compared to the Streamline Upwind/Petrov-Galerkin (SUPG) method.

6.1 Mixed Formulations Treated by Means of BEM-Based FEM

Mixed finite element methods have been instrumental in the development of flexible and accurate approximations of elliptic problems with heterogeneous coefficient on triangular and rectangular grids. The flexibility can even be improved when using polygonal and polyhedral meshes. Such general cells are very desirable in many applications, e.g. flows in heterogeneous porous media as models in hydrology and

© Springer Nature Switzerland AG 2019

S. Weißer, *BEM-based Finite Element Approaches on Polytopal Meshes*,
Lecture Notes in Computational Science and Engineering 130,
https://doi.org/10.1007/978-3-030-20961-2_6

reservoir simulation. Therefore, a variety of approximation and solution methods on general grids, such as mixed finite element methods [120], mimetic finite difference methods [27] and the virtual element methods [29, 30, 42], have been considered, studied, and tested in the last decade. This issue has also been addressed for generalized barycentric coordinates, see [85, 166].

The goal of this section is to introduce a mixed formulation for the BEM-based FEM which has been proposed in [73]. The key idea is to construct a finite dimensional approximation space by implicitly defined basis functions which satisfy certain Neumann boundary value problems on a local, element-by-element-wise level. These problems are treated once more by means of boundary integral formulations which are discretized by boundary element methods.

Since these ideas are applied to the mixed formulation of the problem, we need a suitable discretization of the vector valued Sobolev space

$$\mathbf{H}(\mathrm{div}, \Omega) = \{\mathbf{v} \in \mathbf{L}_2(\Omega) : \mathrm{div}\, \mathbf{v} \in L_2(\Omega)\}$$

on polytopal meshes. This is done by implicitly generating trial functions. A construction of suitable trial function for the mixed FEM on polygonal meshes was done by Kuznetsov and Repin in [120] by using subdivision of the polygonal cell into triangular elements and subsequently generating the test functions locally by mixed FEM. Also similar ideas were implemented in the mixed multiscale finite element method [56, 72]. The novelty in our approach is that instead of treating the local problem by the classical mixed FEM (as in [120]) or by the multiscale FEM (as in [56]) the local problems are treated by means of boundary element methods. Thus, we avoid an additional triangulation of the elements.

6.1.1 Mixed Formulation

We consider the classical model problem of Darcy flow in a porous medium in two-dimensions. Let $\Omega \subset \mathbb{R}^2$ be a convex polygonal domain which is bounded, and let $\mathbf{n}$ be the outer unit normal vector to its boundary $\Gamma = \partial\Omega$. The boundary $\Gamma = \Gamma_D \cup \Gamma_N$ is divided into Γ_D (with non vanishing length) and Γ_N, where Dirichlet and Neumann data is prescribed, respectively. For a given source function $f \in L_2(\Omega)$ and Dirichlet data $g_D \in H^{1/2}(\Gamma_D)$, the boundary value problem for the pressure variable $p \in H^1(\Omega)$ reads

$$\begin{aligned} -\,\mathrm{div}(A\nabla p) &= f \qquad \text{in } \Omega\,, \\ \mathbf{n}\cdot A\nabla p &= 0 \qquad \text{on } \Gamma_N\,, \\ p &= g_D \qquad \text{on } \Gamma_D\,, \end{aligned} \tag{6.1}$$

where the tensor $A \in L_\infty(\Omega)$ represents the permeability of the medium. We assume that $A(\cdot) \in \mathbb{R}^{2\times 2}$ is symmetric, positive definite with

$$0 < a_{\min} \le \frac{\mathbf{v}^\top A(\mathbf{x})\mathbf{v}}{\mathbf{v}^\top \mathbf{v}} \le a_{\max} \quad \forall \mathbf{v} \in \mathbb{R}^2 \setminus \{0\} \quad \text{for almost all } \mathbf{x} \in \Omega$$

for constants $a_{\min}$ and $a_{\max}$, and piecewise constant with respect to the polygonal mesh later on. Vector valued Lebesgue and Sobolev spaces are indicated by bold letters. We further assume that every interior angle at any transient point between the boundary Γ_D and Γ_N is less than π, so that the solution of (6.1) with $A = I$, $f = 0$ and $g_D = 0$ is in the space $H^s(\Omega)$, $s > \frac{3}{2}$, see [87].

Next, a new unknown flux variable $\mathbf{u} = A\nabla p$ is introduced and the boundary value problem is presented as a system of first order differential equations:

$$\begin{aligned} -\operatorname{div} \mathbf{u} &= f && \text{in } \Omega\,, \\ A\nabla p &= \mathbf{u} && \text{in } \Omega\,, \\ \mathbf{n}\cdot\mathbf{u} &= 0 && \text{on } \Gamma_N\,, \\ p &= g_D && \text{on } \Gamma_D\,. \end{aligned} \tag{6.2}$$

This yields the following variational formulation in mixed form, which is actually a saddle point problem:

$$\begin{aligned} &\text{Find } (\mathbf{u}, p) \in \mathbf{H}_N(\operatorname{div}, \Omega) \times L_2(\Omega): \\ &a(\mathbf{u}, \mathbf{v}) + b(\mathbf{v}, p) = (\mathbf{n}\cdot\mathbf{v}, g_D)_{L_2(\Gamma_D)} \quad \forall \mathbf{v} \in \mathbf{H}_N(\operatorname{div}, \Omega)\,, \\ &\qquad\qquad b(\mathbf{u}, q) = -(f, q)_{L_2(\Omega)} \qquad \forall q \in L_2(\Omega)\,, \end{aligned} \tag{6.3}$$

where

$$a(\mathbf{u}, \mathbf{v}) = (A^{-1}\mathbf{u}, \mathbf{v})_{\mathbf{L}_2(\Omega)}\,, \qquad b(\mathbf{v}, q) = (\operatorname{div} \mathbf{v}, q)_{L_2(\Omega)}$$

and

$$\mathbf{H}_N(\operatorname{div}, \Omega) = \{\mathbf{v} \in \mathbf{L}_2(\Omega) : \operatorname{div} \mathbf{v} \in L_2(\Omega) \text{ and } \mathbf{n}\cdot\mathbf{v} = 0 \text{ on } \Gamma_N\}\,.$$

The space $\mathbf{H}(\operatorname{div}, \Omega)$ is equipped with the norm

$$\|\mathbf{v}\|^2_{\mathbf{H}(\operatorname{div},\Omega)} = \|\mathbf{v}\|^2_{\mathbf{L}_2(\Omega)} + \|\operatorname{div} \mathbf{v}\|^2_{L_2(\Omega)}\,.$$

It is easily seen that the bilinear forms $a(\cdot,\cdot)$ and $b(\cdot,\cdot)$ are bounded, i.e.

$$\begin{aligned} |a(\mathbf{u}, \mathbf{v})| &\le \varrho_1 \|\mathbf{u}\|_{\mathbf{H}(\operatorname{div},\Omega)} \|\mathbf{v}\|_{\mathbf{H}(\operatorname{div},\Omega)} && \text{for } \mathbf{u}, \mathbf{v} \in \mathbf{H}(\operatorname{div}, \Omega)\,, \\ |b(\mathbf{v}, q)| &\le \varrho_2 \|\mathbf{v}\|_{\mathbf{H}(\operatorname{div},\Omega)} \|q\|_{L_2(\Omega)} && \text{for } \mathbf{v} \in \mathbf{H}(\operatorname{div}, \Omega), q \in L_2(\Omega)\,, \end{aligned}$$

with some constants $\rho_1, \rho_2 > 0$. Let us set

$$\mathbf{Z} = \{\mathbf{v} \in \mathbf{H}_N(\mathrm{div}, \Omega) : b(\mathbf{v}, q) = 0 \;\; \forall q \in L_2(\Omega)\} .$$

Obviously, we have for $\mathbf{v} \in \mathbf{Z}$ that $\mathrm{div}\, \mathbf{v} = 0$ and hence, the bilinear form $a(\cdot,\cdot)$ is $\mathbf{Z}$-elliptic, i.e., there exists a constant $\alpha > 0$ such that

$$a(\mathbf{v}, \mathbf{v}) \geq \alpha \|\mathbf{v}\|^2_{\mathbf{H}(\mathrm{div},\Omega)} \quad \text{for } \mathbf{v} \in \mathbf{Z} .$$

Furthermore, the form $b(\cdot,\cdot)$ satisfies the so called inf-sup condition, i.e., there exists another constant $\beta > 0$ such that

$$\inf_{q \in L_2(\Omega)} \sup_{\mathbf{v} \in \mathbf{H}_N(\mathrm{div},\Omega)} \frac{b(\mathbf{v}, q)}{\|\mathbf{v}\|_{\mathbf{H}(\mathrm{div},\Omega)} \|q\|_{L_2(\Omega)}} \geq \beta .$$

Consequently, the Babuska–Brezzi theory [43] is applicable and thus, the saddle point problem (6.3) has a unique solution.

Next, we discuss the approximation of the mixed variational formulation (6.3) with the help of BEM-based FEM on polygonal meshes. Therefore, we first need to introduce a $\mathbf{H}(\mathrm{div})$-conforming approximation space.

6.1.2 H(div)-*Conforming Approximation Space*

The construction of an approximation space for $L_2(\Omega)$ is rather easy, later on we use

$$M_h = \{q \in L_2(\Omega) : q|_K = const \;\; \forall K \in \mathcal{K}_h\} \tag{6.4}$$

for this purpose. We concentrate in this section on the definition of a conforming approximation space for $\mathbf{H}(\mathrm{div}, \Omega)$. We consider a regular and stable polygonal mesh $\mathcal{K}_h$ according to Sect. 2.2. The finite dimensional subspace of $\mathbf{H}(\mathrm{div}, \Omega)$ that serves as approximation space is defined through its basis. We restrict ourselves to the lowest order method in which the basis functions are associated with edges only. For $E \in \mathcal{E}_h$, let $\mathbf{n}_E$ be a unit normal vector, which is considered to be fixed in the sequel. Furthermore, let K_1 and K_2 be the two adjacent elements sharing the common edge E with the outer normal vectors $\mathbf{n}_{K_1}$ and $\mathbf{n}_{K_2}$, respectively. The function ϕ_E is defined implicitly as solution of the following local boundary value problem

$$\begin{aligned} \mathrm{div}(A\nabla\phi_E) &= \kappa_E(K)/|K| \quad \text{in } K \in \{K_1, K_2\} , \\ \mathbf{n}_E \cdot A\nabla\phi_E &= \begin{cases} h_E^{-1} & \text{on } E , \\ 0 & \text{on all other edges ,} \end{cases} \end{aligned} \tag{6.5}$$

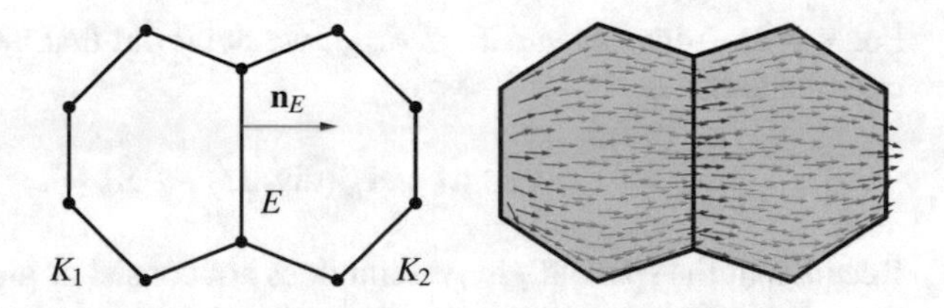

Fig. 6.1 Adjacent elements to E for the definition of ϕ_E (left) and vector field ψ_E (right)

see Fig. 6.1. Here, $\kappa_E(K) = \mathbf{n}_E \cdot \mathbf{n}_K = \pm 1$, such that the solvability condition for the Neumann problem is satisfied and (6.5) has a weak solution $\phi_E \in H^1(\Omega)$ which is unique up to an additive constant. For $E \in \mathscr{E}(K_1) \cap \mathscr{E}(K_2)$, we define

$$\psi_E(\mathbf{x}) = \begin{cases} A\nabla\phi_E(\mathbf{x}) & \text{for } \mathbf{x} \in \overline{K}_1 \cup \overline{K}_2 \,, \\ 0 & \text{else} \,. \end{cases} \tag{6.6}$$

Due to this definition one easily concludes that

$$\|\psi_E\|_{\mathbf{L}_2(K_1 \cup K_2)} = \|\nabla\phi_E\|_{\mathbf{L}_2(K_1 \cup K_2)} \leq c \,, \tag{6.7}$$

cf. also [51]. By construction, ψ_E has continuous normal flux across E and zero normal flux along all other internal edges of Ω so that $\psi_E \in \mathbf{H}(\mathrm{div}, \Omega)$. An edge $E \subset \Gamma_D$ has only one neighbouring element K, and therefore the basis function is constructed in the same way by considering problem (6.5) solely on K.

We set the finite dimensional approximation space as

$$\mathbf{X}_h = \operatorname{span}\{\psi_E : E \in \mathscr{E}_h\} \subset \mathbf{H}(\mathrm{div}, \Omega) \,,$$

and the subspace with vanishing normal traces on Γ_N as

$$\mathbf{X}_{h,N} = \operatorname{span}\{\psi_E : E \in \mathscr{E}_h \setminus \mathscr{E}_{h,N}\} \subset \mathbf{H}_N(\mathrm{div}, \Omega) \,. \tag{6.8}$$

The corresponding vector valued interpolation operator

$$\pi_h : \mathbf{H}(\mathrm{div}, \Omega) \to \mathbf{X}_h$$

is defined by

$$\pi_h \mathbf{v} = \sum_{E \in \mathscr{E}_h} v_E \, \psi_E \,, \tag{6.9}$$

where

$$v_E = \int_E \mathbf{n}_E \cdot \mathbf{v} \, \mathrm{d}s_{\mathbf{x}} \quad \text{for } E \in \mathscr{E}_h \,.$$

For $\mathbf{v} \in \mathbf{H}_N(\mathrm{div}, \Omega)$ and $E \in \mathcal{E}_{h,N}$, we point out that $v_E = 0$. Consequently, the operator satisfies

$$\pi_h : \mathbf{H}_N(\mathrm{div}, \Omega) \to \mathbf{X}_{h,N} .$$

Recall, that the space $\mathbf{X}_h$ in general does not consist of piecewise polynomial functions. The approximation properties of the interpolation operator π_h are established below. First of all, we have the boundedness of this operator.

Lemma 6.1 *Let $\mathcal{K}_h$ be a regular and stable polygonal mesh. The interpolation operator $\pi_h : \mathbf{H}^s(\Omega) \to \mathbf{X}_h$, $s > \frac{1}{2}$, defined by (6.9) is bounded in $\mathbf{H}^s(\Omega)$. Namely, there is a constant $c > 0$ independent of $h = \max\{h_K : K \in \mathcal{K}_h\}$ such that*

$$\|\pi_h \mathbf{v}\|_{\mathbf{L}_2(\Omega)} \le c\|\mathbf{v}\|_{\mathbf{H}^s(\Omega)} \quad \text{for } \mathbf{v} \in \mathbf{H}^s(\Omega) . \tag{6.10}$$

For the restriction of the interpolation operator onto an element $K \in \mathcal{K}_h$ it holds

$$\|\pi_h \mathbf{v}\|_{\mathbf{L}_2(K)} \le c\|\mathbf{v}\|_{\mathbf{H}^s(K)} \quad \text{for } \mathbf{v} \in \mathbf{H}^s(K) .$$

Proof Since π_h is defined locally, it is enough to show that this estimate is valid over each element $K \in \mathcal{K}_h$. Obviously, it holds

$$\pi_h \mathbf{v}|_K = \sum_{E \in \mathcal{E}(K)} v_E \psi_E|_K , \quad v_E = \int_E \mathbf{n}_E \cdot \mathbf{v} \, \mathrm{d}s_{\mathbf{x}} ,$$

and we have

$$\|\pi_h \mathbf{v}\|^2_{\mathbf{L}_2(K)} \le c \sum_{E \in \mathcal{E}(K)} v_E^2 \, \|\psi_E\|^2_{\mathbf{L}_2(K)}$$

with a constant c depending on the number of edges $|\mathcal{E}(K)|$, which is uniformly bounded over all elements due to the stability of the mesh, see Lemma 2.7. By (6.7) we have $\|\psi_E\|_{\mathbf{L}_2(K)} \le c$ and to conclude the proof we need to bound v_E.

We rescale the finite element K to $\widehat{K}$ by using the mapping $\mathbf{x} \mapsto \widehat{\mathbf{x}} = h_K^{-1}\mathbf{x}$, cf. (2.22). Then using the trace inequality [87]

$$\|\widehat{\mathbf{w}}\|_{\mathbf{L}_2(\widehat{E})} \le c\big(\|\widehat{\mathbf{w}}\|_{\mathbf{L}_2(\widehat{K})} + |\widehat{\mathbf{w}}|_{\mathbf{H}^s(\widehat{K})}\big) \quad \text{for } \widehat{\mathbf{w}} \in \mathbf{H}^s(\widehat{K}) , \ s > \tfrac{1}{2}$$

on the scaled element, where $\widehat{E}$ denotes an edge of $\widehat{K}$, we get

$$\begin{aligned} v_E^2 &\le h_E \int_E |\mathbf{n}_E \cdot \mathbf{v}|^2 \, \mathrm{d}s_{\mathbf{x}} \le h_E \int_E |\mathbf{v}|^2 \, \mathrm{d}s_{\mathbf{x}} \le h_E h_K \|\widehat{\mathbf{v}}\|^2_{\mathbf{L}_2(\widehat{E})} \\ &\le c h_E h_K \Big(\|\widehat{\mathbf{v}}\|_{\mathbf{L}_2(\widehat{K})} + |\widehat{\mathbf{v}}|_{\mathbf{H}^s(\widehat{K})} \Big)^2 \le c \Big(\|\mathbf{v}\|^2_{\mathbf{L}_2(K)} + h_K^{2s} \, |\mathbf{v}|^2_{\mathbf{H}^s(K)} \Big) \quad (6.11) \\ &\le c \, \|\mathbf{v}\|^2_{\mathbf{H}^s(K)} \, , \end{aligned}$$

since $h_K \le 1$. Thus, $\|\pi_h \mathbf{v}\|^2_{\mathbf{L}_2(K)} \le c \, \|\mathbf{v}\|^2_{\mathbf{H}^s(K)}$ and after summing for $K \in \mathscr{K}_h$ we get the desired bound. □

Next, we discuss the approximation properties of the interpolation operator π_h.

Lemma 6.2 *Let $\mathscr{K}_h$ be a regular and stable mesh and $\mathbf{v} \in \mathbf{H}^s(\Omega)$, $\frac{1}{2} < s \le 1$. It holds*

$$\|\mathbf{v} - \pi_h \mathbf{v}\|_{\mathbf{H}(\mathrm{div},\Omega)} \le c h^s \, |\mathbf{v}|_{\mathbf{H}^s(\Omega)} + \inf_{q_h \in M_h} \| \operatorname{div} \mathbf{v} - q_h \|_{L_2(\Omega)}$$

with $h = \max\{h_K : K \in \mathscr{K}_h\}$.

Proof On $E \in \mathscr{E}_h$ the interpolant $\pi_h \mathbf{v}$ satisfies

$$\mathbf{n}_E \cdot \pi_h \mathbf{v}\big|_E = h_E^{-1} \int_E \mathbf{n}_E \cdot \mathbf{v} \, \mathrm{d}s_{\mathbf{x}} \, ,$$

and since $\mathbf{n}_K = \kappa_E(K) \mathbf{n}_E$ for $E \in \mathscr{E}(K)$, we have according to the divergence theorem

$$\int_K \operatorname{div} \pi_h \mathbf{v} \, \mathrm{d}\mathbf{x} = \int_{\partial K} \mathbf{n}_K \cdot \pi_h \mathbf{v} \, \mathrm{d}s_{\mathbf{x}} = \int_{\partial K} \mathbf{n}_K \cdot \mathbf{v} \, \mathrm{d}s_{\mathbf{x}} = \int_K \operatorname{div} \mathbf{v} \, \mathrm{d}\mathbf{x} \, .$$

Hence, div $\pi_h \mathbf{v}$ is the L_2-projection of div $\mathbf{v}$ into M_h. Therefore, it is

$$\| \operatorname{div} \mathbf{v} - \operatorname{div} \pi_h \mathbf{v} \|_{L_2(\Omega)} = \inf_{q_h \in M_h} \| \operatorname{div} \mathbf{v} - q_h \|_{L_2(\Omega)} \, ,$$

and we obtain

$$\begin{aligned} \|\mathbf{v} - \pi_h \mathbf{v}\|_{\mathbf{H}(\mathrm{div},\Omega)} &= \Big(\|\mathbf{v} - \pi_h \mathbf{v}\|^2_{\mathbf{L}_2(\Omega)} + \| \operatorname{div}(\mathbf{v} - \pi_h \mathbf{v}) \|^2_{L_2(\Omega)} \Big)^{1/2} \\ &\le \|\mathbf{v} - \pi_h \mathbf{v}\|_{\mathbf{L}_2(\Omega)} + \inf_{q_h \in M_h} \| \operatorname{div} \mathbf{v} - q_h \|_{L_2(\Omega)} \, . \end{aligned}$$

It remains to estimate the error of the projection π_h in the L_2-norm. We consider this term over the scaled element $\widehat{K}$ which is obtained by the mapping $\mathbf{x} \mapsto \widehat{\mathbf{x}} = h_K^{-1} \mathbf{x}$, cf. (2.22). All objects on the scaled element $\widehat{K}$ are indicated by a hat such as the gradient operator $\widehat{\nabla}$ with respect to the variable $\widehat{\mathbf{x}}$. Furthermore, it is $\widehat{\psi}_E(\widehat{\mathbf{x}}) = \psi_E(\mathbf{x})$ and $\psi_{\widehat{E}}$ denotes the basis functions defined on $\widehat{K}$ for the edge $\widehat{E} \in \mathscr{E}(\widehat{K})$ which corresponds to $E \in \mathscr{E}(K)$. First, we show the identity $\widehat{\pi_h \mathbf{v}} = \widehat{\pi}_h \widehat{\mathbf{v}}$. To this end, we observe that

$$\psi_E(\mathbf{x}) = A\nabla\phi_E(\mathbf{x}) = A\nabla\widehat{\phi}_E(h_K^{-1}\mathbf{x}) = h_K^{-1} A\widehat{\nabla}\widehat{\phi}_E(\widehat{\mathbf{x}}) \,. \tag{6.12}$$

Furthermore, $\widehat{\phi}_E$ satisfies

$$\widehat{\operatorname{div}}\big(A\widehat{\nabla}\widehat{\phi}_E\big) = h_K^2 \operatorname{div}\big(A\nabla\phi_E\big) = \frac{\kappa_{\widehat{E}}(\widehat{K})}{|\widehat{K}|} \quad \text{in } \widehat{K} \,,$$

since $|\widehat{K}| = |K|/h_K^2$, and

$$\mathbf{n}_{\widehat{E}} \cdot A\widehat{\nabla}\widehat{\phi}_E = \mathbf{n}_E \cdot h_K A\nabla\phi_E = h_{\widehat{E}}^{-1} \quad \text{on } \widehat{E} \,,$$

since $h_{\widehat{E}} = h_E/h_K$. The basis function $\psi_{\widehat{E}} = A\widehat{\nabla}\phi_{\widehat{E}}$ on the scaled element $\widehat{K}$ is given according to (6.5) and (6.6). Obviously, $\widehat{\phi}_E$ and $\phi_{\widehat{E}}$ are solutions of the same Neumann problem on $\widehat{K}$ and consequently it is

$$\widehat{\phi}_E = \phi_{\widehat{E}} + C$$

for a constant $C \in \mathbb{R}$. Hence, (6.12) yields

$$\widehat{\psi}_E = h_K^{-1}\psi_{\widehat{E}} \,.$$

For the interpolation operator we thus get on each element $K \in \mathscr{K}_h$

$$\widehat{\pi_h \mathbf{v}} = \sum_{E\in\mathscr{E}(K)} v_E\widehat{\psi}_E = \sum_{\widehat{E}\in\mathscr{E}(\widehat{K})} v_E h_K^{-1}\psi_{\widehat{E}} = \sum_{\widehat{E}\in\mathscr{E}(\widehat{K})} v_{\widehat{E}}\psi_{\widehat{E}} = \widehat{\pi}_h\widehat{\mathbf{v}} \,,$$

because of

$$v_E = \int_E \mathbf{n}_E \cdot \mathbf{v}\,\mathrm{d}s_{\mathbf{x}} = \frac{h_E}{h_{\widehat{E}}} \int_{\widehat{E}} \mathbf{n}_{\widehat{E}} \cdot \widehat{\mathbf{v}}\,\mathrm{d}s_{\widehat{\mathbf{x}}} = h_K\, v_{\widehat{E}}$$

due to $h_{\widehat{E}} = h_E/h_K$.

With the help of Lemma 6.1 and exploiting the reverse triangle inequality, we have for $s > \frac{1}{2}$

$$\|\widehat{\mathbf{v}} - \widehat{\pi}_h\widehat{\mathbf{v}}\|_{\mathbf{L}_2(\widehat{K})} \le c\|\widehat{\mathbf{v}}\|_{\mathbf{H}^s(\widehat{K})} \,. \tag{6.13}$$

Next, in order to apply the Bramble–Hilbert Lemma, see Theorem 1.9, to the functional

$$\mathbf{f}(\widehat{\mathbf{v}}) = \|\widehat{\mathbf{v}} - \widehat{\pi}_h\widehat{\mathbf{v}}\|_{\mathbf{L}_2(\widehat{K})} ,$$

we further have to show that $\widehat{\pi}_h\mathbf{d} = \mathbf{d}$ if $\mathbf{d} = (d_1, d_2)^\top \in \mathbb{R}^2$ is a constant vector. By construction, it is $\widehat{\pi}_h\mathbf{d} = \widehat{\nabla}\phi$ over $\widehat{K}$, where ϕ is the solution of

$$-\widehat{\Delta}\phi = 0 \quad \text{in } \widehat{K} \qquad \text{and} \qquad \mathbf{n}_{\widehat{K}} \cdot \widehat{\nabla}\phi = \mathbf{n}_{\widehat{K}} \cdot \mathbf{d} \quad \text{on } \partial\widehat{K} . \tag{6.14}$$

The boundary data for this problem is compatible,

$$\int_{\partial\widehat{K}} \mathbf{n}_{\widehat{K}} \cdot \mathbf{d} \, \mathrm{d}s_{\widehat{\mathbf{x}}} = \int_{\widehat{K}} \widehat{\operatorname{div}}\, \mathbf{d} \, \mathrm{d}\widehat{\mathbf{x}} = 0 ,$$

and therefore the problem has a unique solution up to an additive constant. Obviously, $\phi(\widehat{\mathbf{x}}) = d_1\widehat{x}_1 + d_2\widehat{x}_2 + C$ satisfies (6.14) for $C \in \mathbb{R}$ and so $\widehat{\pi}_h\mathbf{d} = \widehat{\nabla}\phi = \mathbf{d}$.

Finally, the scaling and the application of the Bramble–Hilbert Lemma to the functional $\mathbf{f}$ yields

$$\|\mathbf{v} - \pi_h\mathbf{v}\|_{\mathbf{L}_2(K)} = h_K\|\widehat{\mathbf{v}} - \widehat{\pi}_h\widehat{\mathbf{v}}\|_{\mathbf{L}_2(\widehat{K})} \le c h_K |\widehat{\mathbf{v}}|_{\mathbf{H}^s(\widehat{K})} = c h_K^s |\mathbf{v}|_{\mathbf{H}^s(K)} ,$$

and after summation over all elements we obtain the desired bound. □

Remark 6.3 The constant c in Lemmata 6.1 and 6.2 only depend on the regularity and stability of the mesh. This can be seen as in [51], since the estimates in the proofs, which might incorporate additional dependencies, have only been performed on the scaled element.

6.1.3 Approximation of Mixed Formulation

By the use of the previously introduced spaces, the discrete version of the variational formulation (6.3) reads:

$$\begin{aligned} &\text{Find } (\mathbf{u}_h, p_h) \in \mathbf{X}_{h,N} \times M_h : \\ &a(\mathbf{u}_h, \mathbf{v}_h) + b(\mathbf{v}_h, p_h) = (\mathbf{n} \cdot \mathbf{v}_h, g_D)_{L_2(\Gamma_D)} \quad \forall \mathbf{v}_h \in \mathbf{X}_{h,N} , \\ &\qquad\qquad\quad b(\mathbf{u}_h, q_h) = -(f, q_h)_{L_2(\Omega)} \qquad\quad \forall q_h \in M_h . \end{aligned} \tag{6.15}$$

To prove unique solvability of the discrete problem, we use a fundamental theorem in the mixed finite element analysis, see [43]. This theory relies on the space

$$\mathbf{Z}_h = \{\mathbf{v}_h \in \mathbf{X}_{h,N} : b(\mathbf{v}_h, q_h) = 0 \quad \forall q_h \in M_h\}$$

and the following two assumptions.

A1: There exists a constant $\alpha^* > 0$ such that

$$a(\mathbf{v}_h, \mathbf{v}_h) \geq \alpha^* \|\mathbf{v}_h\|^2_{\mathbf{H}(\mathrm{div},\Omega)} \quad \text{for } \mathbf{v}_h \in \mathbf{Z}_h \,.$$

A2: There exists a constant $\beta^* > 0$ such that

$$\inf_{q_h \in M_h} \sup_{\mathbf{v}_h \in \mathbf{X}_{h,N}} \frac{b(\mathbf{v}_h, q_h)}{\|\mathbf{v}_h\|_{\mathbf{H}(\mathrm{div},\Omega)} \|q_h\|_{L_2(\Omega)}} \geq \beta^* \,.$$

Such assumptions hold in the continuous setting and they are used in order to prove unique solvability of the mixed formulation (6.3). In the discrete case, however, we have to verify these assumptions for the introduced approximation spaces. Afterwards, the continuity of the bilinear forms $a(\cdot,\cdot)$ on $\mathbf{X}_h \times \mathbf{X}_h$ and $b(\cdot,\cdot)$ on $\mathbf{X}_h \times M_h$ as well as A1 and A2 are sufficient for the existence and uniqueness of the solution of the discrete problem (6.15), see [43]. Furthermore, this theory gives an error estimate. Thus, Babuska–Brezzi theory yields the main result of this section.

Theorem 6.4 *The problem* (6.15) *with* $\mathbf{X}_{h,N}$ *defined by* (6.8) *and* M_h *defined by* (6.4) *has a unique solution* $(\mathbf{u}_h, p_h) \in \mathbf{X}_{h,N} \times M_h$. *Furthermore, there exists a constant c depending only on* α^*, β^*, ϱ_1 *and* ϱ_2 *as well as on the mesh regularity and stability such that*

$$\begin{aligned}\|\mathbf{u} - \mathbf{u}_h\|_{\mathbf{H}(\mathrm{div},\Omega)} &+ \|p - p_h\|_{L_2(\Omega)} \\ &\leq c \left\{ \inf_{\mathbf{v}_h \in \mathbf{X}_h} \|\mathbf{u} - \mathbf{v}_h\|_{\mathbf{H}(\mathrm{div},\Omega)} + \inf_{q_h \in M_h} \|p - q_h\|_{L_2(\Omega)} \right\} \,. \end{aligned} \tag{6.16}$$

Proof To show existence and uniqueness we need to verify A1 and A2. Assumption A1 is shown in a straightforward manner. Since div $\mathbf{v}_h$ is constant on each element it follows

$$\mathbf{Z}_h = \{\mathbf{v}_h \in \mathbf{X}_{h,N} : \mathrm{div}\ \mathbf{v}_h = 0 \text{ in } K \in \mathscr{K}_h\} \,, \tag{6.17}$$

and therefore we get for $\mathbf{v}_h \in \mathbf{Z}_h$

$$\begin{aligned} a(\mathbf{v}_h, \mathbf{v}_h) &= \sum_{K \in \mathscr{K}_h} \int_K A^{-1} \mathbf{v}_h \cdot \mathbf{v}_h \, \mathrm{d}\mathbf{x} \\ &\geq a_{\max}^{-1} \sum_{K \in \mathscr{K}_h} \left\{ \|\mathbf{v}_h\|^2_{\mathbf{L}_2(K)} + \|\,\mathrm{div}\ \mathbf{v}_h\|^2_{L_2(K)} \right\} = \alpha^* \|\mathbf{v}_h\|^2_{\mathbf{H}(\mathrm{div},\Omega)} \,. \end{aligned}$$

To verify A2 we use the interpolation operator π_h defined by (6.9). We have shown that π_h satisfies (6.10). In the following we make use of an auxiliary problem.

For given $q_h \in M_h$, we consider ϕ as unique solution of the boundary value problem

$$\begin{aligned} \Delta\phi &= q_h \quad \text{in } \Omega\,, \\ \mathbf{n}\cdot\nabla\phi &= 0 \quad \text{on } \Gamma_N\,, \\ \phi &= 0 \quad \text{on } \Gamma_D\,. \end{aligned} \tag{6.18}$$

Since we have assumed that Ω is convex, it is well known that if either Γ_N or Γ_D is an empty set, then the solution of this problem belongs to $H^2(\Omega)$, see, e.g., [87]. The general case has been studied in details by Bacuta et al. [17] by the use of FEM tools. If all angles between edges with Neumann and Dirichlet data are strictly less than π, then there exists $s > \frac{1}{2}$ such that

$$\|\phi\|_{H^{1+s}(\Omega)} \le c\|q_h\|_{L_2(\Omega)}\,,$$

cf. [17, Theorem 4.1]. Let $\mathbf{w} = \nabla\phi$. Due to the construction, $\mathbf{w}$ has a piecewise constant divergence and the normal trace of $\mathbf{w}$ vanishes on Γ_N. On each $E \in \mathscr{E}_h$ the function $\pi_h\mathbf{w}$ satisfies

$$\mathbf{n}_E \cdot \pi_h\mathbf{w}\big|_E = h_E^{-1}\int_E \mathbf{n}_E\cdot\mathbf{w}\,\mathrm{d}s_{\mathbf{x}}\,,$$

and since $\mathbf{n}_K = \kappa_E(K)\mathbf{n}_E$ for $E \in \mathscr{E}(K)$, we have

$$\int_K \operatorname{div}\,\pi_h\mathbf{w}\,\mathrm{d}\mathbf{x} = \int_{\partial K}\mathbf{n}_K\cdot\pi_h\mathbf{w}\,\mathrm{d}s_{\mathbf{x}} = \int_{\partial K}\mathbf{n}_K\cdot\mathbf{w}\,\mathrm{d}s_{\mathbf{x}} = \int_K \operatorname{div}\,\mathbf{w}\,\mathrm{d}\mathbf{x}\,.$$

Therefore, it is

$$\operatorname{div}\,\mathbf{w} = \operatorname{div}\,\pi_h\mathbf{w} = q_h \quad \text{for } K \in \mathscr{K}_h\,.$$

Making use of the stability of the interpolation operator π_h, see Lemma 6.1, we get

$$\|\pi_h\mathbf{v}\|_{\mathbf{L}_2(\Omega)} \le c\|\mathbf{v}\|_{\mathbf{H}^s(\Omega)} \le c\|\phi\|_{H^{1+s}(\Omega)} \le c\|q_h\|_{L_2(\Omega)}\,, \tag{6.19}$$

where $c > 0$ is a generic constant. Finally, we obtain

$$\begin{aligned} \sup_{\mathbf{v}_h\in\mathbf{X}_{h,N}} \frac{b(\mathbf{v}_h,q_h)}{\|\mathbf{v}_h\|_{\mathbf{H}(\operatorname{div},\Omega)}} &\ge \frac{b(\pi_h\mathbf{w},q_h)}{\|\pi_h\mathbf{w}\|_{\mathbf{H}(\operatorname{div},\Omega)}} = \frac{\|q_h\|^2_{L_2(\Omega)}}{\left(\|\pi_h\mathbf{w}\|^2_{\mathbf{L}_2(\Omega)} + \|\operatorname{div}(\pi_h\mathbf{w})\|^2_{L_2(\Omega)}\right)^{1/2}} \\ &\ge \frac{\|q_h\|^2_{L_2(\Omega)}}{\left(c^2\|q_h\|^2_{L_2(\Omega)} + \|q_h\|^2_{L_2(\Omega)}\right)^{1/2}} \ge \beta^*\|q_h\|_{L_2(\Omega)}\,, \end{aligned}$$

that proves the inf-sup condition.

Following standard arguments of Babuska and Brezzi utilizing A1 and A2, it is easily shown that the discrete problem (6.15) has a unique solution and that the error estimate (6.16) holds, see, e.g., [43]. □

6.1.4 Realization and Numerical Examples

In contrast to (6.2), the following numerical examples are a little bit more general and involve non-homogeneous Neumann data, i.e. $\mathbf{n} \cdot \mathbf{u} = g_N$ on Γ_N for $g_N \in L_2(\Gamma_N)$. As usual, we seek the approximation $\mathbf{u}_h = \mathbf{u}_{h,N} + \mathbf{u}_{h,g_N}$, where $\mathbf{u}_{h,N} \in \mathbf{X}_{h,N}$ has homogeneous Neumann data and $\mathbf{u}_{h,g_N} \in \mathbf{X}_h$ is an extension of the given data g_N in the discrete space, e.g.,

$$\mathbf{u}_{h,g_N} = \sum_{E \in \mathscr{E}_{h,N}} \int_E g_N \,\mathrm{d}s_{\mathbf{x}}\, \psi_E \,.$$

The mixed formulation for the approximation reads:

$$\begin{aligned}
&\text{Find } (\mathbf{u}_{h,N}, p_h) \in \mathbf{X}_{h,N} \times M_h : \\
&a(\mathbf{u}_{h,N}, \mathbf{v}_h) + b(\mathbf{v}_h, p_h) = (\mathbf{n} \cdot \mathbf{v}_h, g_D)_{L_2(\Gamma_D)} - a(\mathbf{u}_{h,g_N}, \mathbf{v}_h) \quad \forall \mathbf{v}_h \in \mathbf{X}_{h,N} \,, \\
&\qquad\qquad b(\mathbf{u}_{h,N}, q_h) = -(f, q_h)_{L_2(\Omega)} - b(\mathbf{u}_{h,g_N}, q_h) \qquad \forall q_h \in M_h \,.
\end{aligned} \tag{6.20}$$

It remains to discuss the computation of the involved terms. Afterwards, the system of linear equations can be set up for the expansion coefficients of the approximations $\mathbf{u}_h$ and p_h in the form

$$\begin{pmatrix} A_h & B_h^\top \\ B_h & 0 \end{pmatrix} \begin{pmatrix} \underline{\mathbf{u}}_h \\ \underline{p}_h \end{pmatrix} = \begin{pmatrix} \underline{r}_1 \\ \underline{r}_2 \end{pmatrix} , \tag{6.21}$$

where A_h and B_h are the matrices given by testing the bilinear forms $a(\cdot, \cdot)$ and $b(\cdot, \cdot)$ with the basis functions of $\mathbf{X}_{h,N}$ and M_h, respectively. The vectors $\underline{r}_1$ and $\underline{r}_2$ contain the corresponding right hand sides of (6.20). The system can be solved with the favourite linear algebra algorithm. Alternatively, one might use the Schur complement. The first equation in (6.21) yields

$$\underline{\mathbf{u}}_h = A_h^{-1} \left(\underline{r}_1 - B_h^\top \underline{p}_h \right) ,$$

and inserting into the second equation of (6.21) gives

$$B_h A_h^{-1} B_h^\top \underline{p}_h = B_h A_h^{-1} \underline{r}_1 - \underline{r}_2$$

for the computation of $\underline{p}_h$.

6.1.4.1 Computational Realization

In this section we address the computational realization of the terms within the mixed formulation (6.20). The integrals $(f, q_h)_{L_2(\Omega)}$ and $(\mathbf{n}\cdot\mathbf{v}_h, g_D)_{L_2(\Gamma_D)}$ are rather standard. The first integral is split into its contribution over each polygonal element and then a quadrature formula is applied over the auxiliary triangulation as in (4.43). For the second integral we recognize that $\mathbf{n}\cdot\mathbf{v}_h$ is constant over each edge of the discretization. Consequently, we split the integral on Γ_D into its contributions over the single edges and apply Gaussian quadrature. We recall that $\operatorname{div}\mathbf{v}_h$ is constant on each element for $\mathbf{v}_h \in \mathbf{X}_h$. Therefore, the entries of B_h have an analytic expression. For $\mathbf{v}_h \in \mathbf{X}_h$ and $q_h \in M_h$, we obtain

$$b(\mathbf{v}_h, q_h) = \sum_{K\in\mathcal{K}_h} |K| \operatorname{div}\mathbf{v}_h\big|_K q_h .$$

In order to treat the bilinear form $a(\cdot,\cdot)$, we apply boundary element techniques. We exploit the definition of the basis functions ψ_E in (6.6) with the help of ϕ_E. Obviously, the function $\mathbf{u}_h \in \mathbf{X}_h$ can be expressed locally over each element $K \in \mathcal{K}_h$ as

$$\mathbf{u}_h = A\nabla\phi_{\mathbf{u}}$$

where $\phi_{\mathbf{u}}$ is the unique solution of

$$\begin{aligned} \operatorname{div}(A\nabla\phi_{\mathbf{u}}) &= f_{\mathbf{u}} \quad \text{in } K ,\\ \mathbf{n}_K\cdot A\nabla\phi_{\mathbf{u}} &= g_{\mathbf{u}} \quad \text{on } \partial K , \end{aligned} \tag{6.22}$$

with a constant $f_{\mathbf{u}}$ and piecewise constant $g_{\mathbf{u}} \in \mathcal{P}^0_{\mathrm{pw,d}}(\partial K)$. Furthermore, the function $\phi_{\mathbf{u}}$ is decomposed into $\phi_{\mathbf{u}} = \phi_{\mathbf{u},0} + \phi_{\mathbf{u},f}$ with

$$\phi_{\mathbf{u},f}(\mathbf{x}) = \tfrac{1}{4} f_{\mathbf{u}}(\mathbf{x}-\bar{\mathbf{x}}_K)^\top A^{-1}(\mathbf{x}-\bar{\mathbf{x}}_K) \subset \mathcal{P}^2(K) ,$$

such that

$$\operatorname{div}(A\nabla\phi_{\mathbf{u},f}) = f_{\mathbf{u}} \quad \text{in } K , \tag{6.23}$$

and hence, $\phi_{\mathbf{u},0}$ is the solution of the Neumann problem

$$\begin{aligned} -\operatorname{div}(A\nabla\phi_{\mathbf{u},0}) &= 0 && \text{in } K ,\\ \mathbf{n}_K\cdot A\nabla\phi_{\mathbf{u},0} &= g_{\mathbf{u}} - \mathbf{n}_K\cdot A\nabla\phi_{\mathbf{u},f} && \text{on } \partial K . \end{aligned} \tag{6.24}$$

The function $\phi_{\mathbf{u},0}$ is unique up to an additive constant. A small exercise shows that $g_{\mathbf{u}} - \mathbf{n}_K\cdot A\nabla\phi_{\mathbf{u},f} \in \mathcal{P}^0_{\mathrm{pw,d}}(\partial K)$, since the gradient of a quadratic function is linear and since the normal component of a linear function along a straight edge is constant. In the case of a scalar valued diffusion coefficient $A \in \mathbb{R}$, we apply the discussed boundary element method from Chap. 4 for the Neumann problem of the Laplace equation. But, there is also a boundary element method available for a general, symmetric and positive definite matrix $A \in \mathbb{R}^{2\times 2}$, see [151]. We comment on this in Sect. 6.2.3.

Consequently, we have the tools to approximate the Dirichlet trace of $\phi_{\mathbf{u},0}$ on ∂K and we utilize the representation formula to evaluate $\phi_{\mathbf{u},0}$ and its derivatives inside the elements. This allows for a very accurate approximation of $\mathbf{u}_h$ inside K. Thus, we have different possibilities to treat the bilinear form $a(\cdot,\cdot)$ as in Sect. 4.5. Either we use a numerical integration scheme over the polygonal elements and evaluate $\mathbf{u}_h$ and $\mathbf{v}_h$ with the help of the representation formula in the quadrature nodes, or we utilize partial integration locally in order to reformulate the volume integrals into boundary integrals. The first strategy is analog to the volume quadrature in (4.43). For the second strategy we write

$$\mathbf{u}_h = A\nabla\phi_{\mathbf{u}} \qquad \text{and} \qquad \mathbf{v}_h = A\nabla\phi_{\mathbf{v}}$$

with

$$\phi_{\mathbf{u}} = \phi_{\mathbf{u},0} + \phi_{\mathbf{u},f} \qquad \text{and} \qquad \phi_{\mathbf{v}} = \phi_{\mathbf{v},0} + \phi_{\mathbf{v},f}$$

as above. This decomposition and the symmetry of A yield

$$\begin{aligned}
a(\mathbf{u}_h,\mathbf{v}_h) &= \sum_{K\in\mathscr{K}_h} \left(A^{-1}\mathbf{u}_h,\mathbf{v}_h\right)_{\mathbf{L}_2(K)} = \sum_{K\in\mathscr{K}_h} (A\nabla\phi_{\mathbf{u}},\nabla\phi_{\mathbf{v}})_{\mathbf{L}_2(K)} \\
&= \sum_{K\in\mathscr{K}_h} \Big\{ \left(A\nabla\phi_{\mathbf{u},0},\nabla\phi_{\mathbf{v},0}\right)_{\mathbf{L}_2(K)} + \left(A\nabla\phi_{\mathbf{u},f},\nabla\phi_{\mathbf{v},f}\right)_{\mathbf{L}_2(K)} \\
&\qquad\qquad + \left(A\nabla\phi_{\mathbf{u},0},\nabla\phi_{\mathbf{v},f}\right)_{\mathbf{L}_2(K)} + \left(\nabla\phi_{\mathbf{u},f},A\nabla\phi_{\mathbf{v},0}\right)_{\mathbf{L}_2(K)} \Big\} \\
&= \sum_{K\in\mathscr{K}_h} \Big\{ I + II + III + IV \Big\} .
\end{aligned}$$

The terms I–IV are treated separately employing integration by parts and the properties (6.23) and (6.24). We obtain

$$III = \left(\mathbf{n}_K\cdot A\nabla\phi_{\mathbf{u},0},\phi_{\mathbf{v},f}\right)_{L_2(\partial K)} \qquad \text{and} \qquad IV = \left(\mathbf{n}_K\cdot A\nabla\phi_{\mathbf{v},0},\phi_{\mathbf{u},f}\right)_{L_2(\partial K)} ,$$

and consequently the terms are given as integrals of piecewise quadratic polynomials over ∂K that are computed analytically. The same arguments yield

$$II = \left(\mathbf{n}_K\cdot A\nabla\phi_{\mathbf{u},f},\phi_{\mathbf{v},f}\right)_{L_2(\partial K)} - \left(f_{\mathbf{u}},\phi_{\mathbf{v},f}\right)_{L_2(K)}$$

with an integral of a piecewise quadratic polynomial over the boundary ∂K and an integral of a quadratic polynomial over K, since $f_{\mathbf{u}}$ is constant. Both integrals are computed analytically, where we apply the divergence theorem to transform the volume integral to a boundary integral. Finally, I has the form

$$I = \left(\mathbf{n}_K \cdot A\nabla\phi_{\mathbf{u},0}, \phi_{\mathbf{v},0}\right)_{L_2(\partial K)}$$

after integration by parts. Here, $\mathbf{n}_K \cdot A\nabla\phi_{\mathbf{u},0}$ is a piecewise constant function on ∂K and $\phi_{\mathbf{v},0}$ is treated by means of boundary element methods as discussed in Chap. 4. For scalar valued diffusion $A \in \mathbb{R}$, we obtain with the notation of trace and boundary integral operators

$$I = \left(A\gamma_1^K\phi_{\mathbf{u},0}, \gamma_0^K\phi_{\mathbf{v},0}\right)_{L_2(\partial K)} = A\left(\gamma_1^K\phi_{\mathbf{u},0}, \mathbf{P}_K\gamma_1^K\phi_{\mathbf{v},0}\right)_{L_2(\partial K)},$$

where $\mathbf{P}_K$ denotes the Poincaré–Steklov operator (4.14), which maps the Neumann to the Dirichlet trace. Hence, I is approximated utilizing the non-symmetric

$$\mathbf{P}_{K,h}^{\mathrm{unsym}} = \mathbf{M}_{K,h}\widetilde{\mathbf{D}}_{K,h}^{-1}\left(\tfrac{1}{2}\mathbf{M}_{K,h}^\top - \mathbf{K}_{K,h}^\top\right)$$

or the symmetric

$$\mathbf{P}_{K,h} = \mathbf{V}_{K,h} + \left(\tfrac{1}{2}\mathbf{M}_{K,h} - \mathbf{K}_{K,h}\right)\widetilde{\mathbf{D}}_{K,h}^{-1}\left(\tfrac{1}{2}\mathbf{M}_{K,h}^\top - \mathbf{K}_{K,h}^\top\right)$$

discretization of $\mathbf{P}_K$, see (4.14) as well as (4.15) and the more detailed discussion in Sect. 4.5. For matrix valued diffusion $A \in \mathbb{R}^{2\times 2}$, the Neumann trace is defined by

$$\gamma_1^K\phi_{\mathbf{u},0} = \mathbf{n}_K \cdot A\nabla\phi_{\mathbf{u},0}$$

for sufficiently regular functions and we can proceed analogously with the BEM.

6.1.4.2 Numerical Examples

To validate our theoretical findings, we give some numerical experiments for the mixed formulation of the BEM-based FEM. In the realization, we set up the matrix A_h with the brute force approach utilizing numerical integration over polygonal elements, where the test and trial functions are evaluated with the help of the representation formula. Furthermore, the system of linear equations (6.21) is solved by means of GMRES [150].

Two model problems are posed on the domain $\Omega = (-1, 1)^2$ and we decompose its boundary into

$$\Gamma_D = \{(x_1, -1)^\top : -1 \le x_1 \le 1\} \quad \text{and} \quad \Gamma_N = \partial\Omega \setminus \Gamma_D\,.$$

In the first example, we choose the data g_D and g_N in such a way that the smooth function $p(\mathbf{x}) = \exp(2\pi(x_1 - 0.3)) \cos(2\pi(x_2 - 0.3))$, $\mathbf{x} \in \mathbb{R}^2$ is the exact solution of

$$-\Delta p = 0 \quad \text{in } \Omega\,, \qquad \mathbf{n}\cdot\nabla p = g_N \quad \text{on } \Gamma_N\,, \qquad p = g_D \quad \text{on } \Gamma_D\,.$$

Thus, $(\mathbf{u}, p)$ with $\mathbf{u} = \nabla p$ solves the corresponding mixed formulation (6.20). For the second example, we take $p(\mathbf{x}) = \sin(\pi x_1)\sin(\pi x_2)$, $\mathbf{x} \in \mathbb{R}^2$ as solution of

$$-\Delta p = f \quad \text{in } \Omega\,, \qquad \mathbf{n}\cdot\nabla p = g_N \quad \text{on } \Gamma_N\,, \qquad p = 0 \quad \text{on } \Gamma_D$$

with corresponding data f and g_N. The BEM-based FEM is applied on a sequence of honeycomb meshes consisting of hexahedral elements with decreasing mesh size h, see Fig. 6.2. We analyse numerically the relative error

$$\frac{\|\mathbf{u} - \mathbf{u}_h\|_{\mathbf{H}(\mathrm{div},\Omega)} + \|p - p_h\|_{L_2(\Omega)}}{\|\mathbf{u}\|_{\mathbf{H}(\mathrm{div},\Omega)} + \|p\|_{L_2(\Omega)}}\,. \tag{6.25}$$

According to Theorem 6.4, the interpolation error in Lemma 6.2 and known approximation properties of the space M_h, cf. Lemma 3.4, we expect linear convergence of the relative error (6.25) with respect to the mesh size $h = \max\{h_K : K \in \mathcal{K}_h\}$. The numerical experiments confirm this fact, see Fig. 6.3. In Fig. 6.4, the approximations p_h and $\mathbf{u}_h$ of the primal and the flux variable are visualized for the second problem.

In the third and final example, we consider a problem with unknown solution. Let $\Omega = (0, 1)^2$ and we prescribe Dirichlet data on the left edge of the square and Neumann data else, such that

$$\Gamma_D = \{(0, x_2)^\top : 0 \le x_2 \le 1\} \quad \text{and} \quad \Gamma_N = \partial\Omega \setminus \Gamma_D\,.$$

We choose the Dirichlet data as

$$g_D(\mathbf{x}) = 1 - x_2 \quad \text{for } \mathbf{x} \in \Gamma_D\,,$$

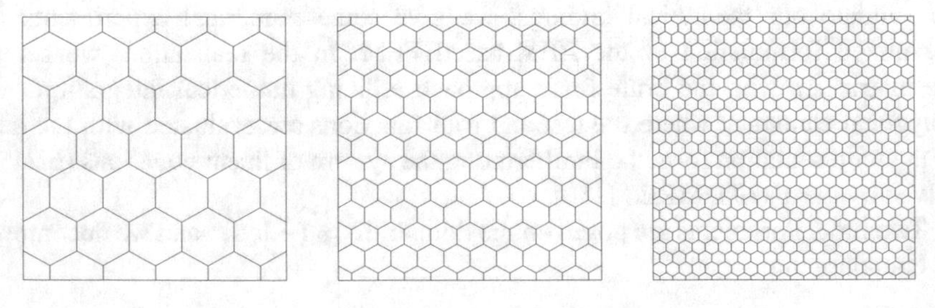

Fig. 6.2 Sequence of honeycomb meshes with hexahedral elements and decreasing mesh size h from left to right

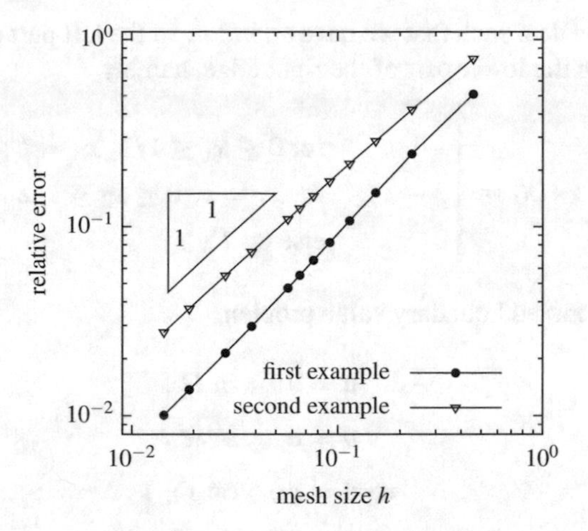

Fig. 6.3 Relative error (6.25) with respect to h in logarithmic scale for first and second example

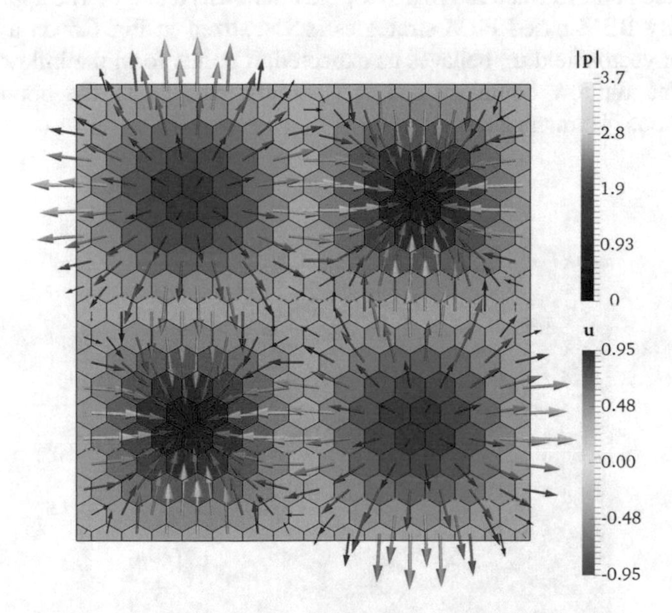

Fig. 6.4 Visualization of the approximation of the second example, the primal variable p_h and the flux unknown $\mathbf{u}_h$

and the Neumann data such that we have an inflow in the left part of the upper edge and an outflow in the lower part of the right edge, namely

$$g_N(\mathbf{x}) = \begin{cases} -3 & \text{for } 0 \leq x_1 \leq 1/2,\ x_2 = 1\,, \\ 3 - 3x_2 & \text{for } x_1 = 1,\ 0 \leq x_2 \leq 1/2\,, \\ 0 & \text{else on } \Gamma_N\,. \end{cases}$$

We consider the mixed boundary value problem

$$\begin{aligned} -\operatorname{div} \mathbf{u} &= 0 && \text{in } \Omega\,, \\ \nabla p &= \mathbf{u} && \text{in } \Omega\,, \\ \mathbf{n} \cdot \mathbf{u} &= g_N && \text{on } \Gamma_N\,, \\ p &= g_D && \text{on } \Gamma_D \end{aligned}$$

in the saddle point formulation (6.20) for the unknowns $\mathbf{u}$ and p. The approximation obtained by BEM-based FEM strategies is visualized in Fig. 6.5 on a polygonal mesh. The vector field $\mathbf{u}_h$ behaves as expected, it points form the inflow boundary towards the outflow boundary and it is almost parallel to the boundary with homogeneous Neumann data.

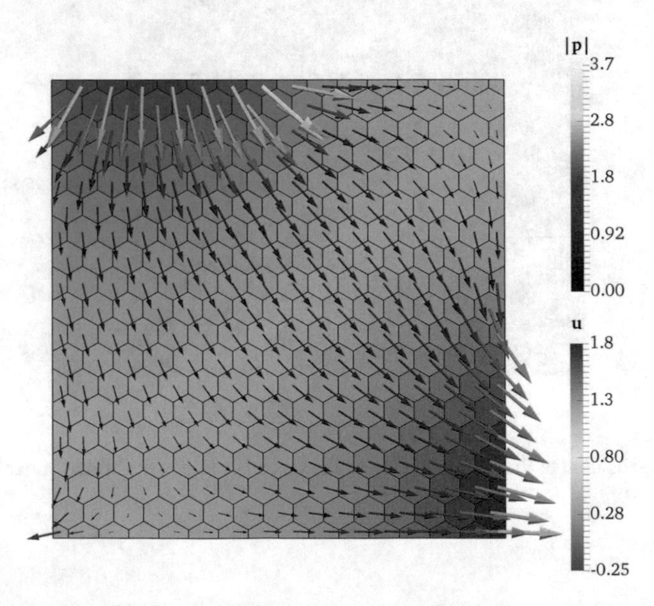

Fig. 6.5 Visualization of the approximation of third example, the primal variable p_h and the flux unknown $\mathbf{u}_h$

6.2 3D Generalization with Application to Convection-Diffusion-Reaction Equation

In this section, we discuss a generalization of a variant of the BEM-based finite element method studied so far. We address the definition of basis functions on meshes with polyhedral elements having polygonal faces. These functions are used to construct an approximation space V_h which can be utilized in the discrete Galerkin formulation of the finite element method. The idea of the BEM-based FEM is to define the basis functions implicitly on each element as local solutions of the underlying differential equation and to treat the local problems by boundary element methods. In the following, we push this idea one step further. As model problem, we consider once more the diffusion equation (2.1) and in addition a general convection-diffusion-reaction equation. In particular, the forthcoming construction of V_h will improve the stability of the discretization method for convection-dominated problems both when compared to a standard FEM and to previous BEM-based FEM approaches. The experiments also show an improved resolution of exponential layers at the outflow boundaries when the proposed method is compared to the Streamline Upwind/Petrov-Galerkin (SUPG) method [48].

6.2.1 *Generalization for Diffusion Problem*

In a first step we consider the generalization to polyhedral elements with polygonal faces for the diffusion problem (2.1). This problem reads

$$\begin{aligned} -\operatorname{div}(a\nabla u) &= f \quad \text{in } \Omega\ , \\ u &= g_D \quad \text{on } \Gamma_D\ , \\ a\nabla u \cdot \mathbf{n} &= g_N \quad \text{on } \Gamma_N\ , \end{aligned}$$

with the assumptions on the data as described in Chap. 2. Section 2.3 gives a detailed construction of basis functions for the two-dimensional case and a simple generalization for the three-dimensional case under the restriction that the polyhedral elements only have triangular faces. These functions are not limited to the diffusion equation, but they have been especially designed for that problem. Here, we first examine the situation for the first order approximation space V_h and give an alternative construction of its basis functions allowing polytopal elements with polygonal faces directly. Afterwards, we present the general space V_h^k yielding k-th order approximations.

If we look again into the two-dimensional case and the definition of the nodal basis functions (2.6), we observe that the values of the basis functions are fixed in the nodes and extended uniquely along the edges by linear functions. This linear extension is nothing else than a harmonic extension along the edge, and thus the

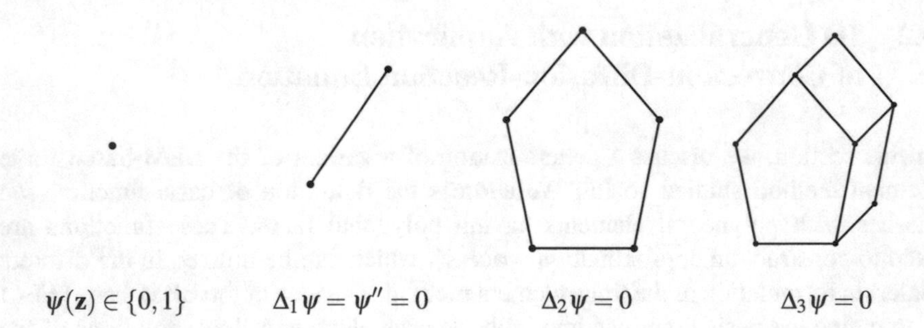

Fig. 6.6 Stepwise construction of basis functions

basis functions are also defined on the edges according to the underlying differential equation. Therefore, we propose a stepwise and hierarchical construction for the basis functions in the case of polyhedral elements with polygonal faces as sketched in Fig. 6.6. This approach has been first proposed in [147]. A similar idea has been used in two-dimensions for the construction of multiscale finite elements in [104].

In order to get a nodal basis of V_h, we declare for each node $\mathbf{z} \in \mathscr{N}_h$ a basis function $\psi_\mathbf{z}$ which is equal to one in $\mathbf{z}$ and zero in all other nodes of the mesh. Denoting the i-dimensional Laplace operator by Δ_i, we define the basis function $\psi_\mathbf{z}$ as unique solution of

$$\begin{aligned} -\Delta_3 \psi_\mathbf{z} &= 0 \quad \text{in } K \quad \text{for all } K \in \mathscr{K}_h\,, \\ -\Delta_2 \psi_\mathbf{z} &= 0 \quad \text{in } F \quad \text{for all } F \in \mathscr{F}_h\,, \\ -\Delta_1 \psi_\mathbf{z} &= 0 \quad \text{in } E \quad \text{for all } E \in \mathscr{E}_h\,, \\ \psi_\mathbf{z}(\mathbf{x}) &= \begin{cases} 1 & \text{for } \mathbf{x} = \mathbf{z}\,, \\ 0 & \text{for } \mathbf{x} \in \mathscr{N}_h \setminus \{\mathbf{z}\}\,, \end{cases} \end{aligned}$$

where the Laplace operators have to be understood in the corresponding linear parameter spaces. The values in the nodes are prescribed. Afterwards, we solve a Dirichlet problem for the Laplace equation on each edge. Then, we use the computed data as Dirichlet datum for the Laplace problem on each face, and finally, we proceed with the Laplace problem on each element, where the solutions on the faces are used as boundary values. In the case of convex faces and elements, these problems are understood in the classical sense and we have $\psi_\mathbf{z} \in C^2(K) \cap C^0(\overline{K})$. In the more general situation of non-convex elements, the weak solution is considered such that we have at least $\psi_\mathbf{z} \in H^1(K)$.

Building the span of these nodal basis functions, we obtain a first order approximation space $V_h = V_h^1$ on general meshes containing polyhedral elements with polygonal faces. In [147], this space has been analysed for its approximation properties and an interpolation operator analog to the one defined in Sect. 2.4 has been studied. The discrete Galerkin formulation for the model problem (2.1) with the generalized approximation space applicable on polyhedral elements with polygonal faces reads as in Sect. 2.5, see (2.28).

Having this hierarchical construction in mind for the definition of nodal basis functions it is clear how to proceed with higher order basis functions. In the two-dimensional setting we enriched the approximation space with element bubble functions which have a polynomial Laplacian, see the motivation in Sect. 2.3.2. Thus, instead of prescribing the Laplace equation on edges, faces and elements, we use the Poisson equation with polynomial right hand side. Consequently, the discrete space V_h^k consists of function that are polynomials along the edges, their restriction onto a face $F \in \mathscr{F}_h$ lies in the two-dimensional approximation space $V_h^k(F)$ defined in Sect. 2.3.3, and they have a polynomial Laplacian inside the three-dimensional element $K \in \mathscr{K}_h$. More precisely, it is

$$V_h^k = \left\{ v \in H^1(\Omega) : \Delta v\big|_K \in \mathscr{P}^{k-2}(K)\ \forall K \in \mathscr{K}_h \text{ and } v\big|_F \in V_h^k(F)\ \forall F \in \mathscr{F}_h \right\}.$$

We easily see that $\mathscr{P}^k(K) \subset V_h^k|_K$, such that polynomials are contained in the approximation space locally. This ensures the approximation properties of the discrete space.

Remark 6.5 If the polyhedral elements have by chance only triangular faces, the approximation space described above is equivalent to the simple generalization from Sect. 2.3.4 for $k = 1$. In the case $k > 1$, however, the defined spaces differ between each other. On each triangular face F it is $\mathscr{P}^k(F) \subsetneq V_h^k(F)$ and whereas the simple generalization thus has $\frac{1}{2}(k-1)(k-2)$ internal degrees of freedom per face the above generalization has $\frac{1}{2}k(k-1)$.

6.2.2 Application to Convection-Diffusion-Reaction Problem

The general convection-diffusion-reaction problem in a bounded Lipschitz domain $\Omega \subset \mathbb{R}^3$ is given by

$$\begin{aligned} \mathrm{L}u = -\operatorname{div}(A\nabla u) + \mathbf{b}\cdot\nabla u + cu &= 0 \quad \text{in } \Omega\,, \\ u &= g_D \quad \text{on } \Gamma\,, \end{aligned} \tag{6.26}$$

where we restrict ourselves to the pure Dirichlet problem for shorter notation. Here $A(\mathbf{x}) \in \mathbb{R}^{3\times 3}$, $\mathbf{b}(\mathbf{x}) \in \mathbb{R}^3$, and $c(\mathbf{x}) \in \mathbb{R}$ are the coefficient functions of the partial differential operator L, and $g_D \in H^{1/2}(\Gamma)$ is the given Dirichlet data. We assume that $A(\cdot)$ is symmetric and uniformly positive with minimal eigenvalue $a_{\min}$, and that $c(\cdot)$ is non-negative. The corresponding Galerkin formulation reads as follows:

$$\begin{aligned} &\text{Find } u \in g_D + H_0^1(\Omega): \\ &\int_\Omega (A\nabla u\cdot\nabla v + \mathbf{b}\cdot\nabla u\, v + cuv)\,\mathrm{d}\mathbf{x} = 0 \qquad \forall v \in H_0^1(\Omega)\,. \end{aligned} \tag{6.27}$$

We require that the coefficients A, $\mathbf{b}$, c are $L^\infty(\Omega)$, and that there exists a unique solution of (6.27). The unique solvability can be ensured under several well known conditions. For example, if $c - \frac{1}{2}\operatorname{div}(\mathbf{b}) \geq 0$, the bilinear form in (6.27) is $H_0^1(\Omega)$-elliptic, which guarantees the existence of a unique solution for the Dirichlet problem. Another sufficient condition is $a_{\min}c > |\mathbf{b}|^2$, see, e.g. [151]. Unique solvability of the variational problem (6.27) can be shown under quite general assumptions using results by Droniou [68].

For the application of the BEM-based FEM, we require that the coefficients $A(\cdot)$, $\mathbf{b}(\cdot)$, and $c(\cdot)$ are piecewise constant with respect to all geometrical objects in the polyhedral mesh $\mathscr{K}_h$. Since this is not the case in general, the coefficients are approximated by piecewise constant ones over the edges, faces and elements of the mesh. If the coefficients are smooth, we take their values in the center of mass of the geometrical objects as constant approximations. This corresponds to a first order approximation of the differential equation. If the coefficients are already piecewise constant with respect to the elements, we obtain their values on the edges and faces by computing averages over neighbouring elements. To simplify notation, we omit new symbols for this approximation. The resulting Dirichlet problem is uniquely solvable according to the before mentioned conditions in [68].

We restrict ourselves to the introduction of the first order approximation space. If the polyhedral elements consists of triangular faces only, we can proceed as for the simple generalization in Sect. 2.3.4. Consequently, the basis functions ψ are defined to be piecewise linear and continuous over the surface triangulation and satisfy the underlying differential equation inside each element, i.e., $\mathrm{L}\,\psi = 0$ in K, $\forall K \in \mathscr{K}_h$. This strategy has been introduced in [96] for the convection-diffusion-reaction equation. We refer to it as the original approach. There is a close relation between this original BEM-based FEM with piecewise linear boundary data and the so-called method of residual-free bubbles [41, 44, 45, 47, 80]. Indeed, it has been shown in [94] that the BEM-based FEM, with exact evaluation of the Steklov–Poincaré operator, is equivalent to the method of residual-free bubbles with exactly computed bubbles. Since the latter has been shown to be a stable method for convection-dominated problems, it seems clear that also the BEM-based FEM should have advantageous stability properties. It should be noted that neither the Steklov–Poincaré operator nor the computation of the residual-free bubbles can be realized exactly in practice.

In this chapter we follow the idea of the previous Sect. 6.2.1 and define the basis functions in a hierarchical fashion as in [99]. Thus, we obtain for each node $\mathbf{z} \in \mathscr{N}_h$ a basis function $\psi_\mathbf{z}$ as unique solution of

$$
\begin{aligned}
\mathrm{L}\,\psi_\mathbf{z} &= 0 \quad \text{in } K \quad \text{for all } K \in \mathscr{K}_h\,, \\
\mathrm{L}_F\,\psi_\mathbf{z} &= 0 \quad \text{in } F \quad \text{for all } F \in \mathscr{F}_h\,, \\
\mathrm{L}_E\,\psi_\mathbf{z} &= 0 \quad \text{in } E \quad \text{for all } E \in \mathscr{E}_h\,, \\
\psi_\mathbf{z}(\mathbf{x}) &= \begin{cases} 1 & \text{for } \mathbf{x} = \mathbf{z}\,, \\ 0 & \text{for } \mathbf{x} \in \mathscr{N}_h \setminus \{\mathbf{z}\}\,. \end{cases}
\end{aligned}
$$

The differential operators L_E and L_F are projections of the differential operator L onto the edge E and the face F, respectively, see below for a precise description. Thus, the functions $\psi_{\mathbf{z}}$ are defined implicitly as local solutions of boundary value problems on edges, faces and elements of the decomposition. Equivalently, one can say that these functions are defined via PDE-harmonic extensions. The nodal data is first extended L_E-harmonically along the edges and afterwards, the data on the edges is extended into the faces with the help of a L_F-harmonic operator and so on.

For the definition of L_E and L_F, let $F \in \mathscr{F}_h$ be a face and $E \in \mathscr{E}_h$ an edge on the boundary of F. By rotation and translation of the coordinate system, we map the face F into the (e_1, e_2)-plane and the edge E onto the e_1-axis of the Euclidean coordinate system (e_1, e_2, e_3) such that one node of E lies in the origin. Thus, we have an orthogonal matrix $B \in \mathbb{R}^{3\times 3}$ and a vector $\mathbf{d} \in \mathbb{R}^3$ such that

$$\mathbf{x} \mapsto \widehat{\mathbf{x}} = B\mathbf{x} + \mathbf{d} \quad \text{and} \quad \widehat{\psi}(\widehat{\mathbf{x}}) = \psi(B^{-1}\widehat{\mathbf{x}} - B^{-1}\mathbf{d}) ,$$

and the differential equation in (6.26) yields

$$-\operatorname{div}(A\nabla\psi) + \mathbf{b}\cdot\nabla\psi + c\psi = -\operatorname{div}_{\widehat{\mathbf{x}}}(BAB^\top\nabla_{\widehat{\mathbf{x}}}\widehat{\psi}) + B\mathbf{b}\cdot\nabla_{\widehat{\mathbf{x}}}\widehat{\psi} + c\widehat{\psi} = 0 . \qquad (6.28)$$

Here, the coefficients $BAB^\top$, $B\mathbf{b}$ and c are constant on F and E, respectively, since A, $\mathbf{b}$ and c are constant approximations on each geometrical object of the original coefficients. Furthermore, we only consider tangential components to define the operators L_F and L_E on the face and edge, respectively. This is equivalent to setting

$$\frac{\partial\widehat{\psi}}{\partial\widehat{x}_3} = \frac{\partial^2\widehat{\psi}}{\partial\widehat{x}_3^2} = 0 \text{ in } F \quad \text{and} \quad \frac{\partial\widehat{\psi}}{\partial\widehat{x}_2} = \frac{\partial\widehat{\psi}}{\partial\widehat{x}_3} = \frac{\partial^2\widehat{\psi}}{\partial\widehat{x}_2^2} = \frac{\partial^2\widehat{\psi}}{\partial\widehat{x}_3^2} = 0 \text{ on } E$$

in (6.28). Therefore, the dependence in (6.28) reduces to two and one coordinate directions such that L_F and L_E are defined as differential operators in two- and one-dimensions using the described coordinate system. Overall, the basis functions are constructed with the help of the convection-diffusion-reaction equation on the edges, faces and elements, where the diffusion matrix and the convection vector are adjusted in a proper way. All appearing one-, two- and three-dimensional boundary value problems are uniquely solvable due to the global properties of $A(\cdot)$, which carry over to $BAB^\top$, and since $c - \frac{1}{2}\operatorname{div}(B\mathbf{b}) = c \geq 0$ on F and E, respectively.

To simplify notation, we omit the coordinate transformation in the following and abbreviate the transformed diffusion matrix $BAB^\top$, the convection vector $B\mathbf{b}$ and the reaction term c to A_F, $\mathbf{b}_F$, c_F and A_E, $\mathbf{b}_E$, c_E on the faces and edges, respectively. Furthermore, we treat the basis functions $\psi_{\mathbf{z}}$ as functions of two or one variable depending on the underlying domain F or E. For example, let us assume that E already lies in the e_1-axis and corresponds to the interval $(0, h_E)$. In this case, $\psi_{\mathbf{z}}$ only depends on x_1 and the scalar valued coefficients A_E, $\mathbf{b}_E$, and c_E along E

and the differential equation reads

$$A_E \psi_{\mathbf{z}}'' + \mathbf{b}_E \psi_{\mathbf{z}}' + c_E \psi_{\mathbf{z}} = 0 \quad \text{in } (0, h_E) , \tag{6.29}$$

with some boundary data $\psi_{\mathbf{z}}(0)$ and $\psi_{\mathbf{z}}(h_E)$ that is 0 or 1 depending on the considered basis function.

Having the basis functions $\psi_{\mathbf{z}}$ at hand, we define the approximation spaces as

$$V_h = \operatorname{span}\{\psi_{\mathbf{z}} : \mathbf{z} \in \mathcal{N}_h\} \quad \text{and} \quad V_{h,D} = V_h \cap H_0^1(\Omega) . \tag{6.30}$$

The discrete Galerkin formulation thus reads:

$$\begin{aligned} &\text{Find } u_h \in g_D + V_{h,D} \subset V_h : \\ &\int_\Omega (A\nabla u_h \cdot \nabla v_h + \mathbf{b} \cdot \nabla u_h \, v_h + c u_h v_h) \, \mathrm{d}\mathbf{x} = 0 \qquad \forall v \in V_{h,D} . \end{aligned} \tag{6.31}$$

Remark 6.6 In order to define a high-order approximation space V_h^k with $k > 1$, we may proceed as in the previous Sect. 6.2.1. Consequently, additional edge, face and element bubble functions are introduced which are defined to satisfy the inhomogeneous convection-diffusion-reaction equation with polynomial right hand side inside the edges, faces and elements, respectively.

6.2.3 Realization of the Basis Functions

Of course, the hierarchically defined basis functions do not have a closed analytical form and they have to be treated numerically. In the following, we discuss this issue in more detail, where we solve the boundary value problems on the edges analytically, the problems on the faces with the help of a 2D FEM and the problems on the elements by means of boundary integral equations. For this purpose, an auxiliary discretization of the boundaries of the elements is needed. We apply the construction of the triangular surface mesh discussed in Sect. 2.2.2, which yields a conforming boundary discretization $\mathcal{T}_l(\partial K)$ of level l. Here, first the faces are discretized by connecting their vertices with the point $\mathbf{z}_F$ and afterwards, the resulting triangles are refined successively by splitting them into four similar triangles. According to this construction, the triangulations on all faces can be glued in a conforming manner to obtain a discretization of the whole boundary ∂K. In particular, the strategy yields for $l \geq 1$ a discretization of each edge in the mesh into line segments, see Fig. 6.7.

The advantage of this line of action is, that the two-dimensional finite element spaces on the faces of the elements fit exactly the approximation spaces utilized in three-dimensional boundary element methods. Thus, a 2D FEM approximation on the faces can directly be used in existing boundary element codes. Alternatively,

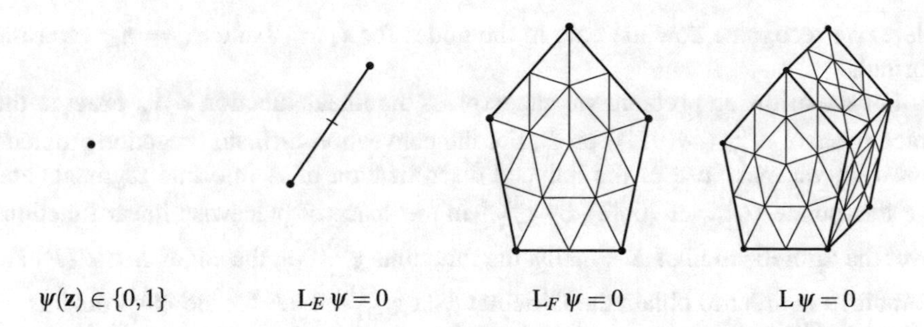

Fig. 6.7 Stepwise approximation of basis functions using the auxiliary discretization with $l = 1$

one might treat the boundary value problems on the faces by a 2D BEM in order to avoid the surface triangulation, but this would result in the need of a 3D BEM on polygonal surface meshes. Hence, we stick with the 2D FEM and 3D BEM strategy that is explained in more detail in the following. Furthermore, we restrict ourselves to $k = 1$.

To be mathematically more precise, we choose a basis function $\psi_{\mathbf{z}}$ and consider its approximation $\psi_{\mathbf{z},l}$ on the edges and faces of $K \in \mathscr{K}_h$ with $\mathbf{z} \in \mathscr{N}(K)$. Here, l refers to the level of the surface triangulation and therefore to the mesh size of the auxiliary discretization. We seek the approximation of $\psi_{\mathbf{z}}\big|_{\partial K}$, namely the Dirichlet data for the three-dimensional problem on K, as

$$g_{l,\partial K}^{(\psi_{\mathbf{z}})} \in \mathscr{P}^1_{\mathrm{pw}}(\mathscr{T}_l(\partial K)) , \quad \text{and set} \quad \psi_{\mathbf{z},l}\big|_{\partial K} = g_{l,\partial K}^{(\psi_{\mathbf{z}})} .$$

The space of piecewise linear polynomials over $\mathscr{T}_l(\partial K)$ has been endowed with a basis Φ_D in Sect. 4.3. We denote by $\underline{g}_{l,\partial K}^{(\psi_{\mathbf{z}})}$ the vector with the expansion coefficients of $g_{l,\partial K}^{(\psi_{\mathbf{z}})}$ in this basis. On all edges $E \in \mathscr{E}(K)$ with $\mathbf{z} \notin \mathscr{N}(E)$ and on all faces $F \in \mathscr{F}(K)$ with $\mathbf{z} \notin \mathscr{N}(F)$, the function $\psi_{\mathbf{z}}$ vanishes and it remains to consider the edges and faces with $\mathbf{z} \in \mathscr{N}(E)$ and $\mathbf{z} \in \mathscr{N}(F)$, respectively.

On Edges

For the pure diffusion problem, the basis functions are obviously linear along the edges with prescribed data in the nodes, which is either zero or one. In the convection-diffusion-reaction regime, however, L_E describes an ordinary differential operator of second order with constant and scalar-valued coefficients, cf. (6.29). Thus the boundary value problems on the edges are solved analytically and $\psi_{\mathbf{z}}$ can be written in closed form on each edge $E \in \mathscr{E}_h$. If $c_E = 0$, for instance, a small exercise shows

$$\psi_{\mathbf{z}}(x_1) = \psi_{\mathbf{z}}(0) + \big(\psi_{\mathbf{z}}(h_E) - \psi_{\mathbf{z}}(0)\big)\frac{1 - \exp\left(\frac{\mathbf{b}_E}{A_E} h_E x_1\right)}{1 - \exp\left(\frac{\mathbf{b}_E}{A_E} h_E\right)} \quad \text{for } x_1 \in [0, h_E] . \tag{6.32}$$

Here, we recognize how the data in the nodes for $x_1 = 0$ and $x_1 = h_E$ enter the formula.

For the diffusion problem we can express the linear function $\psi_{\mathbf{z}}\big|_E$ exact in the trace space of $\mathscr{P}^1_{\mathrm{pw}}(\mathscr{T}_l(\partial K))$ on E. For the convection-diffusion-reaction problem, however, we make use of the induced discretization of E into line segments and we interpolate $\psi_{\mathbf{z}}\big|_E$, cf. (6.32), by $g_{l,E}^{(\psi_{\mathbf{z}})}$ in the space of piecewise linear functions over the discretization of E. Finally, the functions $g_{l,E}^{(\psi_{\mathbf{z}})}$ on the edges $E \in \mathscr{E}(F)$ are combined in order to obtain the Dirichlet data $g_{l,\partial F}^{(\psi_{\mathbf{z}})}$ on ∂F for the 2D problems on the faces of the element.

On Faces

The variational formulation for $\psi_{\mathbf{z}}\big|_F$ reads analog to (6.27). The non-homogeneous Dirichlet data is treated as usual in the Galerkin formulation. Therefore, we interpret $g_{l,\partial F}^{(\psi_{\mathbf{z}})}$ as extension into $\mathscr{P}^1_{\mathrm{pw}}(\mathscr{T}_l(F))$. Furthermore, we denote by $\Phi_{D,F}$ the set of basis functions from Φ_D with support in F, such that

$$\operatorname{span} \Phi_{D,F} = \mathscr{P}^1_{\mathrm{pw}}(\mathscr{T}_l(F)) \cap H^1_0(F)\,.$$

In the case of the pure diffusion problem, the discrete Galerkin formulation for the approximation of the basis functions on the faces $F \in \mathscr{F}(K)$ reads:

$$\text{Find } g_{l,F}^{(\psi_{\mathbf{z}})} \in g_{l,\partial F}^{(\psi_{\mathbf{z}})} + \operatorname{span} \Phi_{D,F} : \qquad \int_F \nabla g_{l,F}^{(\psi_{\mathbf{z}})} \cdot \nabla \varphi \,\mathrm{d}s_{\mathbf{x}} = 0 \quad \forall \varphi \in \Phi_{D,F}\,.$$

We point out that the boundary data on the edges is linear in this case. Hence, it is represented exact in the space of piecewise polynomials. Furthermore, if the faces $F \in \mathscr{F}(K)$ are already triangles, we recover the basis functions discussed in the simple generalization to 3D in Sect. 2.3.4.

In the case of the convection-diffusion-reaction equation we might encounter convection-dominated problems. Consequently, we propose to utilize a stabilized FEM on the faces. We choose the Streamline Upwind/Petrov-Galerkin (SUPG) method [48] such that the discrete formulation for the approximation of the basis functions on the faces reads:

$$\begin{aligned}
&\text{Find } g_{l,F}^{(\psi_{\mathbf{z}})} \in g_{l,\partial F}^{(\psi_{\mathbf{z}})} + \operatorname{span} \Phi_{D,F} : \\
&\int_F (A_F \nabla g_{l,F}^{(\psi_{\mathbf{z}})} \cdot \nabla \varphi + \mathbf{b}_F \cdot \nabla g_{l,F}^{(\psi_{\mathbf{z}})}\, \varphi + c_F g_{l,F}^{(\psi_{\mathbf{z}})}\, \varphi) \,\mathrm{d}s_{\mathbf{x}} \qquad (6.33)\\
&\qquad + \delta_F \int_F (\mathbf{b}_F \cdot \nabla g_{l,F}^{(\psi_{\mathbf{z}})}\, \mathbf{b}_F \cdot \nabla \varphi + c_F g_{l,F}^{(\psi_{\mathbf{z}})}\, \mathbf{b}_F \cdot \nabla \varphi) \,\mathrm{d}s_{\mathbf{x}} = 0 \quad \forall \varphi \in \Phi_{D,F}\,,
\end{aligned}$$

where $\delta_F \geq 0$ is a stabilization parameter which is set to zero in the diffusion-dominated case. The choice of δ_F is discussed in more detail in Sect. 6.2.6. On all faces $F \in \mathscr{F}_h$ with $\mathbf{z} \notin \mathscr{N}(F)$, it is $g_{l,F}^{(\psi_{\mathbf{z}})} = 0$. Here, we point out that the boundary data on the edges is not polynomial in general, cf. (6.32). Thus, $g_{l,\partial F}^{(\psi_{\mathbf{z}})}$ is an approximation of the actual data.

Finally, the functions $g_{l,F}^{(\psi_{\mathbf{z}})}$ on the faces $F \in \mathscr{F}(K)$ are combined in order to obtain the Dirichlet data $g_{l,\partial K}^{(\psi_{\mathbf{z}})} \in \mathscr{P}_{\mathrm{pw}}^1(\mathscr{T}_l(\partial K))$ for the 3D problems on the whole boundary of the element. This construction is well defined since the triangulations of the faces form a conforming discretization of the surface ∂K, cf. Fig. 6.7.

On Elements

After we have computed the Dirichlet traces $g_l^{(\psi_{\mathbf{z}})}$ of all the approximate basis functions $\psi_{\mathbf{z},l}$ on the skeleton of the discretization, i.e., on the boundaries of the polyhedral elements, the three-dimensional local problems are treated by means of boundary integral equations and they are approximated by the boundary element method. For the pure diffusion problem we proceed as discussed in Chap. 4. Consequently, we have the Steklov–Poincaré operator (4.7), which maps the Dirichlet to the Neumann trace, and the representation formula (4.3) for the evaluation of the approximation inside the elements. The approximation of the Steklov–Poincaré operator and the representation formula are given in (4.20) and (4.22), respectively. In particular, the approximation space in the 2D FEM on the faces has been chosen in such a way that $g_{l,\partial K}^{(\psi_{\mathbf{z}})} \in \mathscr{P}_{\mathrm{pw}}^1(\mathscr{T}_l(\partial K))$ for $\psi_{\mathbf{z},l}$ with $\mathbf{z} \in \mathscr{N}(K)$ and $K \in \mathscr{K}_h$. Hence, we can apply directly the results of Sect. 4.3.

The boundary element method is not restricted to the Laplace equation. It generalizes to a large class of problems where the corresponding fundamental solutions are known. This is in particular true for the convection-diffusion-reaction equation. Here, the fundamental solution depends on A_K, $\mathbf{b}_K$ as well as on c_K and consequently on the element $K \in \mathscr{K}_h$. In $\mathbb{R}^3$ and under the assumption $c_K + \|\mathbf{b}_K\|_{A_K^{-1}}^2 \geq 0$, we have

$$U_K^*(\mathbf{x},\mathbf{y}) = \frac{1}{4\pi\sqrt{\det A_K}} \frac{\exp\left(\mathbf{b}_K^\top A_K^{-1}(\mathbf{x}-\mathbf{y}) - \lambda\|\mathbf{x}-\mathbf{y}\|_{A_K^{-1}}\right)}{\|\mathbf{x}-\mathbf{y}\|_{A_K^{-1}}} \quad \text{for } \mathbf{x},\mathbf{y} \in \mathbb{R}^3 ,$$

where

$$\|\mathbf{x}\|_{A_K^{-1}} = \sqrt{\mathbf{x}^\top A_K^{-1}\mathbf{x}} \quad \text{and} \quad \lambda = \sqrt{c_K + \|\mathbf{b}_K\|_{A_K^{-1}}^2} .$$

With the help of $U_K^*(\cdot,\cdot)$, which satisfies

$$\mathrm{L}_{\mathbf{y}}\, U_K^*(\mathbf{x},\mathbf{y}) = \delta_0(\mathbf{y}-\mathbf{x})$$

for the convection-diffusion-reaction operator L, where δ_0 is the Dirac delta distribution, we can formulate the boundary integral operators as in Sect. 4.2. Since L is not a self-adjoint operator, we have to distinguish between the conormal derivative $\gamma_1^K v$, which is given for sufficiently smooth v as

$$(\gamma_1^K v)(\mathbf{x}) = \mathbf{n}_K(\mathbf{x}) \cdot (\gamma_0^K A_K \nabla v)(\mathbf{x}) \quad \text{for } \mathbf{x} \in \partial K ,$$

and the modified conormal derivative

$$\widetilde{\gamma_1^K} v = \gamma_1^K v + (\mathbf{b}_K \cdot \mathbf{n}_K)\gamma_1^K v , \tag{6.34}$$

which is associated with the adjoint problem. The conormal derivative is also called Neumann trace. For $\mathbf{x} \in \partial K$, we have the single-layer potential operator

$$(\mathbf{V}_K \zeta)(\mathbf{x}) = \gamma_0^K \int_{\partial K} U_K^*(\mathbf{x}, \mathbf{y})\zeta(\mathbf{y})\, \mathrm{d}s_{\mathbf{y}} \quad \text{for } \zeta \in H^{-1/2}(\partial K) ,$$

the double-layer potential operator

$$(\mathbf{K}_K \xi)(\mathbf{x}) = \lim_{\varepsilon \to 0} \int_{\mathbf{y} \in \partial K : |\mathbf{y} - \mathbf{x}| \geq \varepsilon} \widetilde{\gamma_{1,\mathbf{y}}^K} U_K^*(\mathbf{x}, \mathbf{y})\xi(\mathbf{y})\, \mathrm{d}s_{\mathbf{y}} \quad \text{for } \xi \in H^{1/2}(\partial K) ,$$

and the adjoint double-layer potential operator

$$(\mathbf{K}_K' \zeta)(\mathbf{x}) = \lim_{\varepsilon \to 0} \int_{\mathbf{y} \in \partial K : |\mathbf{y} - \mathbf{x}| \geq \varepsilon} \gamma_{1,\mathbf{x}}^K U_K^*(\mathbf{x}, \mathbf{y})\zeta(\mathbf{y})\, \mathrm{d}s_{\mathbf{y}} \quad \text{for } \zeta \in H^{-1/2}(\partial K) ,$$

as well as the hypersingular integral operator

$$(\mathbf{D}_K \xi)(\mathbf{x}) = -\gamma_1^K \int_{\partial K} \widetilde{\gamma_{1,\mathbf{y}}^K} U_K^*(\mathbf{x}, \mathbf{y})\xi(\mathbf{y})\, \mathrm{d}s_{\mathbf{y}} \quad \text{for } \xi \in H^{1/2}(\partial K) .$$

These operators have the same mapping properties as the corresponding integral operators for the Laplace operator. We point out that they differ in the fundamental solution $U_K^*(\cdot, \cdot)$ and the use of the modified conormal derivative. As in Chap. 4, we have a representation formula and two representations of the Steklov–Poincaré operator, which maps the Dirichlet to the Neumann trace

$$\gamma_1^K u = \mathbf{S}_K \gamma_0^K u ,$$

in terms of the boundary integral operators:

$$\mathbf{S}_K = \mathbf{V}_K^{-1}(\tfrac{1}{2}\mathbf{I} + \mathbf{K}_K) = \mathbf{D}_K + (\tfrac{1}{2}\mathbf{I} + \mathbf{K}_K')\mathbf{V}_K^{-1}(\tfrac{1}{2}\mathbf{I} + \mathbf{K}_K) , \tag{6.35}$$

provided that $\mathbf{V}_K$ is invertible. The invertibility of the single-layer potential operator $\mathbf{V}_K$ is shown for some special cases, like the Laplace operator or when the material parameters satisfy $a_{\min} c_K > |\mathbf{b}_K|^2$, where $a_{\min}$ is the minimal eigenvalue of A_K, see [151]. For general elliptic operators as in (6.26) with constant coefficients, Costabel [61] has shown that the single-layer potential is a strongly elliptic operator and thus satisfies a Gårding inequality. The discretization of these boundary integral operators follows the line of Sect. 4.3, where the boundary mesh $\mathscr{B}_h$ is chosen to be $\mathscr{T}_l(\partial K)$. Hence, we obtain the corresponding boundary element matrices $\mathbf{V}_{K,l}$, $\mathbf{M}_{K,l}$, $\mathbf{K}_{K,l}$ and so on.

6.2.4 *Fully Discrete Galerkin Formulation*

We consider the convection-diffusion-reaction equation only, since it includes the pure diffusion problem, and we restrict ourselves for shorter notation to Dirichlet boundary conditions and a vanishing source term as in (6.26). Instead of applying the approximation space (6.30) with the implicitly defined basis functions on edges, faces and elements, we use the spaces

$$V_{h,l} = \operatorname{span}\{\psi_{\mathbf{z},l} : \mathbf{z} \in \mathscr{N}_h\} \quad \text{and} \quad V_{h,l,D} = V_{h,l} \cap H_0^1(\Omega)\,,$$

which are spanned by the approximated basis function $\psi_{\mathbf{z},l}$ as described in the previous Sect. 6.2.3. This approximation space is conforming, i.e. $V_{h,l} \subset H^1(\Omega)$, due to the continuity of the functions in $V_{h,l}$ over edges as well as faces and because of the regularity of the local problems defining the basis functions. The discrete Galerkin formulation reads:

$$\text{Find } u_{h,l} \in g_D + V_{h,l,D}: \quad b(u_{h,l}, v_{h,l}) = 0 \quad \forall v_{h,l} \in V_{h,l,D}\,, \tag{6.36}$$

with bilinear form

$$b(u_{h,l}, v_{h,l}) = \int_\Omega \big(A\nabla u_{h,l} \cdot \nabla v_{h,l} + \mathbf{b} \cdot \nabla u_{h,l}\, v_{h,l} + c\, u_{h,l} v_{h,l}\big)\, \mathrm{d}\mathbf{x}\,.$$

For the realization of the bilinear form, we proceed as in Sect. 4.5. Integration by parts and the properties of $V_{h,l}$ yield

$$\begin{aligned} b(u_{h,l}, v_{h,l}) &= \sum_{K \in \mathscr{K}_h} \int_K \big(A_K \nabla u_{h,l} \cdot \nabla v_{h,l} + \mathbf{b}_K \cdot \nabla u_{h,l}\, v_{h,l} + c_K\, u_{h,l} v_{h,l}\big)\, \mathrm{d}\mathbf{x} \\ &= \sum_{K \in \mathscr{K}_h} \int_{\partial K} \gamma_1^K u_{h,l}\, \gamma_0^K v_{h,l}\, \mathrm{d}s_{\mathbf{x}} + \int_K \mathrm{L}\, u_{h,l}\, v_{h,l}\, \mathrm{d}\mathbf{x} \\ &= \sum_{K \in \mathscr{K}_h} \int_{\partial K} \mathbf{S}_K \gamma_0^K u_{h,l}\, \gamma_0^K v_{h,l}\, \mathrm{d}s_{\mathbf{x}}\,. \end{aligned}$$

Next, we replace the Steklov–Poincaré operator by its non-symmetric or symmetric representation, cf. (6.35), and approximate it by means of boundary element methods in analog to Sect. 4.5. Let $\underline{g}^{(u)}_{l,\partial K}$ and $\underline{g}^{(v)}_{l,\partial K}$ be the vectors with the expansion coefficients of $\gamma_0^K u_{h,l}$ and $\gamma_0^K v_{h,l}$ in $\mathscr{P}^1_{\mathrm{pw}}(\mathscr{T}_l(\partial K))$, respectively. These vectors are given as linear combinations of the coefficient vectors $\underline{g}^{(\psi_{\mathbf{z}})}_{l,\partial K}$ from the basis functions $\psi_{\mathbf{z}}$ computed in Sect. 6.2.3 on the faces. Consequently, we obtain for $b(\cdot,\cdot)$ the approximation

$$b_l(u_{h,l}, v_{h,l}) = \sum_{K\in\mathscr{K}_h} (\underline{g}^{(v)}_{l,\partial K})^\top S_{K,l}\, \underline{g}^{(u)}_{l,\partial K}\,,$$

where $S_{K,l} \in \mathbb{R}^{|\mathscr{M}_l(\partial K)|\times|\mathscr{M}_l(\partial K)|}$ is either

$$\mathbf{S}^{\mathrm{unsym}}_{K,l} = \mathbf{M}^\top_{K,l}\mathbf{V}^{-1}_{K,l}\left(\tfrac{1}{2}\mathbf{M}_{K,l} + \mathbf{K}_{K,l}\right),$$

when using the non-symmetric representation, or

$$\mathbf{S}_{K,l} = \mathbf{D}_{K,l} + \left(\tfrac{1}{2}\mathbf{M}^\top_{K,l} + \mathbf{K}^\top_{K,l}\right)\mathbf{V}^{-1}_{K,l}\left(\tfrac{1}{2}\mathbf{M}_{K,l} + \mathbf{K}_{K,l}\right),$$

when using the symmetric representation of the Steklov–Poincaré operator. Here, the matrices in bold letters are the corresponding boundary element matrices for the convection-diffusion-reaction operator defined in Sect. 6.2.3. For this differential operator, however, the hypersingular integral operator $\mathbf{D}_K$ is not self-adjoint and hence, $\mathbf{D}_{K,l}$ is non-symmetric. Consequently, the symmetric representation of the Steklov–Poincaré operator yields a non-symmetric matrix $\mathbf{S}_{K,l}$.

Finally, the fully discrete Galerkin formulation reads:

$$\text{Find } u_{h,l} \in g_D + V_{h,l,D}: \quad b_l(u_{h,l}, v_{h,l}) = 0 \quad \forall v_{h,l} \in V_{h,l,D}\,. \tag{6.37}$$

The assembling of the global FEM matrix is performed as usual by adding up the local element-wise contributions. Therefore, let $D_K \in \mathbb{R}^{|\mathscr{M}_l(\partial K)|\times|\mathscr{N}(K)|}$ be the matrix obtained by gathering the vectors $\underline{g}^{(\psi_{\mathbf{z}})}_{l,\partial K} \in \mathbb{R}^{|\mathscr{M}_l(\partial K)|}$, $\mathbf{z} \in \mathscr{N}(K)$ with the expansion coefficients of $\gamma_0^K \psi_{\mathbf{z},l}$ in $\mathscr{P}^1_{\mathrm{pw}}(\mathscr{T}_l(\partial K))$ computed in Sect. 6.2.3. The matrix

$$D_K^\top S_{K,l} D_K \in \mathbb{R}^{|\mathscr{N}(K)|\times|\mathscr{N}(K)|}$$

with either $S_{K,l} = \mathbf{S}^{\mathrm{unsym}}_{K,l}$ or $S_{K,l} = \mathbf{S}_{K,l}$ serves as local stiffness matrix in the BEM-based FEM simulation. At this point we emphasize that the local auxiliary triangulations $\mathscr{T}_l(\partial K)$ are used only to compute the element stiffness matrices. The level of refinement l chosen for them has no influence on the size of the global FEM system.

6.2.5 Numerical Experiments: Diffusion Problem

In the setup of the local boundary element matrices, we use a semi analytical integration scheme. The inner integral in the Galerkin matrices is evaluated analytically and the outer one is approximated by Gaussian quadrature. For the assembling of the global FEM system matrix we use locally the stiffness matrices resulting from the non-symmetric representation of the Steklov–Poincaré operator, see Sect. 6.2.4. Since this formulation yields non-symmetric local stiffness matrices although the bilinear form is symmetric, we apply a symmetrization in order to retain the symmetry. We write

$$\begin{aligned} b(u_{h,l}, v_{h,l}) &= \sum_{K\in\mathscr{K}_h} \int_K a_K \nabla u_{h,l} \cdot \nabla v_{h,l} \mathrm{d}\mathbf{x} \\ &= \sum_{K\in\mathscr{K}_h} \frac{a_K}{2} \Big(\int_{\partial K} \gamma_1^K u_{h,l}\, \gamma_0^K v_{h,l}\, \mathrm{d}s_{\mathbf{x}} + \int_{\partial K} \gamma_1^K v_{h,l}\, \gamma_0^K u_{h,l}\, \mathrm{d}s_{\mathbf{x}} \Big) , \end{aligned}$$

and use the approximation

$$b_l(u_{h,l}, v_{h,l}) = \sum_{K\in\mathscr{K}_h} \frac{a_K}{2} (\underline{g}_{l,\partial K}^{(v)})^\top \Big(\mathbf{S}_{K,l}^{\text{unsym}} + (\mathbf{S}_{K,l}^{\text{unsym}})^\top \Big) \underline{g}_{l,\partial K}^{(u)} ,$$

which yields locally the symmetric stiffness matrix

$$\frac{a_K}{2} D_K^\top \Big(\mathbf{S}_{K,l}^{\text{unsym}} + (\mathbf{S}_{K,l}^{\text{unsym}})^\top \Big) D_K \in \mathbb{R}^{|\mathscr{N}(K)|\times|\mathscr{N}(K)|} .$$

The symmetric systems of linear equations arising on the faces and in the global FEM system are solved by the conjugate gradient method [90] without any preconditioning. Of course, for larger problems a more efficient solver is of particular interest. It is possible to use FETI-type strategies, for instance. The application of such solvers to the BEM-based FEM has been studied in [94, 97].

The first numerical example in this section is formulated on the unit cube. We utilize Voronoi meshes which are a particular example of polyhedral meshes. In Fig. 6.8, the first three meshes of the sequence are visualized which are used for the convergence experiments. We see that the elements are non-trivial polyhedra with arbitrary polygonal faces. The meshes have been produced by generating random points according to [71] and constructing the corresponding Voronoi diagram in accordance with [70]. It is assumed that the mesh generator provides the points $\mathbf{z}_K$ and $\mathbf{z}_F$ from the Definitions 2.10 and 2.11. However, for convex elements and faces we may use the center of mass instead which is computable.

In Table 6.1, we sketch the number of elements $|\mathscr{K}_h|$ and the number of nodes $|\mathscr{N}_h|$ in the different Voronoi meshes. The proposed strategy approximates the solution by a linear combination of as many basis function as nodes are in

Fig. 6.8 Sequence of Voronoi meshes

Table 6.1 Total number of nodes and elements when working with triangulated surfaces of different mesh levels l

$\lvert\mathcal{K}_h\rvert$	$\lvert\mathcal{N}_h\rvert$	$l = 0$	$l = 1$	$l = 2$
9	46	98	424	1790
76	416	905	4170	18,011
712	4186	9081	42,446	184,170
1316	7850	17,013	79,676	345,903
5606	34,427	74,457	349,663	1,519,143
26,362	164,915	356,189	1,675,171	7,280,603

the mesh. Therefore, the number of degrees of freedom in the BEM-based FEM is $|\mathcal{N}_h|$ minus the number of nodes on the Dirichlet boundary Γ_D. The simple generalization for the first order method from Sect. 2.3.4, initially proposed in [60], needs to triangulate the surfaces of the elements and the number of basis functions corresponds to the total number of nodes after the triangulation. In Table 6.1, this total number of nodes is listed in the case that the faces are triangulated with the level $l = 0, 1, 2$, cf. Fig. 6.9. We recognize that in this situation much more basis functions and thus degrees of freedom are required in the global computations. Roughly speaking, the number of nodes doubles if the coarsest discretization of the faces is used. If a finer triangulation is needed, the number of nodes and thus the number of degrees of freedom increase 10 times for $l = 1$ and even more than forty times for $l = 2$. Since the diameter of the elements are equal in all four situations, the approximation errors of the finite element computations are of the same order for fixed l. However, the constant in [60] might be better than the one obtained for the presented strategy for vanishing right hand side in the differential equation. This is due to the fact that if h is fixed and only l is increased, the method in [60] still converges since it is equivalent to a boundary element domain decomposition approach [106]. The hierarchical construction proposed in this chapter gives for small l comparable approximations while requiring a minimal set of degrees of freedoms.

In the following, we investigate the influence of the face discretization. These triangulations of the faces are utilized to define the approximated basis functions

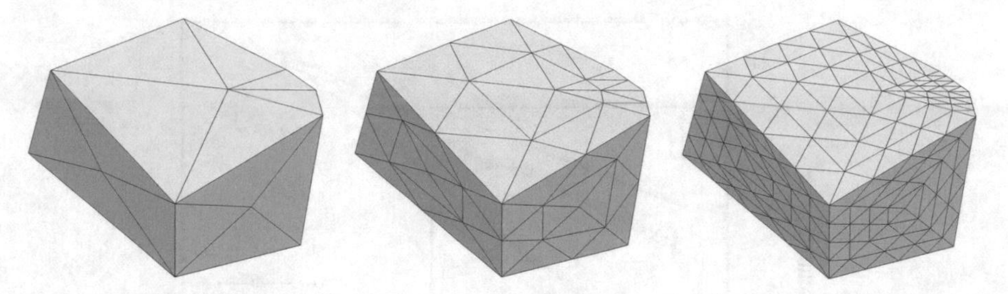

Fig. 6.9 Polyhedral element with surface triangulations of level $l = 0, 1, 2$

Table 6.2 Number of nodes $|\mathscr{M}_l(\partial K)|$ and number of triangles $|\mathscr{T}_l(\partial K)|$ in the surface discretization of the element in Fig. 6.9 for different levels

$\lvert\mathscr{N}(K)\rvert$	l	$\lvert\mathscr{M}_l(\partial K)\rvert$	$\lvert\mathscr{T}_l(\partial K)\rvert$
12	0	20	36
	1	74	144
	2	290	576
	3	1154	2304
	4	4610	9216

$\psi_{\mathbf{z},l}$ on the faces with the help of local, two-dimensional finite element methods. The finer the discretization is chosen the better we approximate the original basis functions $\psi_{\mathbf{z}}$. Even though, the face discretization does not blow up the global system matrix, the computational effort for the local problems increases if the discretization level l is raised. As one example, we pick the element K from Fig. 6.9 and list the number of nodes $|\mathscr{M}_l(\partial K)|$ and the number of triangles $|\mathscr{T}_l(\partial K)|$ in the surface discretization of K for different levels l in Table 6.2. The main tasks in the local problems are the evaluation of the boundary element matrix entries and the inversion of the single-layer potential matrix $\mathbf{V}_{K,l}$, which gives a local complexity of $\mathscr{O}(|\mathscr{T}_l(\partial K)|^3)$.

Next, the rates of convergence are analysed for different values of l. Therefore, consider the Dirichlet boundary value problem

$$-\Delta u = 0 \quad \text{in } \Omega = (0,1)^3\,, \qquad u = g_D \quad \text{on } \Gamma\,,$$

on the sequence of Voronoi meshes, where g_D is chosen such that

$$u(\mathbf{x}) = \exp(2\sqrt{2}\pi(x_1 - 0.3))\cos(2\pi(x_2 - 0.3))\sin(2\pi(x_3 - 0.3)) \qquad (6.38)$$

is the exact solution. The relative errors in the energy and L_2-norm, i.e.

$$\frac{\|u - u_{h,l}\|_b}{\|u\|_b} \quad \text{and} \quad \frac{\|u - u_{h,l}\|_{L_2(\Omega)}}{\|u\|_{L_2(\Omega)}}\,,$$

are given in Fig. 6.10 with respect to $h = \max\{h_K : K \in \mathscr{K}_h\}$ in logarithmic scale for different discretization levels $l = 0, 1, 2$. This example shows that

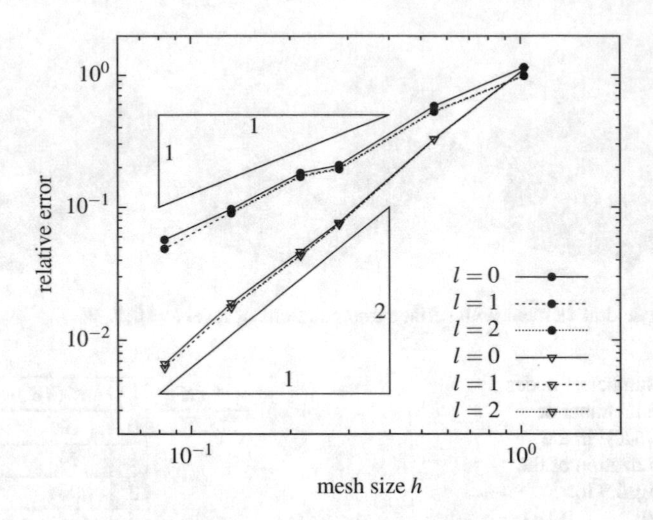

Fig. 6.10 Relative error in $\|\cdot\|_b$ (•) and $\|\cdot\|_{L_2(\Omega)}$ (▽) with respect to h for levels $l = 0, 1, 2$ in the example with solution (6.38)

the discretization level of the faces does not influence the rates of convergence as proofed in [147]. Additionally, Fig. 6.10 indicates that the constant in the error estimate can be chosen to be independent of the level l. The coarsest face discretization with $l = 0$ is sufficient to analyse the convergence rates in the forthcoming numerical experiments. Due to this choice, the local complexity in the two-dimensional finite element method on the faces $F \in \mathscr{F}_h$ and the local boundary element methods on the elements $K \in \mathscr{K}_h$ is rather small. Furthermore, in Fig. 6.10, we recognize linear convergence for the approximation error measured in the energy norm and quadratic convergence if the error is measured in the L_2-norm as expected, see [147].

Finally, we consider the model problem on a L-shaped domain with a singular solution such that $u \notin H^2(\Omega)$, but $u \in H^{5/3}(\Omega)$. Due to the theory of interpolation spaces, see, e.g., [34], we expect a convergence order of $2/3$. With the help of cylindrical coordinates (r, ϕ, x_3), where $r \geq 0$, $\phi \in [\pi/2, 2\pi]$ and $x_3 \in \mathbb{R}$, the function

$$u(r\cos\phi, r\sin\phi, x_3) = r^{2/3}\sin(\tfrac{2}{3}(\phi - \tfrac{\pi}{2})) \in H^{5/3}(\Omega) \tag{6.39}$$

satisfies the Laplace equation in the L-shaped domain

$$\Omega = \big((-1,1)\times(-1,1)\times(0,1)\big) \setminus [0,1]^3$$

with appropriate Dirichlet data. The boundary value problem is solved by means of the BEM-based FEM on a sequence of polyhedral meshes made of polygonal bricks, i.e., the meshes contain as elements prisms having general polygonal ends. In Fig. 6.11, we give the initial mesh of the domain Ω with hanging nodes and

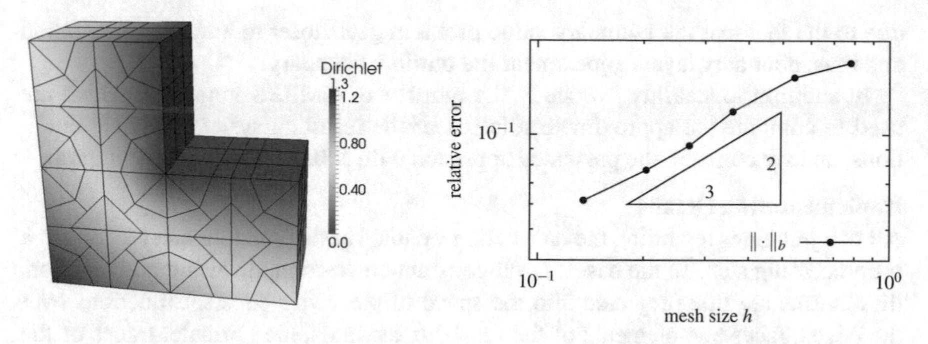

Fig. 6.11 L-shaped domain with polyhedral mesh made of bricks and Dirichlet data (left) and relative error $\|u - u_{h,l}\|_b / \|u\|_b$ with respect to h for $l = 0$ in the example with solution (6.39) (right)

edges. Additionally, we show the relative error $\|u - u_{h,l}\|_b / \|u\|_b$ with respect to $h = \max\{h_K : K \in \mathscr{K}_h\}$ in logarithmic scale. As expected, we obtain the reduced order of convergence for a sequence of uniform refined meshes. To recover the linear convergence in the energy norm for singular solutions, it is necessary to perform adaptive mesh refinement strategies as discussed in Chap. 5.

6.2.6 Numerical Experiments: Convection-Diffusion Problem

In this section, we give some implementation details as well as numerical experiments for the convection-diffusion problem. The computations are done on tetrahedral and polyhedral meshes. For the sake of simplicity, we restrict ourselves to the case of scalar valued diffusion coefficients, i.e., $A = \alpha I$ for some $\alpha > 0$, and a vanishing reaction term $c = 0$. Furthermore, the experiments are carried out with constant and continuously varying convection vector $\mathbf{b}$. Remember, that we have to approximate the coefficients α and $\mathbf{b}$ by constants on each geometrical object for the BEM-based FEM, see Sect. 6.2.2. The method is studied for the case of decreasing diffusion $\alpha \to 0$. Standard numerical schemes like the finite element method become unstable when applied to this type of convection-dominated problems. Typically, the issue manifests itself in the form of spurious oscillations. The critical quantity here is the mesh Péclet number

$$\mathrm{Pe}_K = \frac{h_K |\mathbf{b}_K|}{\alpha_K}, \quad K \in \mathscr{K}_h ,$$

which should be bounded by 2 for standard finite element methods. In the numerical experiments, we give $\mathrm{Pe}_h = \max\{\mathrm{Pe}_K : K \in \mathscr{K}_h\}$. When decreasing the diffusion for fixed h, the mesh Péclet number increases and we expect oscillations. This is due to the fact that the boundary value problem gets closer to a transport equation and thus, boundary layers appear near the outflow boundary.

In addition to stability, we study the number of GMRES iterations, which are used to compute the approximate solution of the resulting system of linear equations, and we compare the presented approach with a 3D SUPG implementation.

Implementation Details

All computations regarding the convection-adapted basis functions can be done in a preprocessing step. In the case of non-constant convection, diffusion and reaction, these terms are first projected into the space of piecewise constant functions over the edges, faces and elements of the mesh. Afterwards, the Dirichlet traces of the basis functions are computed on the edges and faces. Here, an analytic formula is utilized on each edge $E \in \mathscr{E}_h$, and subsequently, the two-dimensional convection-diffusion-reaction problems are treated separately on each face $F \in \mathscr{F}_h$ according to the SUPG formulation (6.33). Let the local Péclet number be defined by

$$\mathrm{Pe}_{F,T} = \frac{h_T |\mathbf{b}_F|}{\alpha_F} \quad \text{for } T \in \mathscr{T}_l(F) .$$

The stabilization parameter δ_F in the SUPG method is chosen to be piecewise constant over the auxiliary triangulation $\mathscr{T}_l(F)$ on each face $F \in \mathscr{F}_h$. The choice

$$\delta_{F,T} = \begin{cases} c_1 h_T / 2 & \text{for } \mathrm{Pe}_{F,T} > 2 , \\ c_2 h_T^2 / \alpha_F & \text{else} , \end{cases}$$

leads to the best possible convergence rate of the discrete solution with respect to the streamline diffusion norm on F, see [149]. However, an 'optimal' choice of the constants c_1 and c_2 is not known. Since we aim to omit additional user defined parameters, the choice

$$\delta_{F,T} = \frac{h_T}{2|\mathbf{b}_F|} \left(\frac{1}{\tan(\mathrm{Pe}_{F,T})} - \frac{1}{\mathrm{Pe}_{F,T}} \right), \tag{6.40}$$

is preferred in the numerical realization, see [111].

The auxiliary triangulations $\mathscr{T}_l(F)$ of level $l \in \mathbb{N}_0$ are constructed as described in Sect. 2.2 and visualized in Figs. 2.4 and 6.9, for example. But, in case of convection-dominated problems on the faces, we decided to move the midpoint $\mathbf{z}_F$ of the mesh, created in $\mathscr{T}_0(F)$, into the direction of the projected convection vector $\mathbf{b}_F$, see Fig. 6.12 (middle). If $\kappa > 0$ is such that $\mathbf{z}_F + \kappa \mathbf{b}_F \in \partial F$, then the translation can be chosen as

$$\mathbf{z}_F \mapsto \mathbf{z}_F + (1 - \vartheta)\kappa \mathbf{b}_F , \quad \text{with transition point} \quad \vartheta = \min\left\{ \frac{1}{2}, \frac{\alpha_F}{|\mathbf{b}_F|} \log(l+1) \right\} .$$

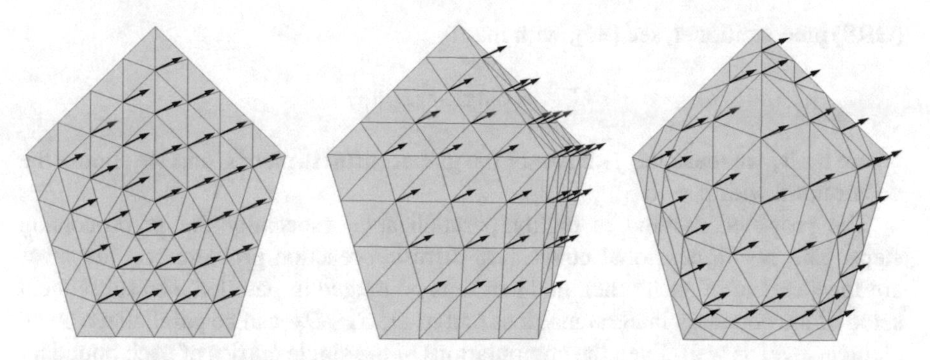

Fig. 6.12 Adaptation of auxiliary triangulation $\mathscr{T}_2(F)$ and projected convection vector: without adaptation (left), by moving the midpoint (middle), as layered mesh (right)

Consequently, the auxiliary meshes get adapted to the local problems. This adaptation is inspired by Shishkin-meshes [156], see also [116, 123, 149], which are graded in such a way that boundary layers are resolved. Another mesh adaptation is to construct layered meshes. This can be achieved as follows. We compute again the point $\mathbf{z}_F + (1 - \vartheta)\kappa\mathbf{b}_F$, but this time, we move the edges created in $\mathscr{T}_1(F)$ that are parallel to ∂F towards the boundary ∂F such that all edges have the same distance to ∂F and one edge lies on the computed point, see Fig. 6.12 (right). In the numerical realization, we set $\vartheta = 0.25$ independent of the local Péclet numbers. Otherwise, the auxiliary triangulations $\mathscr{T}_l(F)$ degenerate in the computations for small $l = 1, 2, 3$ and large Péclet numbers. Furthermore, we only present the results for the first mentioned mesh adaptation technique since the computed values in the experiments differ slightly.

The solutions of the resulting systems of linear equations, coming from the SUPG formulation, with non-symmetric, sparse matrices are approximated with the help of the GMRES method, see [150]. As the stopping criterion, we use the reduction of the norm of the initial residual by a factor of 10^{-10}.

Another preprocessing step is the computation of the matrices arising from the local boundary integral formulations. Here, we use the BEM code developed in the PhD thesis by Hofreither [94], which is based on a fully numerical integration scheme described in [151]. The inversions of the local single-layer potential matrices $\mathbf{V}_{K,l}$ are performed with an efficient LAPACK [6] routine.

The assembling of the global stiffness matrix is performed element-wise as described in Sect. 6.2.4 utilizing the non-symmetric representation of the Steklov–Poincaré operator in the local stiffness matrices. The resulting system of linear equations, which is again sparse and non-symmetric, is solved by GMRES. For the global problem, however, we use the reduction of the norm of the initial residual by a factor of 10^{-6} as the stopping criterion. In our numerical experiments, the GMRES iterations are carried out without preconditioning in general. However, we also implemented a simple diagonal preconditioner, namely a geometric row scaling

(GRS) preconditioner, see [86], with matrix

$$C^{-1} = \mathrm{diag}(1/\|B_j\|_p) ,$$

where by B_j we mean the j-th row of the global stiffness matrix, and we choose the vector norm with $p = 1$.

The proposed method is highly parallelizable, especially the preprocessing steps. The two-dimensional convection-diffusion-reaction problems on the faces are independent of each other, and can thus be treated in parallel. The subsequent setup of the boundary integral matrices and of $D_K^\top S_{K,l} D_K$ can be parallelized on an element level as well. Even the computations of the single entries of each boundary integral matrix are independent of each other.

In the implementation we use another observation to reduce the computational complexity. In the case of constant convection, diffusion and reaction terms, the local boundary integral matrices and the problems on the edges and faces are identical for elements which differ by some translation only. Therefore, we build a lookup table in a preprocessing step such that redundant computations are avoided.

Experiment 1

In the first numerical experiment, a problem with constant convection and diffusion terms is studied. Let $\Omega = (0, 1)^3$, and let us consider the boundary value problem

$$-\alpha \Delta u + \mathbf{b} \cdot \nabla u = 0 \quad \text{in } \Omega , \qquad u = g_D \quad \text{on } \Gamma ,$$

where $\mathbf{b} = (1, 0, 0)^\top$ and $g_D(\mathbf{x}) = x_1 + x_2 + x_3$. The domain Ω is discretized with tetrahedral elements, see Fig. 6.13. The discretization is constructed with the help of a uniform mesh with $8 \times 8 \times 8$ small cubes where each cube is split into 6 tetrahedra. Thus, the mesh consists of 3072 elements, 6528 faces, 4184 edges and 729 nodes

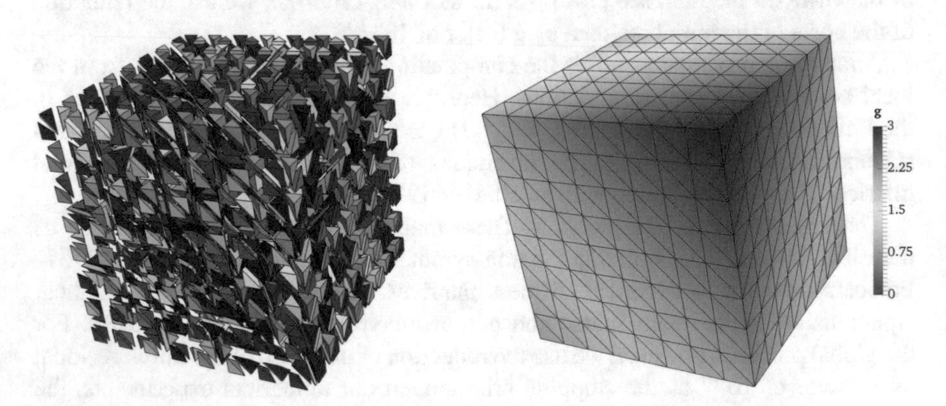

Fig. 6.13 Visualisation of tetrahedral mesh and Dirichlet data for Experiment 1

of which 343 nodes lie in the interior of Ω. Consequently, the number of degrees of freedom in the BEM-based FEM is equal to 343 in this example. The maximal element diameter is $h \approx 0.22$. The mesh is chosen rather coarse, but it is well suited for the study of stability.

Since the convection and diffusion parameters are constant over the whole domain, the lookup table is applied to speed up the computations. Instead of the before mentioned numbers of geometrical object, we only have to treat 6 elements, 12 faces and 7 edges in the preprocessing step, where the traces of the basis functions are computed and the local stiffness matrices are set up.

To handle the Dirichlet boundary condition, we apply pointwise interpolation of the data g_D to obtain an extension into Ω. The interpolant is bounded by 0 from below and by 3 from above on Γ. The convection-diffusion problem satisfies the maximum principle [82, 140], and therefore, we know that $0 \leq u \leq 3$ everywhere for the exact solution. To study stability of the BEM-based FEM, the maximum principle is checked for the approximate solution $u_{h,l} \in V_{h,l}$ obtained by (6.37). Since the basis functions satisfy convection-diffusion problems on the faces and edges and since the maximum principle is also valid there, the maximal values of $u_{h,l}$ should by reached in the nodes of the mesh. However, because of oscillations coming from the SUPG method on the faces, the maximal values might be found at some auxiliary node. Consequently, the maximum principle is tested on the whole skeleton Γ_S.

Table 6.3 gives a comparison of the classical finite element method with continuous piecewise linear basis functions and without stabilization, the original BEM-based FEM proposed in [96] with linear basis functions on the faces and the hierarchical, convection-adapted BEM-based FEM with $l = 2$ discussed in this chapter. The classical FEM satisfies the discrete maximum principle until

Table 6.3 Verifying maximum principle in Experiment 1

		Classic FEM		BEM-based FEM			
				Original [96]		Hierarchical, $l = 2$	
α	Pe_h	$u_{\min}$	$u_{\max}$	$u_{\min}$	$u_{\max}$	$u_{\min}$	$u_{\max}$
1.0×10^{-1}	2	0.00	3.00	0.00	3.00	0.00	3.00
5.0×10^{-2}	4	0.00	3.00	0.00	3.00	0.00	3.00
2.5×10^{-2}	9	0.00	3.00	0.00	3.00	0.00	3.00
1.0×10^{-2}	22	−0.55	3.00	0.00	3.00	−0.01	3.00
5.0×10^{-3}	43	−1.14	3.00	0.00	3.00	−0.01	3.00
2.5×10^{-3}	87	−1.85	3.07	0.00	3.00	−0.01	3.00
1.0×10^{-3}	217			0.00	3.00	−0.01	3.00
5.0×10^{-4}	433			0.00	3.00	−0.01	3.00
2.5×10^{-4}	866			−142.89	399.06	−0.01	3.00
1.0×10^{-4}	2165			−68.85	41.00	−0.01	3.00
5.0×10^{-5}	4330					−0.01	3.08
2.5×10^{-5}	8660					−0.01	14.72

$\alpha = 2.5 \times 10^{-2}$, which corresponds to a Péclet number of 9. The BEM-based strategies, which incorporate the behaviour of the differential operator into the approximation space, are more stable. The method in [96] passes the test up to $\alpha = 5.0 \times 10^{-4}$, which corresponds to $\mathrm{Pe}_h = 433$. In the new, proposed method we might have oscillations occurring in the approximation of the basis functions satisfying convection-dominated problems on the faces. If we neglect these small deviations in the third digit after the decimal point, the proposed method reaches even $\alpha = 1.0 \times 10^{-4}$, i.e. $\mathrm{Pe}_h = 2165$, for $l = 2$ without violation of the maximum principle.

Next, we study the influence of the auxiliary triangulations of the faces on the convection-adapted BEM-based FEM. In Table 6.4, the minimal and maximal values $u_{\min}$ and $u_{\max}$ of the approximate solution are listed for different levels l of the auxiliary meshes. The higher l is chosen, the longer the discrete maximum principle is valid. For $l = 3$, we even have stability until $\alpha = 2.5 \times 10^{-5}$, i.e. $\mathrm{Pe}_h = 8660$. The enhanced stability can be explained by the improved approximations of the boundary value problems on the edges and faces used to construct the basis functions. Obviously, the local oscillations in the construction of basis functions are reduced such that they have less effect to the global approximation.

In Table 6.4, the number of GMRES iterations are given without preconditioning. The GMRES solver for the proposed BEM-based FEM converges faster than for the preceding scheme. For increasing l the convergence slightly improves. Furthermore, the iteration numbers stay bounded without the help of any preconditioning until the maximum principle is violated.

Experiment 2

In this numerical experiment, we compare the convection-adapted BEM-based FEM with a well established method for convection-dominated problems, namely the Streamline Upwind/Petrov–Galerkin (SUPG) finite element method. The three-dimensional SUPG formulation is analogous to (6.33) and the stabilization parameter is chosen according to (6.40). The implementation has been done in the software `FreeFem++`, see [89]. For the comparison, we solve again the problem given in Experiment 1 with the BEM-based FEM and the SUPG method on the coarse tetrahedral discretization. Both approximations have 343 degrees of freedom. Furthermore, a reference solution is computed by the SUPG method on a fine tetrahedral mesh constructed with the help of $128 \times 128 \times 128$ cubes.

Having a closer look at the considered problem, we decompose the boundary of $\Omega = (0, 1)^3$ into the inflow boundary, the outflow boundary and the characteristic boundary which are given by

$$\Gamma_{\mathrm{in}} = \{0\} \times (0,1) \times (0,1)\,, \qquad \Gamma_{\mathrm{out}} = \{1\} \times (0,1) \times (0,1)\,, \qquad \Gamma_{\mathrm{ch}} = \partial\Omega \backslash (\Gamma_{\mathrm{in}} \cup \Gamma_{\mathrm{out}})\,,$$

respectively. It is known, that the solution has an exponential layer near Γ_{out} and a characteristic/parabolic layer near Γ_{ch} in the convection-dominated regime, see [149]. The hierarchical construction of the basis functions for the BEM-based FEM is adapted to the exponential layers but not necessarily to the parabolic

Table 6.4 Verifying maximum principle and comparing GMRES-iterations in Experiment 1 for different mesh levels $l = 1, 2, 3$

		$l = 1$			$l = 2$			$l = 3$		
α	Pe_h	u_{min}	u_{max}	Iter.	u_{min}	u_{max}	Iter.	u_{min}	u_{max}	Iter.
5.0×10^{-3}	43	−0.01	3.00	28	−0.01	3.00	25	0.00	3.00	23
2.5×10^{-3}	87	−0.01	3.00	28	−0.01	3.00	26	0.00	3.00	24
1.0×10^{-3}	217	−0.01	3.00	28	−0.01	3.00	26	0.00	3.00	24
5.0×10^{-4}	433	−0.01	3.00	28	−0.01	3.00	25	0.00	3.00	23
2.5×10^{-4}	866	−0.01	3.00	28	−0.01	3.00	24	0.00	3.00	23
1.0×10^{-4}	2165	−0.01	5.19	30	−0.01	3.00	24	0.00	3.00	23
5.0×10^{-5}	4330	−6.72	169.93	49	−0.01	3.08	25	0.00	3.00	23
2.5×10^{-5}	8660	-4.8×10^{6}	1.3×10^{7}	302	−0.01	14.72	31	0.00	3.00	23
1.0×10^{-5}	21,651				-7.3×10^{3}	3.1×10^{4}	92	0.00	26.11	29
5.0×10^{-6}	43,301							−14.89	36.25	55

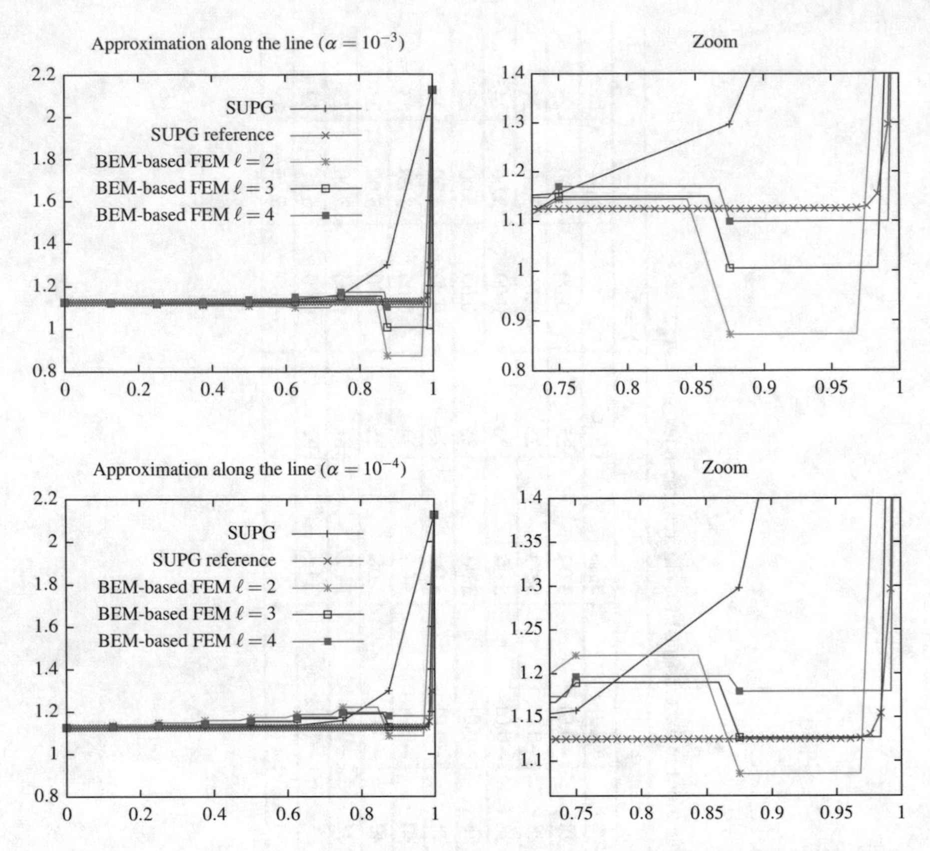

Fig. 6.14 Comparison of convection-adapted BEM-based FEM and SUPG approximation

layer because of the following reason: When we derived the local boundary value problems on the edges, faces and elements, we incorporate the behaviour in the direction of the convection, but we neglect the behaviour orthogonal to the faces, and thus, along the characteristic layer. Consequently, we should study exponential layers to see the advantages of the BEM-based FEM. Therefore, we compare the approximations along the line $s \mapsto (s, 5/8, 1/2)^\top$ for $s \in [0, 1]$, which is far from the characteristic boundary and which is aligned with edges of the discretization.

In Fig. 6.14, we give the approximations of the BEM-based FEM for different levels of the auxiliary triangulations of the faces, the SUPG approximation and the reference solution for $\alpha = 10^{-3}$ ($\mathrm{Pe}_h = 217$) and $\alpha = 10^{-4}$ ($\mathrm{Pe}_h = 2165$). The degrees of freedom are visualized by marks. The SUPG method shows no oscillations, but the layer in the solution is smeared out due to the stabilization. The SUPG approximations for $\alpha = 10^{-3}$ and $\alpha = 10^{-4}$ hardly differ although the layer in the solution changes. The convection-adapted BEM-based FEM has no explicit stabilization, and thus, we recognize some oscillations near the exponential layer. However, the layer is resolved much better with the same number of degrees of freedom. Additionally, we have the possibility to improve the accuracy of the shape functions within the BEM-based FEM by increasing the level l, i.e., by refining the auxiliary triangulations of the faces. Doing this, the oscillations near the exponential layer are reduced and we obtain very accurate solutions for the global problem with only a few degrees of freedom. If we have a closer look at the plots in the right column of Fig. 6.14 with the details near the exponential layer, the curves indicate that the layer of the solution is already smeared out for the reference solution computed with the SUPG method on a very fine mesh.

Experiment 3

In the final numerical experiment, we consider a convection-diffusion problem with non-constant convection vector. In order to compare the experiments, let $\Omega = (0, 1)^3$. We solve

$$-\alpha\Delta u + \mathbf{b}\cdot\nabla u = 0 \quad \text{in } \Omega\,, \qquad u = g_D \quad \text{on } \Gamma\,,$$

where

$$\mathbf{b}(\mathbf{x}) = \frac{0.85}{\sqrt{(1-x_1)^2 + (1-x_3)^2}} \begin{pmatrix} x_3 - 1 \\ 0 \\ 1 - x_1 \end{pmatrix}$$

and g_D is chosen such that it is piecewise bilinear and continuous with $0 \le g_D \le 3$ on one side of the unit cube and zero on all others, see Fig. 6.15. The convection vector $\mathbf{b}$ is scaled in such a way that the Péclet numbers in the computations are comparable with those of Experiment 1. The convection is a rotating field around the upper edge of the unit cube Ω, which lies in the front when looking at Fig. 6.15.

Fig. 6.15 Visualisation of polyhedral mesh and Dirichlet data for Experiment 3

Consequently, we expect that the non-zero Dirichlet data is transported towards the upper side of the cube for low diffusion.

This time, the domain Ω is decomposed into prisms having general polygonal ends, see Fig. 6.15. The polyhedral mesh consists of 350 elements, 1450 faces, 1907 edges and 808 nodes of which 438 nodes lie in the interior of Ω. Thus, the number of degrees of freedom in the BEM-based FEM is equal to 438. The maximal diameter of the elements is $h \approx 0.25$ and the discretization was chosen such that h is approximately the same as in Experiment 1.

In our experiment, the polyhedral mesh has less elements, faces and edges than the tetrahedral discretization. This is beneficial concerning the computations in the preprocessing step. Less local problems have to be solved on edges and faces and there are less boundary element matrices which have to be set up. Furthermore, polyhedral discretizations admit a high flexibility while meshing complex geometries. In Table 6.5, we list the minimal and maximal values of the approximation $u_{h,l}$ on the skeleton for $l = 2$ to verify the discrete maximum principle. Furthermore, the number of GMRES iterations are given with and without GRS preconditioning.

The first observation is that the number of GMRES iterations increases when the diffusion α tends to zero. Thus, the iteration count is not bounded in this experiment. However, this behaviour correlates with the violation of the maximum principle and is therefore the result of inaccuracies. Already with the help of the simple geometric row scaling preconditioner, we overcome the increase of the iteration number.

A more detailed discussion is needed for the discrete maximum principle. In Table 6.5, we observe that this principle is violated in a relatively early stage for $\alpha = 2.5 \times 10^{-2}$, which corresponds to $\mathrm{Pe}_h = 9$. However, the increase of $u_{\max}$ and the decrease of $u_{\min}$ is fairly slow for increasing Péclet number.

Table 6.5 Verifying maximum principle in Experiment 3 for $l = 2$ and number of iterations with/without preconditioning

α	Pe_h	$u_{\min}$	$u_{\max}$	Iter.	Iter. (prec.)
1.0×10^{-1}	2	0.00	3.00	20	20
5.0×10^{-2}	4	0.00	3.00	20	21
2.5×10^{-2}	9	0.00	3.04	20	21
1.0×10^{-2}	22	0.00	3.07	23	22
5.0×10^{-3}	43	−0.01	3.26	29	23
2.5×10^{-3}	86	−0.04	3.37	42	24
1.0×10^{-3}	216	−0.10	3.38	45	23
5.0×10^{-4}	431	−0.13	3.45	48	22
2.5×10^{-4}	863	−0.15	3.51	51	21
1.0×10^{-4}	2157	−0.15	3.53	52	21
5.0×10^{-5}	4313	−0.16	3.57	58	23
2.5×10^{-5}	8627	−0.25	4.38	69	28

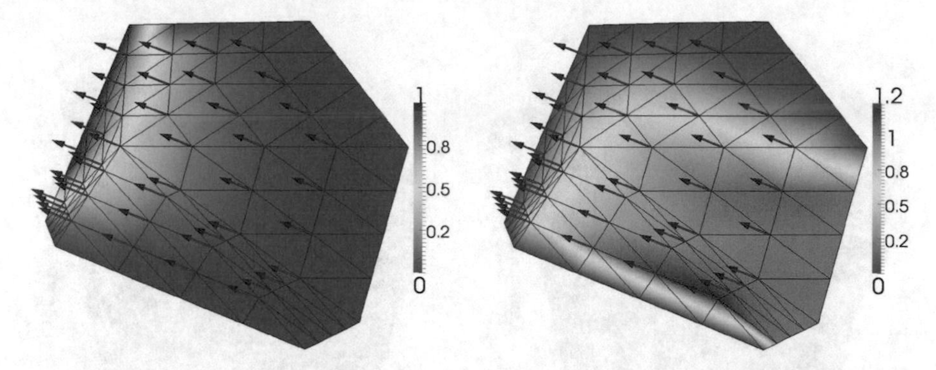

Fig. 6.16 Approximations of basis functions on polygonal face, projected convection vector and auxiliary triangulation with appropriately (left) and not appropriately (right) resolved boundary layer

Here, one has to point out that the computations are done on a polyhedral mesh with a globally continuous approximation $u_{h,l}$. This, by itself, is a current field of research even without dominant convection, see [28]. The geometry of polygonal faces is more complex than the triangles in Experiment 1, and thus, the computations on the faces are more involved.

Figure 6.16 presents the approximation of two different basis functions over the same polygonal face, the auxiliary triangulation and the projected convection vector. We can see how the local mesh has been adapted to the underlying differential operator, namely by moving the node, which lay initially in the center of the polygon, into the direction of the convection. In certain constellations, the boundary layers are not resolved appropriately. In the left picture of Fig. 6.16, the approximation of the basis function is satisfactory. In the right picture, however, oscillations occur in the lower right corner due to the relatively large triangles near the boundary. In many cases these situations are already resolved quite well by the simple mesh adaptation. When we introduced the moving of the auxiliary nodes in the implementation, the numerical results improved. Thus, we expect that a better adaptation of the local meshes, and consequently a better approximation of the local problems, improves the stability of the BEM-based FEM such that we would obtain comparable results to Experiment 1 for the discrete maximum principle.

Finally, in Fig. 6.17, the approximation $u_{h,l}$ is visualized for $l = 2$ and two different values of diffusion $\alpha = 2.5 \times 10^{-2}$ and $\alpha = 5.0 \times 10^{-5}$. The domain Ω has been cut through, such that the approximation is visible on a set of polygonal faces which lie in the interior of the domain. The expected behaviour of the solution can be observed. The Dirichlet data is transported into the interior of the domain along the convection vector. In the case of the convection-dominated problem, oscillations appear near the outflow boundary.

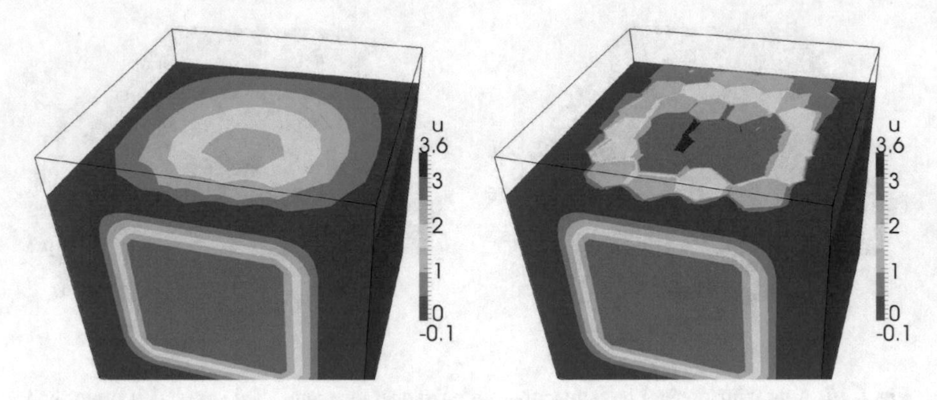

Fig. 6.17 Cut through the domain $\Omega = (0, 3)^3$ and visualisation of the approximation in Experiment 3 for $\alpha = 2.5 \times 10^{-2}$ (left) and $\alpha = 5.0 \times 10^{-5}$ (right)

Conclusion on Convection-Adapted BEM-Based FEM

We have derived convection-adapted BEM-based FEM discretization schemes for convection-diffusion-reaction boundary value problems that considerably extend the range of applicability with respect to the strength of convection. The numerical results have not only confirmed this enhanced stability property of the discretization scheme, but have also indicated faster convergence of the GMRES solver in comparison with the original BEM-based FEM scheme presented in [94, 96]. When compared to the SUPG method, our proposed method shows an improved resolution of the exponential layer at the outflow boundary.

References

1. Adams, R.A.: Sobolev Spaces. Academic, London (1975)
2. Agmon, S.: Lectures on Elliptic Boundary Value Problems. Prepared for publication by B. Frank Jones, Jr. with the assistance of George W. Batten, Jr. Van Nostrand Mathematical Studies, No. 2. D. Van Nostrand Co., Inc., Princeton (1965)
3. Ainsworth, M., Oden, J.T.: A posteriori error estimation in finite element analysis. Comput. Methods Appl. Mech. Eng. **142**(1–2), 1–88 (1997)
4. Ainsworth, M., Oden, T.J.: A posteriori error estimation in finite element analysis. In: Pure and Applied Mathematics. Wiley, New York (2000)
5. Anand, A., Ovall, J.S., Weißer, S.: A Nyström-based finite element method on polygonal elements. Comput. Math. Appl. **75**(11), 3971–3986 (2018)
6. Anderson, E., Bai, Z., Bischof, C., Blackford, S., Demmel, J., Dongarra, J., Du Croz, J., Greenbaum, A., Hammarling, S., McKenney, A., Sorensen, D.: LAPACK Users' Guide, 3rd edn. SIAM, Philadelphia (1999)
7. Antonietti, P.F., Beirão da Veiga, L., Lovadina, C., Verani, M.: Hierarchical a posteriori error estimators for the mimetic discretization of elliptic problems. SIAM J. Numer. Anal. **51**(1), 654–675 (2013)
8. Antonietti, P.F., Beirão da Veiga, L., Verani, M.: A mimetic discretization of elliptic obstacle problems. Math. Comput. **82**(283), 1379–1400 (2013)
9. Antonietti, P.F., Berrone, S., Verani, M., Weißer, S.: The virtual element method on anisotropic polygonal discretizations. In: Radu, F., Kumar, K., Berre, I., Nordbotten, J., Pop, I. (eds.) Numerical Mathematics and Advanced Applications-ENUMATH 2017. Lecture Notes in Computational Science and Engineering, vol. 126, pp. 725–733. Springer, Cham (2019)
10. Apel, T.: Anisotropic finite elements: local estimates and applications. In: Advances in Numerical Mathematics. B. G. Teubner, Stuttgart (1999)
11. Apel, T., Grosman, S., Jimack, P.K., Meyer, A.: A new methodology for anisotropic mesh refinement based upon error gradients. Appl. Numer. Math. **50**(3–4), 329–341 (2004)
12. Arroyo, M., Ortiz, M.: Local maximum-entropy approximation schemes: a seamless bridge between finite elements and meshfree methods. Int. J. Numer. Methods Eng. **65**(13), 2167–2202 (2006)
13. Atkinson, K.E.: The numerical solution of integral equations of the second kind. In: Cambridge Monographs on Applied and Computational Mathematics, vol. 4. Cambridge University Press, Cambridge (1997)
14. Babuška, I., Aziz, A.K.: On the angle condition in the finite element method. SIAM J. Numer. Anal. **13**(2), 214–226 (1976)

© Springer Nature Switzerland AG 2019

S. Weißer, *BEM-based Finite Element Approaches on Polytopal Meshes*,
Lecture Notes in Computational Science and Engineering 130,
https://doi.org/10.1007/978-3-030-20961-2

15. Babuška, I., Rheinboldt, W.C.: A-posteriori error estimates for the finite element method. Int. J. Numer. Methods Eng. **12**(10), 1597–1615 (1978)
16. Babuška, I., Rheinboldt, W.C.: Error estimates for adaptive finite element computations. SIAM J. Numer. Anal. **15**(4), 736–754 (1978)
17. Bacuta, C., Bramble, J.H., Pasciak, J.E.: Using finite element tools in proving shift theorems for elliptic boundary value problems. Numer. Linear Algebra Appl. **10**(1–2), 33–64 (2003). Dedicated to the 60th birthday of Raytcho Lazarov
18. Bangerth, W., Rannacher, R.: Adaptive Finite Element Methods for Differential Equations. Lectures in Mathematics, ETH Zürich. Birkhäuser, Basel (2003)
19. Bank, R.E., Weiser, A.: Some a posteriori error estimators for elliptic partial differential equations. Math. Comp. **44**(170), 283–301 (1985)
20. Bebendorf, M.: Hierarchical matrices. In: Lecture Notes in Computational Science and Engineering, vol. 63. Springer, Berlin (2008)
21. Becker, R., Rannacher, R.: Weighted a posteriori error control in FE methods. In: Bock, H.G., et al. (eds.) ENUMATH'97, pp. 18–22. World Science Publication, Singapore (1995)
22. Becker, R., Rannacher, R.: An optimal control approach to a posteriori error estimation in finite element methods. Acta Numer. **10**, 1–102 (2001)
23. Beirão da Veiga, L., Manzini, G.: Residual a posteriori error estimation for the virtual element method for elliptic problems. ESAIM Math. Model. Numer. Anal. **49**(2), 577–599 (2015)
24. Beirão da Veiga, L., Lipnikov, K., Manzini, G.: Arbitrary-order nodal mimetic discretizations of elliptic problems on polygonal meshes. SIAM J. Numer. Anal. **49**(5), 1737–1760 (2011)
25. Beirão da Veiga, L., Brezzi, F., Cangiani, A., Manzini, G., Marini, L.D., Russo, A.: Basic principles of virtual element methods. Math. Models Methods Appl. Sci. **23**(01), 199–214 (2013)
26. Beirão da Veiga, L., Brezzi, F., Marini, L.: Virtual elements for linear elasticity problems. SIAM J. Numer. Anal. **51**(2), 794–812 (2013)
27. Beirão da Veiga, L., Lipnikov, K., Manzini, G.: The mimetic finite difference method for elliptic problems. In: MS&A. Modeling, Simulation and Applications, vol. 11. Springer, Cham (2014)
28. Beirão da Veiga, L., Brezzi, F., Marini, L.D., Russo, A.: Virtual Element Method for general second-order elliptic problems on polygonal meshes. Math. Models Methods Appl. Sci. **26**(4), 729–750 (2016)
29. Beirão da Veiga, L., Brezzi, F., Marini, L.D., Russo, A.: $H(\mathrm{div})$ and $H(\mathbf{curl})$-conforming virtual element methods. Numer. Math. **133**(2), 303–332 (2016)
30. Beirão da Veiga, L., Brezzi, F., Marini, L.D., Russo, A.: Mixed virtual element methods for general second order elliptic problems on polygonal meshes. ESAIM Math. Model. Numer. Anal. **50**(3), 727–747 (2016)
31. Beirão da Veiga, L., Chernov, A., Mascotto, L., Russo, A.: Exponential convergence of the hp virtual element method in presence of corner singularities. Numer. Math. **138**(3), 581–613 (2018)
32. Beirão da Veiga, L., Lovadina, C., Vacca, G.: Virtual elements for the Navier-Stokes problem on polygonal meshes. SIAM J. Numer. Anal. **56**(3), 1210–1242 (2018)
33. Benedetto, M.F., Berrone, S., Borio, A., Pieraccini, S., Scialò, S.: Order preserving SUPG stabilization for the virtual element formulation of advection-diffusion problems. Comput. Methods Appl. Mech. Eng. **311**, 18–40 (2016)
34. Bergh, J., Löfström, J.: Interpolation Spaces. Springer, Berlin (1976). Grundlehren der Mathematischen Wissenschaften, No. 223
35. Berrone, S., Borio, A.: A residual a posteriori error estimate for the virtual element method. Math. Models Methods Appl. Sci. **27**(8), 1423–1458 (2017)
36. Binev, P., Dahmen, W., DeVore, R.: Adaptive finite element methods with convergence rates. Numer. Math. **97**(2), 219–268 (2004)
37. Braack, M., Ern, A.: A posteriori control of modeling errors and discretization errors. Multiscale Model. Simul. **1**(2), 221–238 (2003)

38. Braess, D.: Finite Elements: Theory, Fast Solvers, and Applications in Elasticity Theory, 3rd edn. Cambridge University Press, Cambridge (2007). Translated from the German by Larry L. Schumaker
39. Braess, D., Schöberl, J.: Equilibrated residual error estimator for edge elements. Math. Comput. **77**(262), 651–672 (2008)
40. Brenner, S.C., Scott, L.R.: The Mathematical Theory of Finite Element Methods. Texts in Applied Mathematics, vol. 15, 2nd edn. Springer, New York (2002)
41. Brezzi, F., Bristeau, M.O., Franca, L.P., Mallet, M., Rogé, G.: A relationship between stabilized finite element methods and the Galerkin method with bubble functions. Comput. Methods Appl. Mech. Eng. **96**(1), 117–129 (1992)
42. Brezzi, F., Falk, R.S., Marini, L.D.: Basic principles of mixed virtual element methods. ESAIM Math. Model. Numer. Anal. **48**(4), 1227–1240 (2014)
43. Brezzi, F., Fortin, M.: Mixed and hybrid finite element methods. In: Springer Series in Computational Mathematics, vol. 15. Springer, New York (1991)
44. Brezzi, F., Franca, L.P., Hughes, T.J.R., Russo, A.: $b = \int g$. Comput. Methods Appl. Mech. Eng. **145**(3–4), 329–339 (1997)
45. Brezzi, F., Hughes, T.J.R., Marini, L.D., Russo, A., Süli, E.: A priori error analysis of residual-free bubbles for advection-diffusion problems. SIAM J. Numer. Anal **36**(6), 1933 1948 (1999)
46. Brezzi, F., Lipnikov, K., Shashkov, M.: Convergence of the mimetic finite difference method for diffusion problems on polyhedral meshes. SIAM J. Numer. Anal. **43**(5), 1872–1896 (2005)
47. Brezzi, F., Marini, D., Russo, A.: Applications of the pseudo residual-free bubbles to the stabilization of convection-diffusion problems. Comput. Methods Appl. Mech. Eng. **166**(1–2), 51–63 (1998)
48. Brooks, A.N., Hughes, T.J.R.: Streamline upwind/Petrov-Galerkin formulations for convection dominated flows with particular emphasis on the incompressible Navier-Stokes equations. Comput. Methods Appl. Mech. Eng. **32**(1–3), 199–259 (1982)
49. Burenkov, V.I.: Sobolev Spaces on Domains, vol. 137. Vieweg+Teubner, Wiesbaden (1998)
50. Cangiani, A., Georgoulis, E.H., Pryer, T., Sutton, O.J.: A posteriori error estimates for the virtual element method. Numer. Math. **137**(4), 857–893 (2017)
51. Cangiani, A., Manzini, G., Sutton, O.J.: Conforming and nonconforming virtual element methods for elliptic problems. IMA J. Numer. Anal. **37**(3), 1317–1354 (2017)
52. Carstensen, C.: Estimation of higher Sobolev norm from lower order approximation. SIAM J. Numer. Anal. **42**(5), 2136–2147 (2005)
53. Carstensen, C., Feischl, M., Page, M., Praetorius, D.: Axioms of adaptivity. Comput. Math. Appl. **67**(6), 1195–1253 (2014)
54. Carstensen, C., Merdon, C.: Estimator competition for poisson problems. J. Comput. Math. **28**(3), 309–330 (2010)
55. Cascon, J.M., Kreuzer, C., Nochetto, R.H., Siebert, K.G.: Quasi-optimal convergence rate for an adaptive finite element method. SIAM J. Numer. Anal. **46**(5), 2524–2550 (2008)
56. Chen, Z., Hou, T.Y.: A mixed multiscale finite element method for elliptic problems with oscillating coefficients. Math. Comput. **72**(242), 541–576 (2003)
57. Chow, P., Cross, M., Pericleous, K.: A natural extension of the conventional finite volume method into polygonal unstructured meshes for CFD application. Appl. Math. Model. **20**(2), 170–183 (1996)
58. Ciarlet, P.G.: The Finite Element Method for Elliptic Problems. North-Holland, Amsterdam (1978)
59. Clément, P.: Approximation by finite element functions using local regularization. ESAIM Math. Model. Numer. Anal. **9**(R-2), 77–84 (1975)
60. Copeland, D., Langer, U., Pusch, D.: From the boundary element domain decomposition methods to local Trefftz finite element methods on polyhedral meshes. In: Domain decomposition methods in science and engineering XVIII. Lecture Notes in Computational Science and Engineering, vol. 70, pp. 315–322. Springer, Berlin (2009)

61. Costabel, M.: Boundary integral operators on Lipschitz domains: elementary results. SIAM J. Math. Anal. **19**(3), 613–626 (1988)
62. Davis, P.J.: On the numerical integration of periodic analytic functions. In: Langer, R.E. (ed) On Numerical Approximation. Proceedings of a Symposium, Madison, April 21–23, 1958, Publication no. 1 of the Mathematics Research Center, U.S. Army, the University of Wisconsin, pp. 45–59. The University of Wisconsin Press, Madison (1959)
63. Demkowicz, L.: Computing with *hp*-adaptive finite elements. Chapman and Hall/CRC Applied Mathematics and Nonlinear Science Series, vol. 1. Chapman and Hall/CRC, Boca Raton (2007)
64. Deuflhard, P., Leinen, P., Yserentant, H.: Concepts of an adaptive hierarchical finite element code. IMPACT Comput. Sci. Eng. **1**, 3–35 (1989)
65. Di Pietro, D.A., Ern, A., Lemaire, S.: An arbitrary-order and compact-stencil discretization of diffusion on general meshes based on local reconstruction operators. Comput. Methods Appl. Math. **14**(4), 461–472 (2014)
66. Dolejší, V., Feistauer, M., Sobotíková, V.: Analysis of the discontinuous Galerkin method for nonlinear convection-diffusion problems. Comput. Methods Appl. Mech. Eng. **194**(25–26), 2709–2733 (2005)
67. Dörfler, W.: A convergent adaptive algorithm for Poisson's equation. SIAM J. Numer. Anal. **33**(3), 1106–1124 (1996)
68. Droniou, J.: Non-coercive linear elliptic problems. Potential Anal. **17**, 181–203 (2002)
69. Dupont, T., Scott, R.: Polynomial approximation of functions in Sobolev spaces. Math. Comp. **34**(150), 441–463 (1980)
70. Ebeida, M.S., Mitchell, S.A.: Uniform random Voronoi meshes. In: Proceedings of the 20th International Meshing Roundtable, pp. 273–290. Springer, Berlin (2012)
71. Ebeida, M.S., Mitchell, S.A., Patney, A., Davidson, A., Owens, J.D.: A simple algorithm for maximal poisson-disk sampling in high dimensions. Comput. Graph. Forum **31** (2012)
72. Efendiev, Y., Hou, T.Y.: Multiscale Finite Element Methods—Theory and Applications. Surveys and Tutorials in the Applied Mathematical Sciences, vol. 4. Springer, New York (2009)
73. Efendiev, Y., Galvis, J., Lazarov, R., Weißer, S.: Mixed FEM for second order elliptic problems on polygonal meshes with BEM-based spaces. In: Lirkov, I., Margenov, S., Waśniewski, J. (eds.) Large-Scale Scientific Computing. Lecture Notes in Computer Science, vol. 8353, pp. 331–338. Springer, Berlin (2014)
74. Eriksson, K., Estep, D., Hansbo, P., Johnson, C.: Introduction to adaptive methods for differential equations. Acta Numer. **4**, 105–158 (1995)
75. Floater, M.S.: Mean value coordinates. Comput. Aided Geom. Des. **20**(1), 19–27 (2003)
76. Floater, M.S.: Generalized barycentric coordinates and applications. Acta Numer. **24**, 161–214 (2015)
77. Floater, M., Hormann, K., Kós, G.: A general construction of barycentric coordinates over convex polygons. Adv. Comput. Math. **24**, 311–331 (2006)
78. Formaggia, L., Perotto, S.: New anisotropic a priori error estimates. Numer. Math. **89**(4), 641–667 (2001)
79. Formaggia, L., Perotto, S.: Anisotropic error estimates for elliptic problems. Numer. Math. **94**(1), 67–92 (2003)
80. Franca, L.P., Nesliturk, A., Stynes, M.: On the stability of residual-free bubbles for convection-diffusion problems and their approximation by a two-level finite element method. Comput. Methods Appl. Mech. Eng. **166**(1–2), 35–49 (1998)
81. Gain, A.L., Talischi, C., Paulino, G.H.: On the virtual element method for three-dimensional linear elasticity problems on arbitrary polyhedral meshes. Comput. Methods Appl. Mech. Eng. **282**, 132–160 (2014)
82. Gilbarg, D., Trudinger, N.S.: Elliptic partial differential equations of second order. In: Classics in Mathematics. Springer, Berlin (2001). Reprint of the 1998 edition
83. Giles, M.B., Süli, E.: Adjoint methods for PDEs: a posteriori error analysis and postprocessing by duality. Acta Numer. **11**, 145–236 (2002)

84. Gillette, A., Rand, A., Bajaj, C.: Error estimates for generalized barycentric interpolation. Adv. Comput. Math. **37**, 417–439 (2012)
85. Gillette, A., Rand, A., Bajaj, C.: Construction of scalar and vector finite element families on polygonal and polyhedral meshes. Comput. Methods Appl. Math. **16**(4), 667–683 (2016)
86. Gordon, D., Gordon, R.: Row scaling as a preconditioner for some nonsymmetric linear systems with discontinuous coefficients. J. Comput. Appl. Math. **234**(12), 3480–3495 (2010)
87. Grisvard, P.: Elliptic problems in nonsmooth domains. In: Monographs and Studies in Mathematics. Pitman Advanced Publication Program, Boston (1985)
88. Gwinner, J., Stephan, E.P.: Advanced Boundary Element Methods. Springer Series in Computational Mathematics, vol. 52. Springer International Publishing, New York (2018)
89. Hecht, F.: New development in FreeFem++. J. Numer. Math. **20**(3–4), 251–265 (2012)
90. Hestenes, M.R., Stiefel, E.: Methods of conjugate gradients for solving linear systems. J. Res. Natl. Bur. Stand. **49**, 409–436 (1952)
91. Hiptmair, R., Moiola, A., Perugia, I.: Plane wave discontinuous Galerkin methods for the 2D Helmholtz equation: analysis of the p-version. SIAM J. Numer. Anal. **49**(1), 264–284 (2011)
92. Hiptmair, R., Moiola, A., Perugia, I.: Error analysis of Trefftz-discontinuous Galerkin methods for the time-harmonic Maxwell equations. Math. Comput. **82**(281), 247–268 (2013)
93. Hofreither, C.: L_2 error estimates for a nonstandard finite element method on polyhedral meshes. J. Numer. Math. **19**(1), 27–39 (2011)
94. Hofreither, C.: A Non-standard Finite Element Method using Boundary Integral Operators. Ph.D. thesis, Johannes Kepler University, Linz (2012)
95. Hofreither, C., Langer, U., Pechstein, C.: Analysis of a non-standard finite element method based on boundary integral operators. Electron. Trans. Numer. Anal. **37**, 413–436 (2010)
96. Hofreither, C., Langer, U., Pechstein, C.: A non-standard finite element method for convection-diffusion-reaction problems on polyhedral meshes. AIP Conf. Proc. **1404**(1), 397–404 (2011)
97. Hofreither, C., Langer, U., Pechstein, C.: FETI solvers for non-standard finite element equations based on boundary integral operators. In: Erhel, J., Gander, M., Halpern, L., Pichot, G., Sassi, T., Widlund, O. (eds.) Domain Decomposition Methods in Science and Engineering XXI. Lecture Notes in Computational Science and Engineering, vol. 98, pp. 731–738. Springer, Cham (2014)
98. Hofreither, C., Langer, U., Pechstein, C.: BEM-based finite element tearing and interconnecting methods. Electron. Trans. Numer. Anal. **44**, 230–249 (2015)
99. Hofreither, C., Langer, U., Weißer, S.: Convection adapted BEM-based FEM. ZAMM Z. Angew. Math. Mech. **96**(12), 1467–1481 (2016)
100. Hörmander, L.: The analysis of linear partial differential operators. I. In: Grundlehren der Mathematischen Wissenschaften [Fundamental Principles of Mathematical Sciences], vol. 256. Springer, Berlin (1983). Distribution theory and Fourier analysis
101. Hormann, K., Floater, M.S.: Mean value coordinates for arbitrary planar polygons. ACM Trans. Graph. **25**(4), 1424–1441 (2006)
102. Hormann, K., Sukumar, N.: Maximum entropy coordinates for arbitrary polytopes. Comput. Graph. Forum **27**(5), 1513–1520 (2008)
103. Hormann, K., Sukumar, N. (eds.): Generalized Barycentric Coordinates in Computer Graphics and Computational Mechanics. CRC Press, Boca Raton (2018)
104. Hou, T.Y., Wu, X.H.: A multiscale finite element method for elliptic problems in composite materials and porous media. J. Comput. Phys. **134**(1), 169–189 (1997)
105. Hsiao, G.C., Wendland, W.L.: A finite element method for some integral equations of the first kind. J. Math. Anal. Appl. **58**(3), 449–481 (1977)
106. Hsiao, G.C., Wendland, W.L.: Domain decomposition in boundary element methods. In: Fourth International Symposium on Domain Decomposition Methods for Partial Differential Equations (Moscow, 1990), pp. 41–49. SIAM, Philadelphia (1991)
107. Hsiao, G.C., Wendland, W.L.: Boundary Integral Equations. Applied Mathematical Sciences, vol. 164. Springer, Berlin (2008)

108. Huang, W.: Mathematical principles of anisotropic mesh adaptation. Commun. Comput. Phys. **1**(2), 276–310 (2006)
109. Huang, W., Kamenski, L., Lang, J.: A new anisotropic mesh adaptation method based upon hierarchical a posteriori error estimates. J. Comput. Phys. **229**(6), 2179–2198 (2010)
110. Jamet, P.: Estimations d'erreur pour des éléments finis droits presque dégénérés. RAIRO Anal. Numér. **10**(R-1), 43–60 (1976)
111. John, V., Knobloch, P.: On spurious oscillations at layers diminishing (SOLD) methods for convection-diffusion equations: Part I-A review. Comput. Methods Appl. Mech. Eng. **196**(17–20), 2197–2215 (2007)
112. Joshi, P., Meyer, M., DeRose, T., Green, B., Sanocki, T.: Harmonic coordinates for character articulation. ACM Trans. Graph. **26**(3), 71.1–71.9 (2007)
113. Karachik, V.V., Antropova, N.A.: On the Solution of the Inhomogeneous Polyharmonic Equation and the inhomogeneous Helmholtz equation. Differ. Equ. **46**(3), 387–399 (2010)
114. Kobayashi, K., Tsuchiya, T.: A priori error estimates for Lagrange interpolation on triangles. Appl. Math. **60**(5), 485–499 (2015)
115. Kobayashi, K., Tsuchiya, T.: Extending Babuška-Aziz's theorem to higher-order Lagrange interpolation. Appl. Math. **61**(2), 121–133 (2016)
116. Kopteva, N., O'Riordan, E.: Shishkin meshes in the numerical solution of singularly perturbed differential equations. Int. J. Numer. Anal. Model. **7**(3), 393–415 (2010)
117. Kress, R.: A Nyström method for boundary integral equations in domains with corners. Numer. Math. **58**(2), 145–161 (1990)
118. Kress, R.: Linear Integral Equations. Applied Mathematical Sciences, vol. 82, 2nd edn. Springer, New York (1999)
119. Kunert, G.: An a posteriori residual error estimator for the finite element method on anisotropic tetrahedral meshes. Numer. Math. **86**(3), 471–490 (2000)
120. Kuznetsov, Y., Repin, S.: New mixed finite element method on polygonal and polyhedral meshes. Russ. J. Numer. Anal. Math. Model. **18**(3), 261–278 (2003)
121. Lax, P.D., Milgram, A.N.: Parabolic equations. In: Contributions to the Theory of Partial Differential Equations, No. 33 in Annals of Mathematics Studies, pp. 167–190. Princeton University Press, Princeton (1954)
122. Lee, J.R.: The law of cosines in a tetrahedron. J. Korea Soc. Math. Educ. Ser. B Pure Appl. Math. **4**(1), 1–6 (1997)
123. Linß, T.: Layer-Adapted Meshes for Reaction-Convection-Diffusion Problem. Lecture Notes in Mathematics, vol. 1985. Springer, Berlin (2010)
124. Lipnikov, K., Manzini, G., Shashkov, M.: Mimetic finite difference method. J. Comput. Phys. **257**(Part B), 1163–1227 (2014)
125. Loseille, A., Alauzet, F.: Continuous mesh framework part I: well-posed continuous interpolation error. SIAM J. Numer. Anal. **49**(1), 38–60 (2011)
126. Martin, S., Kaufmann, P., Botsch, M., Wicke, M., Gross, M.: Polyhedral finite elements using harmonic basis functions. Comput. Graph. Forum **27**(5), 1521–1529 (2008)
127. Maz'ya, V.G.: Boundary integral equations. In: Analysis, IV. Encyclopaedia of Mathematical Sciences, vol. 27, pp. 127–222. Springer, Berlin (1991)
128. McLean, W.C.H.: Strongly Elliptic Systems and Boundary Integral Equations. Cambridge University Press, Cambridge (2000)
129. Morin, P., Nochetto, R.H., Siebert, K.S.: Data oscillation and convergence of adaptive FEM. SIAM J. Numer. Anal. **38**(2), 466–488 (2001)
130. Morin, P., Siebert, K.G., Veeser, A.: A basic convergence result for conforming adaptive finite elements. Math. Models Methods Appl. Sci. **18**(5), 707–737 (2008)
131. Mousavi, S.E., Sukumar, N.: Numerical integration of polynomials and discontinuous functions on irregular convex polygons and polyhedrons. Comput. Mech. **47**, 535–554 (2011)
132. Nochetto, R.H., Veeser, A., Verani, M.: A safeguarded dual weighted residual method. IMA J. Numer. Anal. **29**(1), 126–140 (2009)
133. Nyström, E.J.: Über die praktische auflösung von integralgleichungen mit anwendungen auf randwertaufgaben. Acta Math. **54**(1), 185–204 (1930)

134. Oden, J.T., Prudhomme, S.: On goal-oriented error estimation for elliptic problems: application to the control of pointwise errors. Comput. Methods Appl. Mech. Eng. **176**, 313–331 (1999)
135. Ovall, J.S., Reynolds, S.: A high-order method for evaluating derivatives of harmonic function in planar domains. SIAM J. Sci. Comput. **40**(3), A1915–A1935 (2018)
136. Payne, L.E., Weinberger, H.F.: An optimal Poincaré inequality for convex domains. Arch. Ration. Mech. Anal. **5**, 286–292 (1960)
137. Peraire, J., Patera, A.T.: Bounds for linear-functional outputs of coercive partial differential equations: local indicators and adaptive refinement. In: Advances in adaptive computational methods in mechanics (Cachan, 1997). Studies in Applied Mechanics, vol. 47, pp. 199–216. Elsevier Science B. V., Amsterdam (1998)
138. Perugia, I., Pietra, P., Russo, A.: A plane wave virtual element method for the Helmholtz problem. ESAIM Math. Model. Numer. Anal. **50**(3), 783–808 (2016)
139. Petzoldt, M.: A posteriori error estimators for elliptic equations with discontinuous coefficients. Adv. Comput. Math. **16**(1), 47–75 (2002)
140. Protter, M.H., Weinberger, H.F.: Maximum principles in differential equations. Springer, New York (1984). Corrected reprint of the 1967 original
141. Rannacher, R., Suttmeier, F.T.: A feed-back approach to error control in finite element methods: application to linear elasticity. Comput. Mech. **19**(5), 434–446 (1997)
142. Repin, S.: A Posteriori Estimates for Partial Differential Equations. Radon Series on Computational and Applied Mathematics, vol. 4. Walter de Gruyter GmbH and Co. KG, Berlin (2008)
143. Richter, T., Wick, T.: Variational localizations of the dual weighted residual estimator. J. Comput. Appl. Math. **279**, 192–208 (2015)
144. Rjasanow, S., Steinbach, O.: The fast solution of boundary integral equations. In: Mathematical and Analytical Techniques with Applications to Engineering. Springer, New York (2007)
145. Rjasanow, S., Weißer, S.: Developments in BEM-based finite element methods on polygonal and polyhedral meshes. In: ECCOMAS 2012-European Congress on Computational Methods in Applied Sciences and Engineering, e-Book Full Papers, pp. 1421–1431 (2012)
146. Rjasanow, S., Weißer, S.: Higher order BEM-based FEM on polygonal meshes. SIAM J. Numer. Anal. **50**(5), 2357–2378 (2012)
147. Rjasanow, S., Weißer, S.: FEM with Trefftz trial functions on polyhedral elements. J. Comput. Appl. Math. **263**, 202–217 (2014)
148. Rjasanow, S., Weißer, S.: ACA improvement by surface segmentation. In: Apel, T., Langer, U., Meyer, A., Steinbach, O. (eds.) Advanced Finite Element Methods with Applications-Proceedings of the 30th Chemnitz FEM Symposium 2017. Lecture Notes in Computational Science and Engineering, vol. 128, pp. X, 205. Springer International Publishing, New York (2019)
149. Roos, H.G., Stynes, M., Tobiska, L.: Robust Numerical Methods for Singularly Perturbed Differential Equations. Springer Series in Computational Mathematics, vol. 24, 2nd edn. Springer, Berlin (2008)
150. Saad, Y., Schultz, M.H.: GMRES: a generalized minimal residual algorithm for solving nonsymmetric linear systems. SIAM J. Sci. Statist. Comput. **7**(3), 856–869 (1986)
151. Sauter, S.A., Schwab, C.: Boundary Element Methods. Springer Series in Computational Mathematics, vol. 39. Springer, Berlin (2011)
152. Schneider, R.: A review of anisotropic refinement methods for triangular meshes in FEM. In: Apel, T., Steinbach, O. (eds.) Advanced Finite Element Methods and Applications, pp. 133–152. Springer, Berlin (2013)
153. Scott, R.: Optimal L^∞ estimates for the finite element method on irregular meshes. Math. Comput. **30**(136), 681–697 (1976)
154. Scott, L.R., Zhang, S.: Finite element interpolation of nonsmooth functions satisfying boundary conditions. Math. Comput. **54**(190), 483–493 (1990)

155. Shewchuk, J.R.: Triangle: Engineering a 2D quality mesh generator and Delaunay triangulator. In: Lin, M.C., Manocha, D. (eds.) Applied Computational Geometry Towards Geometric Engineering, pp. 203–222. Springer, Berlin (1996)
156. Shishkin, G.I.: A difference scheme for a singularly perturbed equation of parabolic type with a discontinuous initial condition. Dokl. Akad. Nauk SSSR **300**(5), 1066–1070 (1988)
157. Si, H.: TetGen, a Delaunay-based quality tetrahedral mesh generator. ACM Trans. Math. Softw. **41**(2), 36 (2015). Article 11
158. Sloan, I.H.: Error analysis of boundary integral methods. Acta Numerica **1**, 287–339 (1992)
159. Steinbach, O.: Numerical Approximation Methods for Elliptic Boundary Value Problems: Finite and Boundary Elements. Springer, New York (2007)
160. Steinbach, O., Wendland, W.L.: On C. Neumann's method for second-order elliptic systems in domains with non-smooth boundaries. J. Math. Anal. Appl. **262**(2), 733–748 (2001)
161. Stephan, E.P., Suri, M.: The h-p version of the boundary element method on polygonal domains with quasiuniform meshes. RAIRO Modél. Math. Anal. Numér. **25**(6), 783–807 (1991)
162. Strang, G.: Variational crimes in the finite element method. In: The mathematical foundations of the finite element method with applications to partial differential equations (Proceedings of Symposia in University of Maryland, Baltimore, MD, 1972), pp. 689–710. Academic, New York (1972)
163. Sukumar, N.: Construction of polygonal interpolants: a maximum entropy approach. Int. J. Numer. Methods Eng. **61**(12), 2159–2181 (2004)
164. Sukumar, N., Tabarraei, A.: Conforming polygonal finite elements. Int. J. Numer. Methods Eng. **61**(12), 2045–2066 (2004)
165. Tabarraei, A., Sukumar, N.: Application of polygonal finite elements in linear elasticity. Int. J. Comput. Methods **3**(4), 503–520 (2006)
166. Talischi, C.: A family of $H(\mathrm{div})$ finite element approximations on polygonal meshes. SIAM J. Sci. Comput. **37**(2), A1067–A1088 (2015)
167. Talischi, C., Paulino, G.H., Pereira, A., Menezes, I.F.M.: PolyMesher: a general-purpose mesh generator for polygonal elements written in Matlab. Struct. Multidiscip. Optim. **45**(3), 309–328 (2012)
168. Trefftz, E.: Ein Gegenstück zum Ritzschen Verfahren. In: Proceedings of the 2nd International Congress of Technical Mechanics, pp. 131–137. Orell Fussli, Zürich (1926)
169. Veeser, A., Verfürth, R.: Poincaré constants for finite element stars. IMA J. Numer. Anal. **32**(1), 30–47 (2012)
170. Verfürth, R.: A Posteriori Error Estimation Techniques for Finite Element Methods. Numerical Mathematics and Scientific Computation. Oxford University Press, Oxford (2013)
171. von Petersdorff, T., Stephan, E.P.: Regularity of mixed boundary value problems in $\mathbf{R}^3$ and boundary element methods on graded meshes. Math. Methods Appl. Sci. **12**(3), 229–249 (1990)
172. Wachspress, E.L.: A Rational Finite Element Basis. Academic, London (1975)
173. Wang, J., Ye, X.: A weak Galerkin finite element method for second-order elliptic problems. J. Comput. Appl. Math. **241**, 103–115 (2013)
174. Weißer, S.: Residual error estimate for BEM-based FEM on polygonal meshes. Numer. Math. **118**(4), 765–788 (2011)
175. Weißer, S.: Arbitrary order Trefftz-like basis functions on polygonal meshes and realization in BEM-based FEM. Comput. Math. Appl. **67**(7), 1390–1406 (2014)
176. Weißer, S.: BEM-based finite element method with prospects to time dependent problems. In: Oñate, E., Oliver, J., Huerta, A. (eds.) Proceedings of the Jointly Organized WCCM XI, ECCM V, ECFD VI, Barcelona, July 2014, pp. 4420–4427. International Center for Numerical Methods in Engineeering (CIMNE) (2014)
177. Weißer, S.: Higher Order Trefftz-like Finite Element Method on Meshes with L-shaped Elements. In: Steinmann, P., Leugering, G. (eds.) Special Issue: 85th Annual Meeting of the International Association of Applied Mathematics and Mechanics (GAMM), Erlangen 2014. PAMM, vol. 14, pp. 31–34. Wiley-VCH, Weinheim (2014)

178. Weißer, S.: Residual based error estimate for higher order trefftz-like trial functions on adaptively refined polygonal meshes. In: Abdulle, A., Deparis, S., Kressner, D., Nobile, F., Picasso, M. (eds.) Numerical Mathematics and Advanced Applications-ENUMATH 2013. Lecture Notes in Computational Science and Engineering, vol. 103, pp. 233–241. Springer International Publishing, New York (2015)
179. Weißer, S.: Adaptive BEM-based FEM on polygonal meshes from virtual element methods. In: ECCOMAS Congress 2016-Proceedings of the 7th European Congress on Computational Methods in Applied Sciences and Engineering, vol. 2, pp. 2930–2940 (2016)
180. Weißer, S.: Residual based error estimate and quasi-interpolation on polygonal meshes for high order BEM-based FEM. Comput. Math. Appl. **73**(2), 187–202 (2017)
181. Weißer, S.: Anisotropic polygonal and polyhedral discretizations in finite element analysis. ESAIM Math. Model. Numer. Anal. **53**(2), 475–501 (2019)
182. Weißer, S., Wick, T.: The dual-weighted residual estimator realized on polygonal meshes. Comput. Methods Appl. Math. **18**(4), 753–776 (2018)
183. Zienkiewicz, O.C., Zhu, J.Z.: A simple error estimator and adaptive procedure for practical engineering analysis. Int. J. Numer. Methods Eng. **24**(2), 337–357 (1987)

Index

© Springer Nature Switzerland AG 2019

S. Weißer, *BEM-based Finite Element Approaches on Polytopal Meshes*,
Lecture Notes in Computational Science and Engineering 130,
https://doi.org/10.1007/978-3-030-20961-2

Editorial Policy

1. Volumes in the following three categories will be published in LNCSE:

i) Research monographs
ii) Tutorials
iii) Conference proceedings

Those considering a book which might be suitable for the series are strongly advised to contact the publisher or the series editors at an early stage.

2. Categories i) and ii). Tutorials are lecture notes typically arising via summer schools or similar events, which are used to teach graduate students. These categories will be emphasized by Lecture Notes in Computational Science and Engineering. **Submissions by interdisciplinary teams of authors are encouraged.** The goal is to report new developments – quickly, informally, and in a way that will make them accessible to non-specialists. In the evaluation of submissions timeliness of the work is an important criterion. Texts should be well-rounded, well-written and reasonably self-contained. In most cases the work will contain results of others as well as those of the author(s). In each case the author(s) should provide sufficient motivation, examples, and applications. In this respect, Ph.D. theses will usually be deemed unsuitable for the Lecture Notes series. Proposals for volumes in these categories should be submitted either to one of the series editors or to Springer-Verlag, Heidelberg, and will be refereed. A provisional judgement on the acceptability of a project can be based on partial information about the work: a detailed outline describing the contents of each chapter, the estimated length, a bibliography, and one or two sample chapters – or a first draft. A final decision whether to accept will rest on an evaluation of the completed work which should include

– at least 100 pages of text;
– a table of contents;
– an informative introduction perhaps with some historical remarks which should be accessible to readers unfamiliar with the topic treated;
– a subject index.

3. Category iii). Conference proceedings will be considered for publication provided that they are both of exceptional interest and devoted to a single topic. One (or more) expert participants will act as the scientific editor(s) of the volume. They select the papers which are suitable for inclusion and have them individually refereed as for a journal. Papers not closely related to the central topic are to be excluded. Organizers should contact the Editor for CSE at Springer at the planning stage, see *Addresses* below.

In exceptional cases some other multi-author-volumes may be considered in this category.

4. Only works in English will be considered. For evaluation purposes, manuscripts may be submitted in print or electronic form, in the latter case, preferably as pdf- or zipped ps-files. Authors are requested to use the LaTeX style files available from Springer at http://www.springer.com/gp/authors-editors/book-authors-editors/manuscript-preparation/5636 (Click on LaTeX Template → monographs or contributed books).

For categories ii) and iii) we strongly recommend that all contributions in a volume be written in the same LaTeX version, preferably LaTeX 2e. Electronic material can be included if appropriate. Please contact the publisher.

Careful preparation of the manuscripts will help keep production time short besides ensuring satisfactory appearance of the finished book in print and online.

5. The following terms and conditions hold. Categories i), ii) and iii):

Authors receive 50 free copies of their book. No royalty is paid.
Volume editors receive a total of 50 free copies of their volume to be shared with authors, but no royalties.

Authors and volume editors are entitled to a discount of 40 % on the price of Springer books purchased for their personal use, if ordering directly from Springer.

6. Springer secures the copyright for each volume.

Addresses:

Timothy J. Barth
NASA Ames Research
Center NAS Division
Moffett Field, CA 94035, USA
barth@nas.nasa.gov

Michael Griebel
Institut für Numerische Simulation
der Universität Bonn
Wegelerstr. 6
53115 Bonn, Germany
griebel@ins.uni-bonn.de

David E. Keyes
Mathematical and Computer Sciences
and Engineering
King Abdullah University of Science
and Technology
P. O. Box 55455
Jeddah 21534, Saudi Arabia
david.keyes@kaust.edu.sa

and

Department of Applied Physics
and Applied Mathematics
Columbia University
500 W. 120th Street
New York, NY 10027, USA
kd2112@columbia.edu

Risto M. Nieminen
Department of Applied Physics
Aalto University School of Science
and Technology
00076 Aalto, Finland
risto.nieminen@aalto.fi

Dirk Roose
Department of Computer Science
Katholieke Universiteit Leuven
Celestijnenlaan 200A
3001 Leuven-Heverlee, Belgium
dirk.roose@cs.kuleuven.be

Tamar Schlick
Department of Chemistry
and Courant Institute
of Mathematical Sciences
New York University
251 Mercer Street
New York, NY 10012, USA
schlick@nyu.edu

Editor for Computational Science
and Engineering at Springer:

Martin Peters
Springer-Verlag
Mathematics Editorial IV
Tiergartenstrasse 17
69121 Heidelberg, Germany
martin.peters@springer.com

Lecture Notes in Computational Science and Engineering

1. D. Funaro, *Spectral Elements for Transport-Dominated Equations.*
2. H.P. Langtangen, *Computational Partial Differential Equations.* Numerical Methods and Diffpack Programming.
3. W. Hackbusch, G. Wittum (eds.), *Multigrid Methods V*
4. P. Deuflhard, J. Hermans, B. Leimkuhler, A.E. Mark, S. Reich, R.D. Skeel (eds.), *Computational Molecular Dynamics: Challenges, Methods, Ideas.*
5. D. Kröner, M. Ohlberger, C. Rohde (eds.), *An Introduction to Recent Developments in Theory and Numerics for Conservation Laws.*
6. S. Turek, *Efficient Solvers for Incompressible Flow Problems.* An Algorithmic and Computational Approach.
7. R. von Schwenn, ***Multi Body System SIMulation.*** Numerical Methods, Algorithms, and Software.
8. H.-J. Bungartz, F. Durst, C. Zenger (eds.), *High Performance Scientific and Engineering Computing.*
9. T.J. Barth, H. Deconinck (eds.), *High-Order Methods for Computational Physics.*
10. H.P. Langtangen, A.M. Bruaset, E. Quak (eds.), *Advances in Software Tools for Scientific Computing.*
11. B. Cockburn, G.E. Karniadakis, C.-W. Shu (eds.), *Discontinuous Galerkin Methods.* Theory, Computation and Applications.
12. U. van Rienen, *Numerical Methods in Computational Electrodynamics.* Linear Systems in Practical Applications.
13. B. Engquist, L. Johnsson, M. Hammill, F. Short (eds.), *Simulation and Visualization on the Grid.*
14. E. Dick, K. Riemslagh, J. Vierendeels (eds.), *Multigrid Methods VI.*
15. A. Frommer, T. Lippert, B. Medeke, K. Schilling (eds.), *Numerical Challenges in Lattice Quantum Chromodynamics.*
16. J. Lang, *Adaptive Multilevel Solution of Nonlinear Parabolic PDE Systems.* Theory, Algorithm, and Applications.
17. B.I. Wohlmuth, *Discretization Methods and Iterative Solvers Based on Domain Decomposition.*
18. U. van Rienen, M. Günther, D. Hecht (eds.), *Scientific Computing in Electrical Engineering.*
19. I. Babuška, P.G. Ciarlet, T. Miyoshi (eds.), *Mathematical Modeling and Numerical Simulation in Continuum Mechanics.*
20. T.J. Barth, T. Chan, R. Haimes (eds.), *Multiscale and Multiresolution Methods.* Theory and Applications.
21. M. Breuer, F. Durst, C. Zenger (eds.), *High Performance Scientific and Engineering Computing.*
22. K. Urban, *Wavelets in Numerical Simulation.* Problem Adapted Construction and Applications.
23. L.F. Pavarino, A. Toselli (eds.), *Recent Developments in Domain Decomposition Methods.*

24. T. Schlick, H.H. Gan (eds.), *Computational Methods for Macromolecules: Challenges and Applications.*

25. T.J. Barth, H. Deconinck (eds.), *Error Estimation and Adaptive Discretization Methods in Computational Fluid Dynamics.*

26. M. Griebel, M.A. Schweitzer (eds.), *Meshfree Methods for Partial Differential Equations.*

27. S. Müller, *Adaptive Multiscale Schemes for Conservation Laws.*

28. C. Carstensen, S. Funken, W. Hackbusch, R.H.W. Hoppe, P. Monk (eds.), *Computational Electromagnetics.*

29. M.A. Schweitzer, *A Parallel Multilevel Partition of Unity Method for Elliptic Partial Differential Equations.*

30. T. Biegler, O. Ghattas, M. Heinkenschloss, B. van Bloemen Waanders (eds.), *Large-Scale PDE-Constrained Optimization.*

31. M. Ainsworth, P. Davies, D. Duncan, P. Martin, B. Rynne (eds.), *Topics in Computational Wave Propagation.* Direct and Inverse Problems.

32. H. Emmerich, B. Nestler, M. Schreckenberg (eds.), *Interface and Transport Dynamics.* Computational Modelling.

33. H.P. Langtangen, A. Tveito (eds.), *Advanced Topics in Computational Partial Differential Equations.* Numerical Methods and Diffpack Programming.

34. V John, *Large Eddy Simulation of Turbulent Incompressible Flows.* Analytical and Numerical Results for a Class of LES Models.

35. E. Bänsch (ed.), *Challenges in Scientific Computing- CISC 2002.*

36. B.N. Khoromskij, G. Wittum, *Numerical Solution of Elliptic Differential Equations by Reduction to the Interface.*

37. A. Iske, *Multiresolution Methods in Scattered Data Modelling.*

38. S.-I. Niculescu, K. Gu (eds.), *Advances in Time-Delay Systems.*

39. S. Attinger, P. Koumoutsakos (eds.), *Multiscale Modelling and Simulation.*

40. R. Kornhuber, R. Hoppe, J. Périaux, O. Pironneau, O. Wildlund, J. Xu (eds.), *Domain Decomposition Methods in Science and Engineering.*

41. T. Plewa, T. Linde, V.G. Weirs (eds.), *Adaptive Mesh Refinement – Theory and Applications.*

42. A. Schmidt, K.G. Siebert, *Design of Adaptive Finite Element Software.* The Finite Element Toolbox ALBERTA.

43. M. Griebel, M.A. Schweitzer (eds.), *Meshfree Methods for Partial Differential Equations II.*

44. B. Engquist, P. Lötstedt, O. Runborg (eds.), *Multiscale Methods in Science and Engineering.*

45. P. Benner, V. Mehrmann, D.C. Sorensen (eds.), *Dimension Reduction of Large-Scale Systems.*

46. D. Kressner, *Numerical Methods for General and Structured Eigenvalue Problems.*

47. A. Boriçi, A. Frommer, B. Joó, A. Kennedy, B. Pendleton (eds.), *QCD and Numerical Analysis III.*

48. F. Graziani (ed.), *Computational Methods in Transport.*

49. B. Leimkuhler, C. Chipot, R. Elber, A. Laaksonen, A. Mark, T. Schlick, C. Schütte, R. Skeel (eds.), *New Algorithms for Macromolecular Simulation.*

50. M. Bücker, G. Corliss, P. Hovland, U. Naumann, B. Norris (eds.), *Automatic Differentiation: Applications, Theory, and Implementations.*

51. A.M. Bruaset, A. Tveito (eds.), *Numerical Solution of Partial Differential Equations on Parallel Computers.*

52. K.H. Hoffmann, A. Meyer (eds.), *Parallel Algorithms and Cluster Computing.*

53. H.-J. Bungartz, M. Schäfer (eds.), *Fluid-Structure Interaction.*

54. J. Behrens, *Adaptive Atmospheric Modeling.*

55. O. Widlund, D. Keyes (eds.), *Domain Decomposition Methods in Science and Engineering XVI.*

56. S. Kassinos, C. Langer, G. Iaccarino, P. Moin (eds.), *Complex Effects in Large Eddy Simulations.*

57. M. Griebel, M.A Schweitzer (eds.), *Meshfree Methods for Partial Differential Equations III.*

58. A.N. Gorban, B. Kégl, D.C. Wunsch, A. Zinovyev (eds.), *Principal Manifolds for Data Visualization and Dimension Reduction.*

59. H. Ammari (ed.), *Modeling and Computations in Electromagnetics: A Volume Dedicated to Jean-Claude Nédélec.*

60. U. Langer, M. Discacciati, D. Keyes, O. Widlund, W. Zulehner (eds.), *Domain Decomposition Methods in Science and Engineering XVII.*

61. T. Mathew, *Domain Decomposition Methods* for *the Numerical Solution of Partial Differential Equations.*

62. F. Graziani (ed.), *Computational Methods in Transport: Verification and Validation.*

63. M. Bebendorf, *Hierarchical Matrices.* A Means to Efficiently Solve Elliptic Boundary Value Problems.

64. C.H. Bischof, H.M. Bücker, P. Hovland, U. Naumann, J. Utke (eds.), *Advances in Automatic* Differentiation.

65. M. Griebel, M.A. Schweitzer (eds.), *Meshfree Methods for Partial Differential Equations IV.*

66. B. Engquist, P. Lötstedt, O. Runborg (eds.), *Multiscale Modeling and Simulation in Science.*

67. I.H. Tuncer, Ü. Gülcat, D.R. Emerson, K. Matsuno (eds.), *Parallel Computational Fluid Dynamics 2007.*

68. S. Yip, T. Diaz de la Rubia (eds.), *Scientific Modeling and Simulations.*

69. A. Hegarty, N. Kopteva, E. O'Riordan, M. Stynes (eds.), *BAIL 2008 – Boundary and Interior Layers.*

70. M. Bercovier, M.J. Gander, R. Kornhuber, O. Widlund (eds.), *Domain Decomposition Methods in Science and Engineering XVIII.*

71. B. Koren, C. Vuik (eds.), *Advanced Computational Methods in Science and Engineering.*

72. M. Peters (ed.), *Computational Fluid Dynamics for Sport Simulation.*

73. H.-J. Bungartz, M. Mehl, M. Schäfer (eds.), *Fluid Structure Interaction II - Modelling, Simulation, Optimization.*

74. D. Tromeur-Dervout, G. Brenner, D.R. Emerson, J. Erhel (eds.), *Parallel Computational Fluid Dynamics 2008.*

75. A.N. Gorban, D. Roose (eds.), *Coping with Complexity: Model Reduction and Data Analysis.*

76. J.S. Hesthaven, E.M. Rønquist (eds.), *Spectral and High Order Methods for Partial Differential Equations.*
77. M. Holtz, *Sparse Grid Quadrature in High Dimensions with Applications in Finance and Insurance.*
78. Y. Huang, R. Kornhuber, O.Widlund, J. Xu (eds.), *Domain Decomposition Methods in Science and Engineering XIX.*
79. M. Griebel, M.A. Schweitzer (eds.), *Meshfree Methods for Partial Differential Equations V.*
80. P.H. Lauritzen, C. Jablonowski, M.A. Taylor, R.D. Nair (eds.), *Numerical Techniques for Global Atmospheric Models.*
81. C. Clavero, J.L. Gracia, F.J. Lisbona (eds.), *BAIL 2010 – Boundary and Interior Layers, Computational and Asymptotic Methods.*
82. B. Engquist, O. Runborg, Y.R. Tsai (eds.), *Numerical Analysis and Multiscale Computations.*
83. I.G. Graham, T.Y. Hou, O. Lakkis, R. Scheichl (eds.), *Numerical Analysis of Multiscale Problems.*
84. A. Logg, K.-A. Mardal, G. Wells (eds.), *Automated Solution of Differential Equations by the Finite Element Method.*
85. J. Blowey, M. Jensen (eds.), *Frontiers in Numerical Analysis - Durham 2010.*
86. O. Kolditz, U.-J. Gorke, H. Shao, W. Wang (eds.), *Thermo-Hydro-Mechanical-Chemical Processes in Fractured Porous Media- Benchmarks and Examples.*
87. S. Forth, P. Hovland, E. Phipps, J. Utke, A. Walther (eds.), *Recent Advances in Algorithmic Differentiation.*
88. J. Garcke, M. Griebel (eds.), *Sparse Grids and Applications.*
89. M. Griebel, M.A. Schweitzer (eds.), *Meshfree Methods for Partial Differential Equations VI.*
90. C. Pechstein, *Finite and Boundary Element Tearing and Interconnecting Solvers for Multiscale Problems.*
91. R. Bank, M. Holst, O. Widlund, J. Xu (eds.), *Domain Decomposition Methods in Science and Engineering XX.*
92. H. Bijl, D. Lucor, S. Mishra, C. Schwab (eds.), *Uncertainty Quantification in Computational Fluid Dynamics.*
93. M. Bader, H.-J. Bungartz, T. Weinzierl (eds.), *Advanced Computing.*
94. M. Ehrhardt, T. Koprucki (eds.), *Advanced Mathematical Models and Numerical Techniques for Multi-Band Effective Mass Approximations.*
95. M. Azaïez, H. El Fekih, J.S. Hesthaven (eds.), *Spectral and High Order Methods for Partial Differential Equations ICOSAHOM 2012.*
96. F. Graziani, M.P. Desjarlais, R. Redmer, S.B. Trickey (eds.), *Frontiers and Challenges in Warm Dense Matter.*
97. J. Garcke, D. Pflüger (eds.), *Sparse Grids and Applications – Munich 2012.*
98. J. Erhel, M. Gander, L. Halpern, G. Pichot, T. Sassi, O. Widlund (eds.), *Domain Decomposition Methods in Science and Engineering XXI.*
99. R. Abgrall, H. Beaugendre, P.M. Congedo, C. Dobrzynski, V. Perrier, M. Ricchiuto (eds.), *High Order Nonlinear Numerical Methods for Evolutionary PDEs - HONOM 2013.*
100. M. Griebel, M.A. Schweitzer (eds.), *Meshfree Methods for Partial Differential Equations VII.*

101. R. Hoppe (ed.), *Optimization with PDE Constraints - OPTPDE 2014*.

102. S. Dahlke, W. Dahmen, M. Griebel, W. Hackbusch, K. Ritter, R. Schneider, C. Schwab, H. Yserentant (eds.), *Extraction of Quantifiable Information from Complex Systems*.

103. A. Abdulle, S. Deparis, D. Kressner, F. Nobile, M. Picasso (eds.), *Numerical Mathematics and Advanced Applications - ENUMATH 2013*.

104. T. Dickopf, M.J. Gander, L. Halpern, R. Krause, L.F. Pavarino (eds.), *Domain Decomposition Methods in Science and Engineering XXII*.

105. M. Mehl, M. Bischoff, M. Schäfer (eds.), *Recent Trends in Computational Engineering - CE2014*. Optimization, Uncertainty, Parallel Algorithms, Coupled and Complex Problems.

106. R.M. Kirby, M. Berzins, J.S. Hesthaven (eds.), *Spectral and High Order Methods for Partial Differential Equations - ICOSAHOM'14*.

107. B. Jüttler, B. Simeon (eds.), *Isogeometric Analysis and Applications 2014*.

108. P. Knobloch (ed.), *Boundary and Interior Layers, Computational and Asymptotic Methods – BAIL 2014*.

109. J. Garcke, D. Pflüger (eds.), *Sparse Grids and Applications – Stuttgart 2014*.

110. H. P. Langtangen, *Finite Difference Computing with Exponential Decay Models*.

111. A. Tveito, G.T. Lines, *Computing Characterizations of Drugs for Ion Channels and Receptors Using Markov Models*.

112. B. Karazösen, M. Manguoğlu, M. Tezer-Sezgin, S. Göktepe, Ö. Uğur (eds.), *Numerical Mathematics and Advanced Applications - ENUMATH 2015*.

113. H.-J. Bungartz, P. Neumann, W.E. Nagel (eds.), *Software for Exascale Computing - SPPEXA 2013–2015*.

114. G.R. Barrenechea, F. Brezzi, A. Cangiani, E.H. Georgoulis (eds.), *Building Bridges: Connections and Challenges in Modern Approaches to Numerical Partial Differential Equations*.

115. M. Griebel, M.A. Schweitzer (eds.), *Meshfree Methods for Partial Differential Equations VIII*.

116. C.-O. Lee, X.-C. Cai, D.E. Keyes, H.H. Kim, A. Klawonn, E.-J. Park, O.B. Widlund (eds.), *Domain Decomposition Methods in Science and Engineering XXIII*.

117. T. Sakurai, S.-L. Zhang, T. Imamura, Y. Yamamoto, Y. Kuramashi, T. Hoshi (eds.), *Eigenvalue Problems: Algorithms, Software and Applications in Petascale Computing*. EPASA 2015, Tsukuba, Japan, September 2015.

118. T. Richter (ed.), *Fluid-structure Interactions*. Models, Analysis and Finite Elements.

119. M.L. Bittencourt, N.A. Dumont, J.S. Hesthaven (eds.), *Spectral and High Order Methods for Partial Differential Equations ICOSAHOM2016*. Selected Papers from the ICOSAHOM Conference, June 27-July 1, 2016, Rio de Janeiro, Brazil.

120. Z. Huang, M. Stynes, Z. Zhang (eds.), *Boundary and Interior Layers, Computational and Asymptotic Methods BAIL 2016*.

121. S.P.A. Bordas, E.N. Burman, M.G. Larson, M.A. Olshanskii (eds.), *Geometrically Unfitted Finite Element Methods and Applications*. Proceedings of the UCL Workshop 2016.

122. A. Gerisch, R. Penta, J. Lang (eds.), *Multiscale Models in Mechano and Tumor Biology*. Modeling, Homogenization, and Applications.

123. J. Garcke, D. Pflüger, C.G. Webster, G. Zhang (eds.), *Sparse Grids and Applications - Miami 2016*.

124. M. Schäfer, M. Behr, M. Mehl, B. Wohlmuth (eds.), *Recent Advances in Computational Engineering*. Proceedings of the 4th International Conference on Computational Engineering (ICCE 2017) in Darmstadt.

125. P.E. Bjørstad, S.C. Brenner, L. Halpern, R. Kornhuber, H.H. Kim, T. Rahman, O.B. Widlund (eds.), *Domain Decomposition Methods in Science and Engineering XXIV*. 24th International Conference on Domain Decomposition Methods, Svalbard, Norway, February 6–10, 2017.

126. F.A. Radu, K. Kumar, I. Berre, J.M. Nordbotten, I.S. Pop (eds.), *Numerical Mathematics and Advanced Applications – ENUMATH 2017*.

127. X. Roca, A. Loseille (eds.), *27th International Meshing Roundtable*.

128. Th. Apel, U. Langer, A. Meyer, O. Steinbach (eds.), *Advanced Finite Element Methods with Applications*. Selected Papers from the 30th Chemnitz Finite Element Symposium 2017.

129. M. Griebel, M.A. Schweitzer (eds.), *Meshfree Methods for Partial Differencial Equations IX*.

130. S. Weißer, *BEM-based Finite Element Approaches on Polytopal Meshes*.

131. V.A. Garanzha, L. Kamenski, H. Si (eds.), *Numerical Geometry, Grid Generation and Scientific Computing*. Proceedings of the 9th International Conference, NUMGRIG2018/Voronoi 150, Celebrating the 150th Anniversary of G. F. Voronoi, Moscow, Russia, December 2018.

For further information on these books please have a look at our mathematics catalogue at the following URL: `www.springer.com/series/3527`

Monographs in Computational Science and Engineering

1. J. Sundnes, G.T. Lines, X. Cai, B.F. Nielsen, K.-A. Mardal, A. Tveito, *Computing the Electrical Activity in the Heart.*

For further information on this book, please have a look at our mathematics catalogue at the following URL: www.springer.com/series/7417

Texts in Computational Science and Engineering

1. H. P. Langtangen, *Computational Partial Differential Equations.* Numerical Methods and Diffpack Programming. 2nd Edition
2. A. Quarteroni, F. Saleri, P. Gervasio, *Scientific Computing with MATLAB and Octave.* 4th Edition
3. H. P. Langtangen, *Python Scripting for Computational Science.* 3rd Edition
4. H. Gardner, G. Manduchi, *Design Patterns for e-Science.*
5. M. Griebel, S. Knapek, G. Zumbusch, *Numerical Simulation in Molecular Dynamics.*
6. H. P. Langtangen, *A Primer on Scientific Programming with Python.* 5th Edition
7. A. Tveito, H. P. Langtangen, B. F. Nielsen, X. Cai, *Elements of Scientific Computing.*
8. B. Gustafsson, *Fundamentals of Scientific Computing.*
9. M. Bader, *Space-Filling Curves.*
10. M. Larson, F. Bengzon, *The Finite Element Method: Theory, Implementation and Applications.*
11. W. Gander, M. Gander, F. Kwok, *Scientific Computing: An Introduction using Maple and MATLAB.*
12. P. Deuflhard, S. Röblitz, *A Guide to Numerical Modelling in Systems Biology.*
13. M. H. Holmes, *Introduction to Scientific Computing and Data Analysis.*
14. S. Linge, H. P. Langtangen, *Programming for Computations* - A Gentle Introduction to Numerical Simulations with MATLAB /Octave.
15. S. Linge, H. P. Langtangen, *Programming for Computations* - A Gentle Introduction to Numerical Simulations with Python.
16. H.P. Langtangen, S. Linge, *Finite Difference Computing with PDEs - A* Modern Software Approach.
17. B. Gustafsson, *Scientific Computing from a Historical Perspective.*
18. J. A. Trangenstein, *Scientific Computing.* Volume I - Linear and Nonlinear Equations.

19. J. A. Trangenstein, *Scientific Computing*. Volume II - Eigenvalues and Optimization.

20. J. A. Trangenstein, *Scientific Computing*. Volume III - Approximation and Integration.

For further information on these books please have a look at our mathematics catalogue at the following URL: www.springer.com/series/5151

Printed in the United States
By Bookmasters